TRANSFER MATRIX METHOD

MATHEMATICS AND ITS APPLICATIONS
(EAST EUROPEAN SERIES)

TRANSFER MATRIX METHOD

ALEXANDER TESÁR

Institute of Construction and Architecture,
Slovak Academy of Sciences, Bratislava,
Czechoslovakia

ĽUDOVÍT FILLO

Faculty of Civil Engineering,
Slovak Technical University, Bratislava,
Czechoslovakia

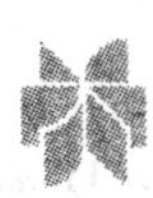

KLUWER ACADEMIC PUBLISHERS
DORDRECHT / BOSTON / LONDON

Library of Congress Cataloging-in-Publication Data

Tesár Alexander, Fillo Ľudovít.
Transfer matrix method.
(Mathematics and its applications. East European series).
Enlarged and revised translation of: Metóda prenosových matíc.
Bibliography: p.
Includes index.

1. Structures, Theory of—Matrix methods.
I. Fillo, Ľudovít. II. Title. III. Series:

Mathematics and its applications (Kluwer Academic
Publishers). East European series.

TA642.T4713 1988 642.1′7 87-23352

ISBN 90-277-2590-X

Scientific Editor
Prof. Ing. Vladimír Křístek, DrSc.

Distributors for the U.S.A. and Canada
Kluwer Academic Publishers,
101 Philip Drive, Norwell, MA 02061 U.S.A.

Distributors for East European socialist countries, Democratic People's Republic of Korea, People's Republic of China, People's Republic of Mongolia, Republic of Cuba, Socialist Republic of Vietnam
Veda, Publishing House of the Slovak Academy of Sciences,
814 30 Bratislava, Klemensova 19, Czechoslovakia

Distributors for all remaining countries
Kluwer Academic Publishers Group,
P.O. Box 322, 3300 AH Dordrecht, The Netherlands

Original title: Metóda prenosových matíc.
First edition published in 1985 by Veda, Bratislava.
First English edition published in 1988 by Veda, Bratislava,
in co-edition with Kluwer Academic Publishers, Dordrecht,
The Netherlands.

Printed in Czechoslovakia.

CONTENTS

SERIES EDITOR'S PREFACE

Approach your problems from the right end and begin with the answers. The one day, perhaps you will find the final question.

'The Hermit Clad in Crane Feathers' in R. van Gulik's *The Chinese Maze Murders*.

It isn't that they can't see the solution. It is that they can't see the problem.

G. K. Chesterton. *The Scandal of Father Brown* 'The Point of a Pin'.

Growing specialization and diversification have brought a host of monographs and textbooks on increasingly specialized topics. However, the 'tree' of knowledge of mathematics and related fields does not grow only by putting forth new branches. It also happens, quite often in fact, that branches which were thought to be completely disparate are suddenly seen to be related.

Further, the kind and level of sophistication of mathematics applied in various sciences has changed drastically in recent years: measure theory is used (non-trivially) in regional and theoretical economics; algebraic geometry interacts with physics; the Minkowsky lemma, coding theory and the structure of water meet one another in packing and covering theory; quantum fields, crystal defects and mathematical programming profit from homotopy theory, Lie algebras are relevant to filtering; and prediction and electrical engineering can use Stein spaces. And in addition to this there are such new emerging subdisciplines as 'experimental mathematics', 'CFD', 'completely integrable systems', 'chaos, synergetics and large-scale order', which are almost impossible to fit into the existing classification schemes. They draw upon widely different sections of mathematics. This programme, Mathematics and Its Applications, is devoted to new emerging (sub)disciplines and to such (new) interrelations as exempli gratia:

— a central concept which plays an important role in several different mathematical and/or scientific specialized areas;
— new applications of the results and ideas from one area of scientific endeavour into another;
— influences which the results, problems and concepts of one field of enquiry have and have had on the development of another.

The Mathematics and Its Applications programme tries to make available a careful selection of books which fit the philosophy outlined above. With such books, which are stimulating rather than definitive, intriguing rather than encyclopaedic, we hope to contribute something towards better communication among the practitioners in diversified fields.

The word 'transfer matrix', which occurs in the title of this book has (at least) three rather different meanings in mathematics: transfer matrices occur in system/control theory to describe the relation between inputs and outputs of a (linear) system, they play an important role in lattice models in statistical mechanics, and they are a most important tool in the structural analysis of all kinds of mechanical structures both in static and dynamic settings. This last one is the one relevant for this book. Thus here we are concerned with the analysis of the load-bearing capacities and the visco-elastic and vibratory properties of rods, beams, shells, plates, pipes, frames, grids, chains, and other mechanical structures, and, most importantly, with calculating these things quickly and reliably. Now the premier numerical tools for this appear to be 'finite element methods'; thus the AMS classification scheme mentions in section 73K (Structural Mechanics) right below each other: 'finite element methods' and 'other numerical methods'. And of course the importance of finite-element methods is by no means limited to structural mechanics. Cf. e.g. the book by Cuvelier, Segal, and Van Steenhoven, Finite Element Methods and the Navier—Stokes equations (in this series) for an account of its role in fluid dynamics. Yet finite-element methods by themselves in structural dynamics quickly lead to problems that are large, too large even for modern computers. At the same time as finite-element methods developed, transfer-matrix techniques, which are exact solution methods, arose. It is precisely the combination of the two: 'finite element' and 'transfer matrix' which has now developed into a very powerful tool used all over the world. And that is more precisely what this book is about.

The unreasonable effectiveness of mathematics in science ...

Eugene Wigner

Well, if you know of a better 'ole, go to it.

Bruce Bairnsfather

What is now proved was once only imagined.

William Blake

As long as algebra and geometry proceeded along separate paths, their advance was slow and their applications limited.

But when these sciences joined company they drew from each other fresh vitality and thenceforward marched on at a rapid pace towards perfection.

Joseph Louis Lagrange

Bussum, September 1987 Michiel Hazewinkel

INTRODUCTION

In recent years progressive structural systems have obtained increased importance due to new developments in industry and technology, especially in nuclear, aviation, off-shore, military and bridge engineering. Economic considerations and functional requirements of such structures have resulted in a drive for more daring designs with better utilization of all reserves in load-bearing capacity of structural systems and applied materials. The safety of such structures cannot be estimated unless the generalized spatial analysis of motion, taking account of the complete static and dynamic behaviour, is realized.

The analysis of nonlinear demeanour of structures has become the focus of intense efforts because of the pressing problems of reactor safety, vehicle crashworthiness, disaster prevention designs of bridges, missiles or planes. Considerable work has been directed towards improving the computational efficiency of certain components of finite static and dynamic processes, such as temporal integration of discrete equations of motion, the solution of nonlinear algebraic equations, optimization and condensation of numerical techniques, etc. In analysis, the shortcut and simplified methods of the past have been replaced by general techniques, such as matrix and finite element methods, which permit the analysis of the most complex structures to whatever degree of modelling desired. Material and geometric nonlinearities, time-dependent behaviour, dynamic and thermal loads can now be treated for an essentially unlimited range of structures, geometries and exploitation conditions.

Structural synthesis and optimization are becoming integral parts of the structural design process, if not as production tools using actual cost factors, then at least as tools for evaluating alternate design concepts on a consistent basis.

Finite element methods of structural analysis present a general approach for solution of such problems. These methods are valuable tools for the analyses of complex structural problems of modern linear and nonlinear mechanics.

Finite element methods provide a concept for the approximate solution of a wide class of engineering problems by reducing the analysis to the solution of large systems of algebraic equations. Substructuring procedures are used to reduce the number of unknowns in a finite element analysis, which requires the inversion of sizeable matrices to reduce the order of the final matrix equation.

However, the solution of most large-scale linear and especially nonlinear transient structural problems is still not economically feasible when applying such concepts. This raised the question as to whether the use of vector processing computer systems might result in bringing down the computational times for advanced analyses of large structural systems within the acceptable practical range.

There were formulated the requirements for development of problem-oriented computational algorithms of the discrete streamline type, exploiting the vector processing capacity of modern computers. The increased speed and expanded storage capacity of modern computers, especially of the vector type, coupled with the newest advances and trends in numerical methods and programming techniques have significantly improved the engineers' ability to solve such problems.

During the period of development of the finite element techniques the transfer matrix method of finite structural analysis also advanced significantly. In contrast to the finite element method, the transfer matrix method is a solution procedure working with discrete mesh simulated by movable finite elements. The analysed discrete simulation is modelled during the motion of elements over the structure. The transfer matrix method has the streamline character suitable for vector processing computers. From such point of view the transfer matrix method is a more efficient procedure in comparison with finite element method. There exists a significant difference in the number of unknowns that must be detemined in order to complete the analysis. Since these two numerical methods have different strengths and deficiencies, the advantages may be gained by combining both concepts for effective utilization of modern computer techniques. The CPU-time required for solving a direct finite element method increases exponentially with the number of elements, while the analysis time for a combined finite element versus transfer matrix approach (FETM-method) remains essentially independent of the number of units. Since a stiffness formulation is used for the development of such a typical unit, the inversion of a sizeable matrix is required to eliminate the displacements at interior nodes before converting the governing equations for the periodic unit into transmission form.

This book presents the basic principles in the theory and application of transfer matrix and FETM-methods for solution of linear and nonlinear problems of progressive subtle structural systems exposed to the action of static and dynamic loads. The basic philosophy of the methods adopted in this book is based on the idea of breaking up a complicated structural system into component parts with simple elastic and dynamic properties that can be readily expressed in a matrix form. These component matrices are considered as movable finite blocks that, when fitted together according to a set of predetermined rules, express the static and dynamic, linear as well as nonlinear properties of the entire system. The matrix and vector formulations of these rules, embodied in present methods, are superbly suited for streamline computer analyses using the standard or vector computer techniques or packages. Present concepts are significant first of all in nonlinear limit state analysis of subtle thinwalled structural configurations. For such structures is typical the general variability of physical parameters in nonlinear regions of deformed configurations as well as in various time steps of their response. Nonlinear solutions and particularly the discrete dynamic response analysis of progressive structures are very expensive. Studying reliable mehods by which the computational costs can be reduced compared with direct integration of the nonlinear finite element equations of motion may thus be of considerable importance.

The problems analysed in this book are dealt with in five chapters and in enclosed Catalogue of Transfer Matrices. After the introduction, the attention in the first chapter is focused on the generalized analysis of motion regarding all linear and nonlinear effects and their interactions that are present in limit analysis of modern structural systems. Analysed are the influences of geometric and material nonlinearities, effects of linear and nonlinear material and structural damping as well as the linear and nonlinear behaviour of structures in resonance, critical or postcritical limit regions of exploitation. Linear and static analyses represent special simplified cases of above generalized concepts.

In the second chapter the theoretical and numerical backgrounds of the discrete variant of the transfer matrix method for the solution of linear problems are presented, together with corresponding computer algorithms. The transfer matrices are derived for one- and multidimensional elements, for bars, beams, plates, shells, grid and frame systems, for thinwalled members as well as for other typical members of the up-to-date structural analysis. The corresponding nodal matrices are also derived for static and dynamic analyses of such members when using the discrete variant of the transfer matrix method. The matrices of initial parameters and boundary conditions for simulation of various boundary and intermediate conditions of studied structures are developed. All presented matrices are accepted as element blocks for the modelling of various structural simulations.

The third chapter deals with the FETM-method as a problem-oriented combination of finite element and transfer matrix techniques adopted for solution of generalized nonlinear problems. The concept of hybrid substructures of finitization in space and time is utilized. The mathematical principles of such concept are analysed. The determination of transfer hypermatrices from stiffness matrices of the finite element method is described. The global hypermatrices of spatial interaction of individual members for simulation of multidimensional problems are determined. The hypermatrices for typical macroelements of spatial grid simulations are derived. The set-up of such microelements submits the macroelements of the algorithm of the FETM-method for linear and nonlinear analyses of structural simulations. The incorporation of the pseudo-force techniques into algorithms of the FETM-method is presented when solving the nonlinear problems of slender structural simulations. Furthermore, attention is paid to the development of methods important in condensation and total vectorization techniques of the programmatory code. Mixed time discretization is performed using the combined Newmark—Wilson method of direct numerical integration in time, incorporated into linear and nonlinear algorithms of the FETM-method. The nonlinear interactions occurring in advanced nonlinear regions are studied, e.g., the elastic-plastic, damping, pre- and post-buckling nonlinear interrelations in the static and dynamic response of slender structures. Resulting structural synthesis and optimization techniques are based on the analogies with evolution processes in biology. Such modern synthesis concepts present one of the last achievements in this branch.

In the fourth chapter of this book the application of the developed theoretical and numerical concepts is presented. Real thinwalled beams, plate and shell structures are dealt with. Ultimate and limit state analyses in stability, post-buckling and resonance regions yield the comprehension on the safety and optimization of such structures. Theoretical, numerical and experimental solutions of present problems are performed.

Further research topics and future orientations in the advanced numerical analyses of the modern structural configurations are covered in the fifth chapter.

The Catalogue of Transfer Matrices which has been added contains the matrix expressions for solutions of all studied problems dealt with in this book. The Catalogue (CTM) also contains the basic computer program for solving the problems presented in this book when applying the transfer matrix method (TMM).

All symbols are defined in the text when they first appear. The symbols that are introduced in one section but not referred to later are not included. Vectors and matrices are represented by bold-type characters. Some symbols may have two or more meanings in different chapters; these are clearly defined when used thus avoiding confusion.

The successful application of the research involving the transfer matrix and FETM-methods generally results in producing of a computer program to be used in the practical solution of engineering problems. In order for the program to be an effective tool, it must be based on theories and techniques from three different disciplines. Firstly, the approximations used to develop the properties of the various elements must be based on sound fundamental principles of mechanics. Secondly, the numerical methods selected for spatial integration, solution of the sets of equations, evaluation of eigenvalues and step-by-step mixed time-integration techniques must be accurate and efficient. Thirdly, the computer implementation of the numerical techniques used must be approached with great care if the number of numerical operations is to be minimized, high speed and low speed storage units are to be used effectively and the resulting program is to be reasonably machine independent.

The purpose of this book is to present the necessary background in all three of these areas.

Alexander Tesár

Chapter 1

Generalized Analysis of Motion

1.1 General Concepts

Displacements of a material particle of arbitrary structure are considered as a family of mappings from one region in space to another. The momentary configuration of the structure is completely defined by the locations of its material particles at the given time. Variations of configurations are assumed to be continuous and new boundaries will not arise during deformation. The path of a material point is followed through the various configurations of the structure and each position is defined in relation to a reference position.

Assuming the coordinates x, y, z and corresponding displacements u, v, w, the *Green strain tensor* is defined as

$$E_{xx} = \frac{\partial u}{\partial x} + \frac{1}{2}\left[\left(\frac{\partial u}{\partial x}\right)^2 + \left(\frac{\partial v}{\partial x}\right)^2 + \left(\frac{\partial w}{\partial x}\right)^2\right], \tag{1.1}$$

$$E_{xy} = \frac{1}{2}\left(\frac{\partial v}{\partial x} + \frac{\partial u}{\partial y} + \frac{\partial u}{\partial x}\frac{\partial u}{\partial y} + \frac{\partial v}{\partial x}\frac{\partial v}{\partial y} + \frac{\partial w}{\partial x}\frac{\partial w}{\partial y}\right), \tag{1.2}$$

$\vdots$

etc.

To establish the constitutive equations with Green strain tensor, a *stress tensor* with the same reference is needed. A symmetric one will be advantageous in applications. The 2nd Piola—Kirchhoff stress tensor denoted by S_{ij} has the desired properties. Using *Gauss' theorem* for the surface integral yields the general equilibrium equation for the deformed configuration expressed by the *2nd Piola—Kirchhoff stress tensor* as

$$S_{ij} = g(E_{ij}), \tag{1.3}$$

where g is a single valued function of the Green strain tensor E_{ij}.

Consider a structure with volume, surface area and mass density in an initial configuration denoted by V, S and ϱ_0, respectively. The body forces per unit mass are denoted $F_{0,i}$ and surface tractions are specified by force components T_i.

The structure in equilibrium is subjected to a virtual displacement δu_i, which is kinematically consistent with present boundary conditions. The balance of work in the structure may then be written as

$$\int_V S_{ij}\delta E_{ij}\,\mathrm{d}V + \int_S T_i\delta u_i\,\mathrm{d}S + \int_V P_i\delta u_i\,\mathrm{d}V = 0\,, \tag{1.4}$$

where

$$P_i = \varrho_0 F_{0,i}\,. \tag{1.5}$$

Expression (1.4) states that among all kinematically admissible displacement fields u_i the actual one renders the value of the total potential energy stationary.

The incremental form of the variational principle for two configurations of analysed structure is given by

$$\int_V S_{ij}^{(1)}\delta E_{ij}^{(1)}\,\mathrm{d}V + \int_S T_i^{(1)}\delta u_i^{(1)}\,\mathrm{d}S + \int_V P_i^{(1)}\delta u_i^{(1)}\,\mathrm{d}V = 0\,, \tag{1.6}$$

$$\int_V S_{ij}^{(2)}\delta E_{ij}^{(2)}\,\mathrm{d}V + \int_S T_i^{(2)}\delta u_i^{(2)}\,\mathrm{d}S + \int_V P_i^{(2)}\delta u_i^{(2)}\,\mathrm{d}V = 0\,, \tag{1.7}$$

where superscripts (1) and (2) denote the two configurations. The components of surface tractions and body forces refer to the same reference configuration and may therefore be subtracted directly to give

$$\Delta T_i = T_i^{(2)} - T_i^{(1)}\,, \tag{1.8}$$

$$\Delta P_i = P_i^{(2)} - P_i^{(1)}\,. \tag{1.9}$$

The variations of the two displacement fields are chosen to be the same

$$\delta u_i = \delta u_i^{(1)} = \delta u_i^{(2)}\,. \tag{1.10}$$

An incremental form of the virtual work equations is then obtained by subtracting equations (1.7) and (1.6) giving

$$\int_V (S_{ij}^{(2)}\delta E_{ij}^{(2)} - S_{ij}^{(1)}\delta E_{ij}^{(1)})\,\mathrm{d}V + \int_S \Delta T_i\delta u_i\,\mathrm{d}S + \int_V \Delta P_i\delta u_i\,\mathrm{d}V = 0 \tag{1.11}$$

and considering the virtual variations of both analysed configurations.

The neglection of the higher order strain energy terms can be done only if the two configurations are sufficiently close to each other. Equation (1.11) will then give configuration (2) from the known configuration (1) and known load increments.

1.2. Analysis of Motion

When the work done by inertial and damping forces over virtual displacements δu_i is added to equation (1.4), the virtual work principle for dynamic problems can be written as

$$\int_V S_{ij}\delta E_{ij}\,\mathrm{d}V + \int_V \varrho\ddot{u}_i\delta u_i\,\mathrm{d}V + \int_V \mathbf{C}\dot{u}_i\delta u_i\,\mathrm{d}V +$$

$$+ \int_S T_i\delta u_i\,\mathrm{d}S + \int_V P_i\delta u_i\,\mathrm{d}V = 0\,, \tag{1.12}$$

where the dots express derivatives with respect to time, ϱ and $\mathbf{C}$ are the mass and damping terms, respectively. The body forces and surface tractions are now time-dependent.

The generalized analysis of motion defines the vibrating structure as a *dynamic transfer system* characterized by *output signal* $y(t)$ as a dynamic structural response to *input (inducing) signal* $x(t)$ in accordance with Figure 1.

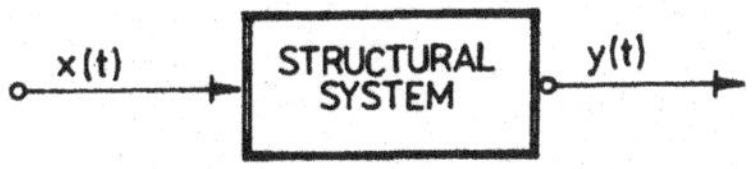

x(t) STRUCTURAL SYSTEM y(t)

Fig. 1. Dynamic transfer system.

Physical properties of a transfer system are cumulated in the *complex transfer function* $F(\omega)$ defined as a *system response* to a harmonic unit signal $x(t)$ with frequency ω

$$x(t) = 1\,\mathrm{e}^{\mathrm{i}\omega t}\,. \tag{1.13}$$

The response output $y(t)$ to an arbitrary input signal $x(t)$ can be determined by the application of a Duhamel integral of the form

$$y(t) = \int_0^\infty x(t-\tau)G(\tau)\,\mathrm{d}\tau\,, \tag{1.14}$$

where $G(\tau)$ denotes the system response to a Dirac delta impulse with spectral parameter τ. Thus, the distribution of an input signal into individual impulses is realized. The structural response is determined by integration of output signals corresponding to each of inputs. The system transfer function $F(\omega)$ is given by the Fourier transform of the system response.

Regarding the *power spectral densities* of input and output signals $S_x(\omega)$ and $S_y(\omega)$, the present relations can be expressed in a generalized deterministic-stochastic form

$$S_y(\omega) = |F(\omega)|^2 S_x(\omega)\,, \tag{1.15}$$

where $|F(\omega)|^2$ is defined as the corresponding *power spectral density factor* of the studied structure. The physical interpretation of this concept is illustrated in Figure 2.

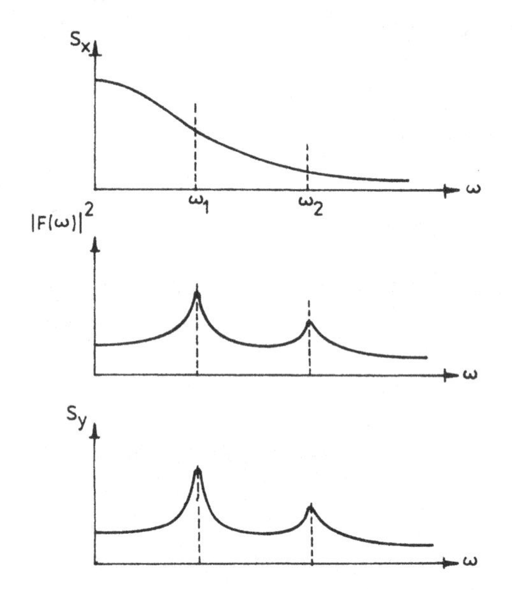

Fig. 2. Physical simulation of power spectral density.

1.3. Incremental Form of Equations of Motion

An incremental form of the equations of motion is readily obtained by considering the dynamic equilibrium at two configurations a time step Δt apart. The increments of external loading then balance dynamic equilibrium at time $t + \Delta t$ as

$$\mathbf{M}_t \Delta \ddot{\boldsymbol{r}}_t + \mathbf{C}_t \Delta \dot{\boldsymbol{r}}_t + \mathbf{K}_t \Delta \boldsymbol{r}_t = \boldsymbol{R}_{t+\Delta t} - (\boldsymbol{V}_t^{\mathrm{I}} + \boldsymbol{V}_t^{\mathrm{D}} + \boldsymbol{V}_t^{\mathrm{S}}), \tag{1.16}$$

where

— inertia forces $\boldsymbol{V}_t^{\mathrm{I}} = \mathbf{M}_t \ddot{\boldsymbol{r}}_t$, (1.17)
— damping forces $\boldsymbol{V}_t^{\mathrm{D}} = \mathbf{C}_t \dot{\boldsymbol{r}}_t$, (1.18)
— elastic forces $\boldsymbol{V}_t^{\mathrm{S}} = \mathbf{K}_t \boldsymbol{r}_t$. (1.19)

The vectors of nodal point accelerations and velocities are given as time derivatives of the vector of nodal deformations $\boldsymbol{r}$. The mass matrix $\mathbf{M}$, damping matrix $\mathbf{C}$ and stiffness matrix $\mathbf{K}$ are constructed of element matrices established in incremental fashion directly for applied discrete simulation. The subscript t denotes current time and $\boldsymbol{R}$ is the vector of external loads. If the structure is in equilibrium at time t, the right-hand side of equation (1.16) will be identical to the increment in external loads over the time increment Δt.

Increments in nodal displacements, velocities and accelerations have thus been expressed by external load increments and known physical property matrices. If however, these matrices change during the time step, then equation (1.16) is only an approximately true. The variability of incremental stiffness during time step Δt may be considered in accordance with Figure 3. The vector of residual forces given by

$$\Delta \boldsymbol{V}_{t+\Delta t} = \boldsymbol{R}_{t+\Delta t} - (\boldsymbol{V}^{\mathrm{I}}_{t+\Delta t} + \boldsymbol{V}^{\mathrm{D}}_{t+\Delta t} + \boldsymbol{V}^{\mathrm{S}}_{t+\Delta t}), \tag{1.20}$$

is a measure of how close to equilibrium at time $t + \Delta t$ the solution has been increased by the approximate expression (1.16).

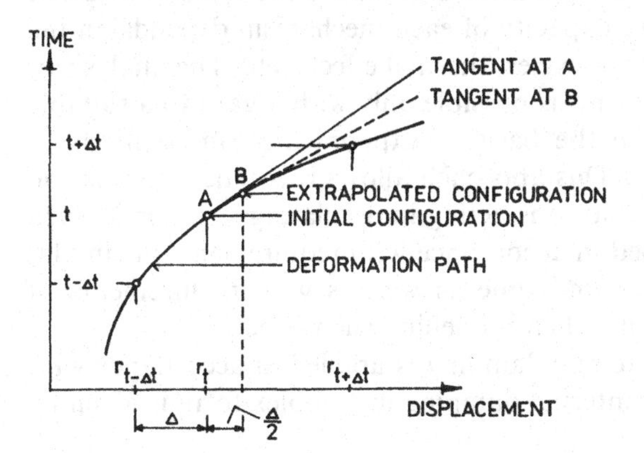

Fig. 3. Alternative incremental stiffness approach.

In the next sections attention will be paid to the formulation of physical property matrices in linear as well as nonlinear incremental fashion.

1.4. Mass Matrix

The mass matrix defined in equation (1.16) is called a consistent mass matrix which in terms of energetic relations means that the representation of kinetic energy is consistent with that of potential energy. The total element mass for translational degrees of freedom is lumped with equal amounts to each node. For the rotational degrees of freedom the mass is evaluated either by using the consistent mass expression assuming zero displacements and equal rotations about correspoding axes at both ends of the element, or by neglecting the values which are small compared to translational masses. Numerical calculations have demonstrated that the actual numerical values of rotational masses are of minor importance if the element discretization is reasonable with respect to significant nodes. In fact, rotational masses are often neglected and the corresponding degrees of freedom are eliminated by static condensation.

1.5. Damping Matrix

The linear and nonlinear resonance response of structures is influenced primarily by the effects of linear and nonlinear material and structural damping and also by the dissipative capacity of environment. The equivalent damping covering the total energy dissipation can be simulated by several concepts which express the relations between damping forces and strain velocities. Energy transformations due to material damping can be modelled by theoretical mechanisms which are simulating magnetostriction effects, local plastic deformations, nonstationary thermic variations connected with intercrystalline friction and motion of atomic groups in the crystalline lattice of the material. The dissipative capacity of each mechanism depends on the frequency, stress level, number of cycles, thermal effects, etc. The analysis of interaction of all these mechanisms is possible only with regard to a suitable material model constructed on the basis of experimental observations of individual damping influences. This approach allows theoretical simulation of complicated dissipative mechanisms acting in oscillating structures. Such an approach must be developed in accordance with requirements of simulation of real structure in space and time as well as with requirements of accuracy of numerical solutions when modelling such effects.

The complex theory of hysteretic damping is applied in accordance with the references [83, 92]. Linear internal damping in complex form is given by the relation

$$\sigma = E_0(u + \mathrm{i}v)\,\varepsilon, \tag{1.21}$$

with stress and strain denoted by σ and ε, respectively. E_0 is the complex modulus of elasticity if the real part of the strain ε converges to zero and parameters u and v are real functions of the damping factor η_s which are expressed as

$$u = \frac{1 - \dfrac{\eta_s^2}{4}}{1 + \dfrac{\eta_s^2}{4}}, \quad v = \frac{\eta_s}{1 + \dfrac{\eta_s^2}{4}}. \tag{1.22}$$

The damping factor η_s and the logarithmic decrement of damping δ_f are coupled by ratio

$$\eta_s = \delta_f/\pi. \tag{1.23}$$

If the stress is varying sinusoidally, the strain varies in time with equivalent frequency ω and with phase shift α. If the stress is written as

$$\sigma = \sigma_0 \mathrm{e}^{\mathrm{i}\omega t}, \tag{1.24}$$

where σ_0 is the stress amplitude, then corresponding strain may be given by

$$\varepsilon = \varepsilon_0 e^{i(\omega t - \alpha)} . \tag{1.25}$$

The complex modulus of elasticity is defined by

$$E = E_1 + iE_2 = \sigma/\varepsilon = \sigma_0 e^{i\alpha}/\varepsilon_0 , \tag{1.26}$$

with real and imaginary components given by

$$E_1 = \frac{\sigma_0}{\varepsilon_0} \cos\alpha , \tag{1.27}$$

$$E_2 = \frac{\sigma_0}{\varepsilon_0} \sin\alpha . \tag{1.28}$$

The damping factor η_s is given by

$$\eta_s = \operatorname{tg}\alpha = E_2/E_1 \tag{1.29}$$

and is incorporated into the complex modulus of elasticity given by

$$E = E_0(1 + i\eta_s) . \tag{1.30}$$

Similarly, shear modulus of elasticity is given by

$$G = G_0(1 + i\eta_e) , \tag{1.31}$$

in which η_e is the shear damping factor.

Structural damping can be approximated by complex spring characteristics

$$K_B = K_0(1 + i\eta_B) , \tag{1.32}$$

formulated as nodal couplings of elastic or stiff supports or of other types of structural discontinuities. Parameters K_0 and η_B are defined as real spring constant and factor of structural damping, respectively. The structural damping behaves in interaction with aforementioned influences of material damping.

The specific work D of total damping can be expressed as a function of the stress σ

$$D = J\sigma^n , \tag{1.33}$$

with experimentally given parameters J and n [47]. The total work of damping is determined by integration of individual specific works of damping. The variability of stress causes that each element of structure has its own hysteretic curve contributing to the total damping. The work of damping in the element volume V_g is given by

$$D_g = \int_0^{V_g} D\,\mathrm{d}V = \int_0^{\sigma_{max}} D \frac{\mathrm{d}V}{\mathrm{d}\sigma}\mathrm{d}\sigma. \tag{1.34}$$

The maximum stress σ_{max} corresponds to the maximum work of damping D_{max} and D_g can be written as

$$D_g = D_{max} V_g \beta_1, \tag{1.35}$$

with nondimensional parameter β_1

$$\beta_1 = \int_0^1 \left(\frac{D}{D_{max}}\right) \frac{\mathrm{d}\left(\frac{V}{V_g}\right)}{\mathrm{d}\left(\frac{\sigma}{\sigma_{max}}\right)} \mathrm{d}\left(\frac{\sigma}{\sigma_{max}}\right). \tag{1.36}$$

The energy cumulated in analysed volume can be written as

$$U_g = \int_0^{V_g} \frac{\sigma^2}{E}\mathrm{d}V = \frac{1}{2} V_g \frac{\sigma_{max}^2}{E}\beta_2, \tag{1.37}$$

with nondimensional parameter β_2

$$\beta_2 = \int_0^1 \left(\frac{\sigma}{\sigma_{max}}\right)^2 \frac{\mathrm{d}\left(\frac{V}{V_g}\right)}{\mathrm{d}\left(\frac{\sigma}{\sigma_{max}}\right)} \mathrm{d}\left(\frac{\sigma}{\sigma_{max}}\right). \tag{1.38}$$

The factor of damping for the analysed volume element is then given by the relation

$$\eta_s = \frac{D_g}{2\pi U_g} = \frac{E D_{max} \beta_1}{\pi \sigma_{max}^2 \beta_2}, \tag{1.39}$$

and is incorporated into the complex modulus of elasticity in accordance with the stresses in individual elements of the applied discrete simulation.

In the case of linear damping there may be assumed the ratio $\beta_1/\beta_2 = 1$. The resonance analysis of nonlinear damping is based on the evaluation of the parameters β_1 and β_2 for given geometric and stress relations, using the equations (1.36) and (1.38). An iterative scheme is used for the determination of the damping factors in each element of utilized discretization mesh. In the first iteration the stresses are determined which yield the values of parameters $\beta_1^{(1)}$ and $\beta_2^{(1)}$ as well as the values of the preliminary damping factor $\eta_s^{(1)}$ or of corresponding logarithmic decrement of damping $\delta_f^{(1)}$. These parameters are the basis for further iteration steps determining the modified incremental stresses and parameters $\beta_1^{(i)}$, $\beta_2^{(i)}$, $\eta_s^{(i)}$ or $\delta_f^{(i)}$. The calculation continues until the applied convergence criterion, such as for example

$$\left|\frac{\eta_s^{(i+1)}}{\eta_s^{(i)}}\right| - 1 \leqslant 0.01\,, \tag{1.40}$$

is satisfied. The functions β_1, β_2 and δ_f are plotted in Figures 4 and 5 for the illustrative case of profile IPE 30 made of steel with allowable stress limit 210 MPa, related to the normal stresses due to transversal and axial loads. The graphs illustrate the typical nonlinear behaviour of the logarithmic decrement of damping when obtaining the total stress level over 150 or 200 MPa for steel material.

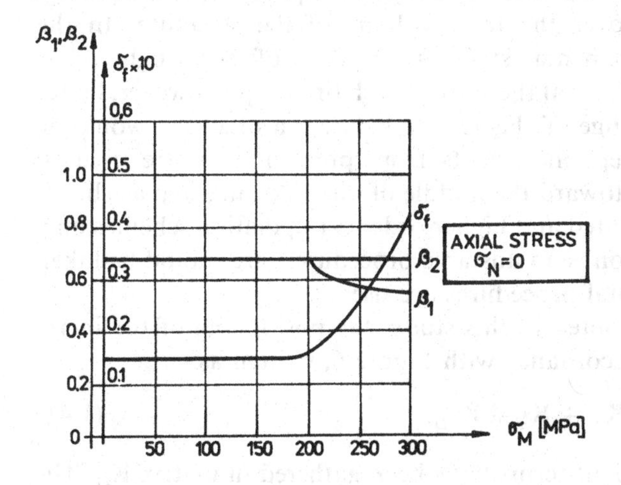

Fig. 4. Parameters β_1, β_2 and δ_f as functions of normal (σ_M) and axial (σ_N) stresses for IPE-30 profile produced of steel having allowable normal stress 290 MPa.

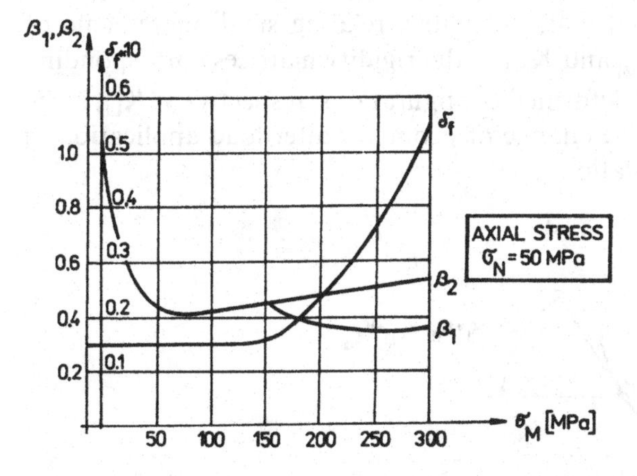

Fig. 5. Parameters β_1, β_2 and δ_f as functions of normal and axial stresses for IPE-30 profile produced of steel having allowable normal stress limit 290 MPa.

This concept allows the determination of the damping parameters for each discrete element for real stress levels of studied structural region. The damping parameters determine the matrices of linear and nonlinear damping for the structural simulation investigated.

1.6. Stiffness Matrix

The elements of the incremental stiffness matrix are expressed as the sum of the corresponding elements of the usual small displacement stiffness matrix and the terms which are the functions of displacements. For the nonlinear formulation applied here, each displacement increment must be so small that equation (1.12) is valid for extending the solution to a new configuration. The incremental stiffness for a complete structure is obtained by summing up the contributions from all individual elements, which is equivalent to integrating the incremental stiffness over the total volume of the structure. In the response analysis the incremental stiffness matrix is often thought of as identical to the tangent stiffness at the start of each time step. However, better approximations of the change of elastic forces during a time step would be expected if the displacement increments from previous step are used to extrapolate displacements toward the middle of the following step as shown in Figure 3. More sophisticated and higher order extrapolation schemes may be used, but as extrapolation is an unstable procedure, care should be taken especially if pure incremental procedures are used.

For the formulation adopted in this study, the positioning of the incremental stiffness $\mathbf{K}_\mathrm{I}$ is, in accordance with Figure 6, written as

$$\mathbf{K}_\mathrm{I} = \mathbf{K}_\mathrm{L} + \mathbf{K}_\mathrm{G} = \mathbf{K}_\mathrm{L} + \mathbf{K}_{\sigma_0} + \mathbf{K}_\sigma + \mathbf{K}_\mathrm{LD}, \tag{1.41}$$

where all geometry-dependent terms have been gathered in matrix $\mathbf{K}_\mathrm{G}$. The matrix $\mathbf{K}_\mathrm{L}$ is the incremental stiffness matrix relating small increments of loads and displacements. $\mathbf{K}_{\sigma_0}$ and $\mathbf{K}_\sigma$ are the rigidity matrices corresponding to stress states of initial and deformed configurations, respectively. $\mathbf{K}_\mathrm{LD}$ is the stiffness matrix expressing the change of geometry after load application in a large displacement formulation.

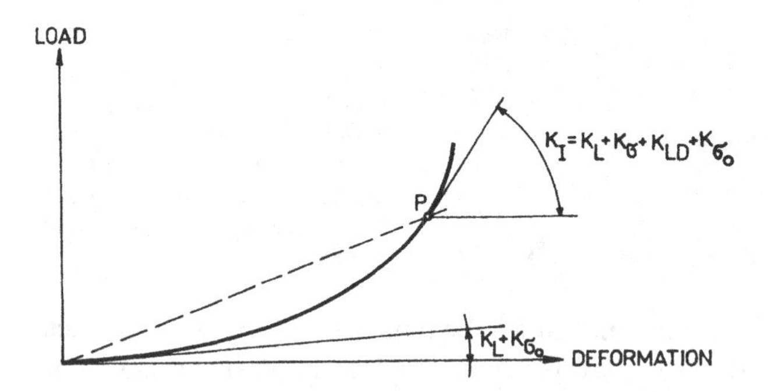

Fig. 6. Stiffness concepts.

1.7. Numerical Solution Techniques

The nonlinearities in structures occur in two different forms: the geometric nonlinearities caused by finite displacements, and the material nonlinearities arising from nonlinear constitutive equations. Usually these two sources of nonlinearities occur simultaneously in structural problems.

The best method for solving the nonlinear equations depends on the actual physical behaviour of the structure under consideration. The choice of the method to be used is much more difficult than in the case of linear problems. Whereas in linear analyses the solution is always unique, it is not necessarily the case in nonlinear problems. Accordingly, insight into the nature of the structural problem is essential in addition to good knowledge of numerical analysis. Often, a combined technique using several methods must be used to obtain a good accuracy and reliability at a relatively low cost.

The methods can be subdivided into four groups according to their mathematical basis for solution. These groups may be characterized as follows:

— energy minimization,
— direct iteration methods,
— incremental methods,
— combined methods.

The first group covers methods that employ some nonlinear programming scheme for a direct search for the extreme values of the potential energy. One of the most efficient minimization techniques is the *Fletcher—Powell method* [21] which employs a very fast approximation to the inverse gradient (*inverse Jacobian matrix*). However, the energy minimization approach is not reliable when used for problems in which structural instability occurs due to the existence of local minima.

The second group comprises methods that utilize some direct iteration technique applied to the equilibrium equations. The lowest order iteration method available is *functional iteration* [39] which implies successive substitutions of the displacement vector into the inverted equilibrium equation. The method is illustrated for a one degree of freedom system in Figure 7a. In the *chord method* as shown in Figure 7b, a *Newton type* iteration is performed maintaining a constant preselected gradient such as the initial linear stiffness of the system.

A higher order iteration scheme is the *Newton—Raphson iteration* which utilizes the gradient of the stiffness relation in accordance with Figure 7c. *Quasi-Newton methods* based on the approximations to the gradient can also be used; in fact they are highly recommendable in most cases in which they do not impede convergence. The method of false position and *Aitkens δ^2-process* [39] have also been used by some investigators.

The third group comprises methods treating the incremental form of the equilibrium equations as a first-order differential equation which is solved as an initial value problem by applying load increments.Numerous numerical methods capable of solving this problem have been described, see for example references [8, 29] and [30]. Such methods are characterized as single step or multistep, implicit or explicit, and they are based on quadrature, series expansions or finite differences. The most important features of the various methods from a numerical point of view are their stability (related to round-off errors) and truncation errors (resulting from omission of higher order terms). For practical applications their efficiency or time consumption per

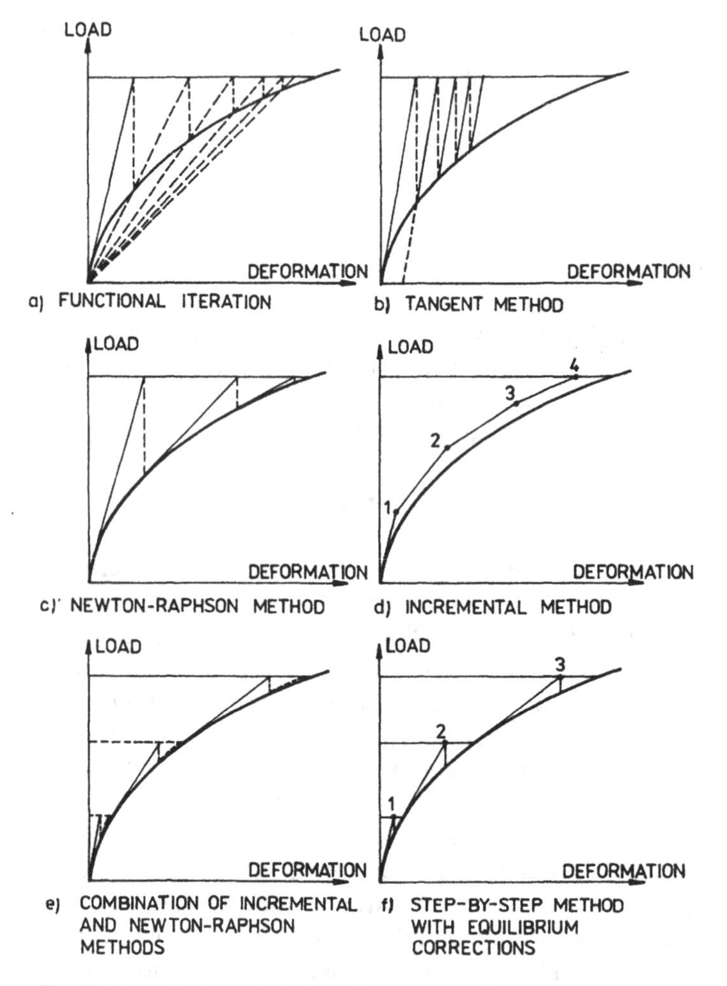

Fig. 7. Numerical methods for solving of nonlinear problems.

step is also a highly critical quantity. The simplest incremental scheme is the *Euler—Cauchy method* (simple step-by-step) as shown in Figure 7d.

The fourth group of methods comprises those procedures which in some way are combinations of groups 2 and 3. A noteworthy method of this category is a step-by-step procedure combined with Newton iteration at certain load intervals (see Figure 7e). A somewhat similar method uses residual force (equilibrium) corrections combined with load incrementation. This method is shown in Figure 7f.

For strongly nonlinear problems the true equilibrium path should be followed closely. The most reliable procedure will then be to combine an incremental method with equilibrium iterations. For more moderate nonlinearities, the equilibrium path may also be traced with sufficient accuracy by a pure step-by-step procedure. For problems that are not path-dependent, the methods from the first two groups may be preferable if the nonlinearities are not too strong for obtaining an acceptable rate of convergence.

For a discretized system the true path corresponds to a specific displacement vector function of given loading. The incremental stiffness for some point on the path is a function of all elements of the associated displacement vector. Any other displacement vector that does not correspond to a point on the true path yields a hypersurface that is different from any slope at the true path. For example, for thin plates or shells for which the membrane stiffness is high compared with the bending stiffness, even a small perturbation of the displacement vector has a tremendous influence on the incremental stiffness. Only a small perturbation in the in-plane displacement components in the nonlinear range significantly changes the stiffness associated with transverse displacements. For such cases the rate of convergence is usually very slow, particularly when the incremental stiffness matrix is kept constant during the iteration process. Similar difficulties have been experienced when applying various nonlinear programming techniques to an optimization problem. Following a narrow valley towards an extremum may result in slow convergence of the Newton—Raphson or other numerical techniques. A method based on the principle of extrapolating errors to zero has proved to be useful in such cases. Such an approach can be used directly in connection with the slowly converging Newton—Raphson iteration. Figure 8 shows how the out-of-balance force associated with one arbitrary displacement component is used for finding a new displacement value for which such a force is presumed to be zero. The following formula is easily derived using Figure 8

$$\Delta r_{j+1} = \Delta r^*_{j+1} \frac{\Delta R_j}{\Delta R_j - \Delta R^*_{j+1}}. \tag{1.42}$$

The values marked with an asterisk represent intermediate values that are improved by using equation (1.42). This equation should be used for all displacement components. The extrapolation method itself is generally not stable and should be used only as a "single shot" adjustment in connection with a stable type of iteration.

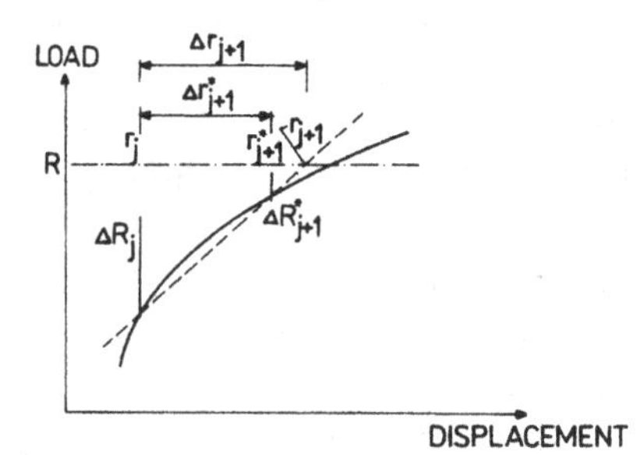

Fig. 8. Method for finding improved displacement values.

1.8. Convergence and Accuracy

When an iteration technique is used for solving the nonlinear problems, a convergence criterion capable of determining when the procedure has converged to the desired degree of accuracy is needed. Principally, three groups of such convergence criteria are available — force criteria, displacement criteria and stress criteria. For thinwalled plate or shell problems the displacement criteria are preferable. According to this convergence criterion, an error vector is defined as follows

$$\varepsilon_{ij} = \left(\frac{\Delta r_1}{r_{1,\mathrm{ref}}}, \frac{\Delta r_2}{r_{2,\mathrm{ref}}} \cdots \frac{\Delta r_N}{r_{N,\mathrm{ref}}} \right)^{\mathrm{T}}_{(j)}, \tag{1.43}$$

where Δr_1, Δr_2, ..., etc. are the changes of displacement components during cycle j, and N is the total number of nonzero components. Each component is scaled by a reference quantity to obtain a nondimensional measure. These reference quantities are generally not chosen to be equal to the corresponding total components. The changes of transversal displacements are scaled by the largest of the transversal components, the changes in rotations are scaled by the largest rotation and similarily with the in-plane displacements. Three alternative norms can be suggested to measure the value of the error vector:

— the *modified absolute norm*

$$|\varepsilon|_1 = \frac{1}{N} \sum_{k=1}^{N} \frac{\Delta r_k}{r_{k,\mathrm{ref}}}, \tag{1.44}$$

— the *modified Euclidean (spectral) norm*

$$|\varepsilon|_2 = \sqrt{\frac{1}{N} \sum_{k=1}^{N} \left| \frac{\Delta r_k}{r_{k,\mathrm{ref}}} \right|^2}, \tag{1.45}$$

— the *maximum (uniform) norm*

$$|\varepsilon|_\infty = \max \left| \frac{\Delta r_k}{r_{k,\mathrm{ref}}} \right|, \qquad (1.46)$$

where r_k is the k-th component of the vector of displacements. These definitions yield convergence criteria that have a direct physical meaning.

Two first norms are modified by dividing by N to get quantities that are independent of the total number of displacement components. Error bounds on the displacement vector indicate the accuracy both of displacements and stresses. The three different norms follow each other in a parallel manner. Therefore, it has no great importance which particular norm is being chosen. However, the maximum norm is probably the safest measure of convergence since it gives a specific error bound on all displacement components; the other norms yield more of an *average error bound.* The practical range of $|\varepsilon|$ lies approximately between 10^{-2} and 10^{-1}. Application of a fixed number of iteration cycles will not work satisfactorily since some systems converge rapidly and others very slowly; even the same structure may completely change its character when a load is applied. In-plane forces, transversal forces and moments may all be of a different order. The variable number of iteration cycles is used when dealing with various problems. The linearity of the convergence criterion given by the maximum norm (1.46) makes it possible during the iteration process to estimate how many more cycles are necessary to satisfy a given convergence criterion.

1.9. Automatic Incrementation Procedures

The incremental load factors may be given as input to the computer program, which may require a considerable amount of data preparation unless the size of the increment is kept constant throughout the entire load regime. However, in order to account for the inherent truncation errors in the iteration schemes, the load steps should be varied according to the local curvature of the load-displacement curve. Since in general the actual structural response is not known in advance, the need for an automatic scheme for determining the size of the load increment is obvious.

A method is presented for controlling the step lengths so that the difference in the displacement norm between the actual load-deformation curve and the initial tangent is the same for each load step. The method is easily explained using Figure 9, which illustrates the relation between a load factor λ and a *modified Euclidean norm* of displacement increments between consecutive equilibrium configurations. Since the actual curve is not known in advance, a second degree parabola is used as extrapolation curve, and the constant truncation error is measured as shown in Figure 9. The parabola passes through A and B on the actual curve with a common tangent direction at point B. The approximation is therefore based on the true initial tangent.

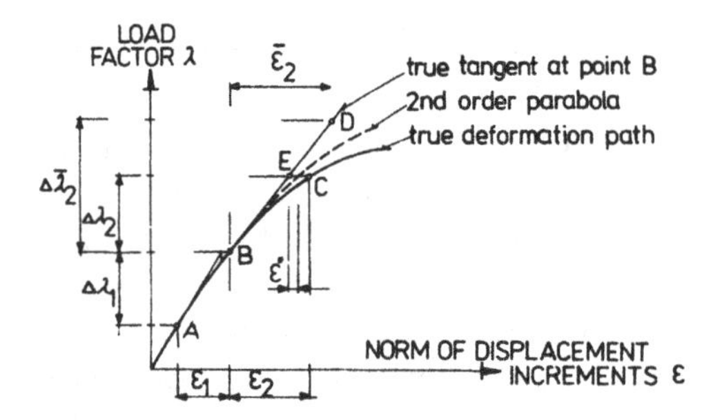

Fig. 9. Automatic load incrementation.

To start the procedure for the unloaded structure, the point A at the origin of the first load step has to be prescribed. The value of the displacement norm may either be given as input or computed as the truncation error for the prescribed load step.

The automatic incrementation procedure for the second and following load steps will be:

— calculate displacement increments $\Delta \boldsymbol{r}_2 = \mathbf{K}_I^{-1} \Delta \bar{\boldsymbol{R}}_2$ for an arbitrary load increment $\Delta \boldsymbol{R}_2$ corresponding to an increment $\Delta \bar{\lambda}_2$ in the load factor. $\mathbf{K}_I$ is the incremental stiffness matrix with corresponding tangent slope γ expressed as

$$\gamma = \Delta \bar{\lambda}_2 / \bar{\varepsilon}_2 , \tag{1.47}$$

— use the second degree parabola through A and B and the predefined value of the constant truncation error ε^* to calculate the increment of the load parameter $\Delta \lambda_2$

$$\Delta \lambda_2 = \Delta \lambda_1 \sqrt{\frac{\gamma \varepsilon^*}{\lambda_1 - \gamma \varepsilon_1}} \leqslant \Delta \lambda_{max} , \tag{1.48}$$

where $\Delta \lambda_{max}$ is a prescribed limit related $\Delta \lambda_1$,

— scale $\Delta \bar{\boldsymbol{R}}_2$ and $\Delta \boldsymbol{r}_2$ to find point E on the tangent, corresponding to the vectors of load and displacement increments

$$\Delta \boldsymbol{R}_2 = \frac{\Delta \lambda_2}{\Delta \bar{\lambda}_2} \Delta \bar{\boldsymbol{R}}_2 , \quad \Delta \boldsymbol{r}_{2.0} = \frac{\Delta \lambda_2}{\Delta \bar{\lambda}_2} = \Delta \bar{\boldsymbol{r}}_2 , \tag{1.49}$$

where the subscript zero indicates the first approximation before the equilibrium iterations,

— perform the equilibrium iterations of load level represented by $\Delta \boldsymbol{R}_2$ to reach the point C on the load-deformation curve with repetition of procedure for consecutive load steps.

The nonlinear response of multidegrees of freedom systems is commonly illustrated by load-displacement diagrams for a so-called representative degree of freedom. Such diagrams are sometimes inadequate in that they do not give all information about the actual mode of deformation sustained by the structure.

For characterizing the over-all behaviour of multidegrees of freedom systems the parameter termed as the *"current stiffness parameter"* has been suggested in [8]. The parameter is defined for load increment number i

$$S_{p,i} = \frac{\dfrac{\Delta r_1^T}{\Delta p_1} R_{ref}}{\dfrac{\Delta r_i^T}{\Delta p_i} R_{ref}} = \frac{\Delta p_i \Delta r_1^T}{\Delta p_1 \Delta r_i^T}, \tag{1.50}$$

where p_1 or p_i are loading parameters. Observe that the current stiffness parameter has the initial value of 1 for any nonlinear problem. A "softening" system is characterized by a value of S_p less than 1 while a system with gradually increasing stiffness will have S_p greater than 1.

Graphical representations of the current stiffness parameter are sketched

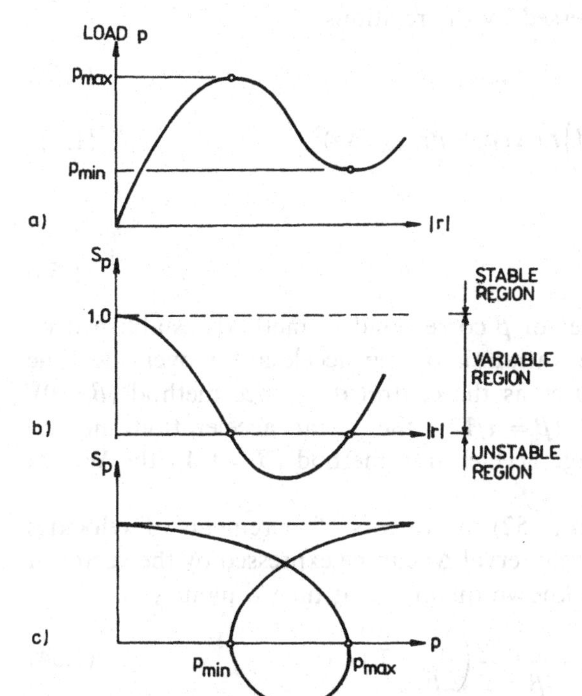

Fig. 10. Current stiffness parameter for snap-trough problem.

in Figure 10. As an illustrative example a snap-through problem is considered (Figure 10a). The variation of S_p as a function of the displacement norm $|r|$ is shown in Figure 10b. The unstable region is characterized by a negative value of S_p. The points of intersection with the $|r|$-axis facilitate the identification of the extremum points. A similar plot of S_p versus the load parameter p (Figure 10c) may be useful to accurately estimate the value of the load parameter at these points. The above properties of the current stiffness parameter imply that S_p can also be used to control the solution algorithm for limit point behaviour.

1.10. Time Integration of Incremental Equations of Motion

The incremental schemes developed for direct time integration of the equations of motion are based on the Newmark β-family of integration operators [56], including the Wilson Θ-modification [119]. Thereby a series of operators with different characteristics is available using the same computer code. The constant average acceleration method corresponding to $\beta = 1/4$ is perhaps the most commonly used method for time integration in structural dynamics problems.

With the Θ-modification included, the vectors of displacements and velocities at time $t + \Delta\tau$ are expressed by the relations

$$\dot{\boldsymbol{r}}_{t+\Delta\tau} = \dot{\boldsymbol{r}}_t + (1 - \gamma)\ddot{\boldsymbol{r}}_t\Delta\tau + \gamma\ddot{\boldsymbol{r}}_{t+\Delta\tau}\Delta\tau, \tag{1.51}$$

$$\boldsymbol{r}_{t+\Delta\tau} = \boldsymbol{r}_t + \dot{\boldsymbol{r}}_t\Delta\tau + \left(\frac{1}{2} - \beta\right)\ddot{\boldsymbol{r}}_t(\Delta\tau)^2 + \beta\ddot{\boldsymbol{r}}_{t+\Delta\tau}(\Delta\tau)^2, \tag{1.52}$$

where

$$\Delta\tau = \Theta\Delta t, \quad \Theta \geqslant 1.0. \tag{1.53}$$

For $\gamma = 1/2$ different values of β correspond to methods established by various assumptions for the variation of the acceleration over the time interval Δt, defining procedures as the central difference method ($\beta = 0$), the Fox—Goodwin method ($\beta = 1/12$), the linear acceleration method ($\beta = 1/6$), the constant average acceleration method ($\beta = 1/4$), the Wilson Θ-method ($\beta = 1/6$), etc.

From equations (1.51) and (1.52) the vectors of increments of velocities and accelerations over the time interval $\Delta\tau$ can be expressed by the vector of displacement increments and known quantities at time t, namely

$$\Delta\dot{\boldsymbol{r}}_t = \dot{\boldsymbol{r}}_{t+\Delta\tau} - \dot{\boldsymbol{r}}_t = \frac{\gamma}{\beta\Delta\tau}\Delta\boldsymbol{r}_t - \frac{\gamma}{\beta}\dot{\boldsymbol{r}}_t - \left(\frac{\gamma}{2\beta} - 1\right)\Delta\tau\ddot{\boldsymbol{r}}_t, \tag{1.54}$$

$$\Delta \ddot{r}_t = \ddot{r}_{t+\Delta\tau} - \ddot{r}_t = \frac{1}{\beta(\Delta\tau)^2}\Delta r_t - \frac{1}{\beta\Delta\tau}\dot{r}_t - \frac{1}{2\beta}\ddot{r}_t. \tag{1.55}$$

By rewriting expression (1.16) for dynamic equilibrium at time $t + \Delta\tau$ and inserting equations (1.54) and (1.55), there will result an incremental matrix equation in displacement increments over the time interval $\Theta\Delta t$

$$\mathbf{K}_t^{\text{eff}}\Delta r_t = \Delta R_t^{\text{eff}}, \tag{1.56}$$

where effective stiffness and load increment vectors are defined by the relations

$$\mathbf{K}_t^{\text{eff}} = \frac{1}{\beta(\Delta\tau)^2}\mathbf{M}_t + \frac{\gamma}{\beta\Delta\tau}\mathbf{C}_t + \mathbf{K}_{\text{I},t}, \tag{1.57}$$

$$\Delta R_t^{\text{eff}} = \Delta R_t + \mathbf{M}_t\left(\frac{1}{\beta\Delta\tau}\dot{r}_t + \frac{1}{2\beta}\ddot{r}_t\right) + \mathbf{C}_t\left[\frac{\gamma}{\beta}\dot{r}_t + \left(\frac{\gamma}{2\beta} - 1\right)\Delta\tau\ddot{r}_t\right]. \tag{1.58}$$

The Wilson Θ-method interpolates the accelerations at time $t + \Delta t$ between the values at times $t + \Delta\tau$

$$\ddot{r}_{t+\Delta t} = \ddot{r}_t + \frac{1}{\Theta}(\ddot{r}_{t+\Delta\tau} - \ddot{r}_t). \tag{1.59}$$

The integration procedure may now be defined by the following sequence of operations:

— $\mathbf{K}_t^{\text{eff}}$ and ΔR_t^{eff} from equations (1.57) and (1.58),
— Δr_t from equation (1.56),
— $\ddot{r}_{t+\Delta\tau}$ from equation (1.55),
— $\ddot{r}_{t+\Delta t}$ from equation (1.59) (redundant for $\Theta = 1{,}0$),
— $\dot{r}_{t+\Delta t}$ from equation (1.51), for $\Delta\tau = \Delta t$,
— $r_{t+\Delta t}$ from equation (1.52), ($\Theta < 1.0$),
— $r_{t+\Delta t} = r_t + \Delta r_t$

Equilibrium iterations may be performed according to the modified equation (1.16) via

$$({}^{i-1}V^{\text{I}}_{t+\Delta\tau} + {}^{i}V^{\text{I}}_{t+\Delta\tau}) + ({}^{i-1}V^{\text{D}}_{t+\Delta\tau} + {}^{i}V^{\text{D}}_{t+\Delta\tau}) + + ({}^{i-1}V^{\text{S}}_{t+\Delta\tau} + {}^{i}V^{\text{S}}_{t+\Delta\tau}) = R_{t+\Delta\tau}. \tag{1.60}$$

The sums of forces in each pair of brackets represent the total intertia, damping and elastic forces, respectively, after iteration number i. Introducing additional displacement increments ${}^{i}\Delta_r$ by

$${}^{i}\Delta r_t = {}^{i-1}\Delta r_t - {}^{i}\Delta_r, \tag{1.61}$$

into equation (1.55), the inertia force is given as

$$ {}^{i}\Delta \boldsymbol{V}^{\mathrm{I}}_{t+\Delta\tau} = \frac{\gamma}{\beta(\Delta\tau)^2} \mathbf{M}_t\, {}^{i}\boldsymbol{\Delta}_r . \tag{1.62} $$

Correspondingly, equation (1.55) yields

$$ {}^{i}\Delta \boldsymbol{V}^{\mathrm{D}}_{t+\Delta\tau} = \frac{\gamma}{\beta\Delta\tau} \mathbf{C}_t\, {}^{i}\boldsymbol{\Delta}_r . \tag{1.63} $$

The dynamic equilibrium at time $t + \Delta\tau$ given by equation (1.60) can then be rewritten in the form

$$ \left(\frac{1}{\beta(\Delta\tau)^2}\mathbf{M}_t + \frac{\gamma}{\beta\Delta\tau}\mathbf{C}_t + {}^{i}\mathbf{K}_{\mathrm{I},t}\right){}^{i}\boldsymbol{\Delta}_r = $$

$$ = \boldsymbol{R}_{t+\Delta\tau} - ({}^{i-1}\boldsymbol{V}^{\mathrm{I}}_{t+\Delta\tau} + {}^{i-1}\boldsymbol{V}^{\mathrm{D}}_{t+\Delta\tau} + {}^{i-1}\boldsymbol{V}^{S}_{t+\Delta\tau}), \tag{1.64} $$

where the internal forces at the start of iteration step are given by

$$ {}^{i-1}\boldsymbol{V}^{\mathrm{I}}_{t+\Delta\tau} = \mathbf{M}_t\, {}^{i-1}\ddot{\boldsymbol{r}}_{t+\Delta\tau}, \tag{1.65} $$

$$ {}^{i-1}\boldsymbol{V}^{\mathrm{D}}_{t+\Delta\tau} = \mathbf{C}_t\, {}^{i-1}\dot{\boldsymbol{r}}_{t+\Delta\tau}, \tag{1.66} $$

$$ {}^{i-1}\boldsymbol{V}^{\mathrm{S}}_{t+\Delta\tau} = \mathbf{K}_t\, {}^{i-1}\boldsymbol{r}_{t+\Delta\tau}. \tag{1.67} $$

The incremental stiffness may be recomputed for each iteration step corresponding to the Newton—Raphson procedure or may be kept constant during the iterations. Equilibrium is reached when the modified Euclidean norm of ${}^{i}\boldsymbol{\Delta}_r$ becomes less than a predefined value.

The sequence of equilibrium iterations enters the itegration procedure and the iteration loop number i is then defined by the following subsequent calculations:

— ${}^{i-1}\dot{\boldsymbol{r}}_{t+\Delta\tau}$ from equation (1.54),
— ${}^{i-1}\boldsymbol{r}_{t+\Delta\tau} = \boldsymbol{r}_t + {}^{i-1}\Delta\boldsymbol{r}_t$,
— ${}^{i-1}\boldsymbol{V}^{\mathrm{I}}_{t+\Delta\tau}$, ${}^{i-1}\boldsymbol{V}^{\mathrm{D}}_{t+\Delta\tau}$, ${}^{i-1}\boldsymbol{V}^{\mathrm{S}}_{t+\Delta\tau}$ from equations (1.65), (1.66) and (1.67),
— ${}^{i}\boldsymbol{\Delta}_r$ from equation (1.64) regarding the corresponding Euclidean norm,
— ${}^{i}\Delta\boldsymbol{r}_t$ and ${}^{i}\ddot{\boldsymbol{r}}_{t+\Delta\tau}$ from equations (1.62) and (1.55).

In case of positive equilibrium value of the norm of parameter ${}^{i}\boldsymbol{\Delta}_r$ the integration procedure continues by step 4. In other case the realization of further iteration cycles is inevitable [70].

The obtainable accuracy of the present time-integration methods will depend on the loading and physical properties of studied structural simulation and particularly on the length of applied time step. The integration is considered as a transformation of displacements, velocities and accelerations over the time interval Δt. The transformation is defined by the matrices **A** and **L** in the relation

$$ \boldsymbol{h}_{t+\Delta\tau} = \mathbf{A}\boldsymbol{h}_t + \mathbf{L}\boldsymbol{p}_{t+\Delta\tau}, \tag{1.68} $$

where displacements, velocities and accelerations have been assembled in the vectors $\boldsymbol{h}_t$ and $\boldsymbol{h}_{t+\Delta t}$ and the matrices $\mathbf{A}$ and $\mathbf{L}$ denote the integration approximation and load operators, respectively. The vector $\boldsymbol{p}_{t+\Delta t}$ represents the loading at time $t + \Theta\Delta t$. The approximation operator is defined to be stable if the amplitude values of $\boldsymbol{h}_n$ do not grow unboundedly as the number of time steps n is increased for arbitrary initial vector $\boldsymbol{h}_0$. This is assured when the spectral radius of the approximation operator satisfies relation

$$\varrho(\mathbf{A}) = 1 . \tag{1.69}$$

The spectral radius is defined by

$$\varrho(\mathbf{A}) = \max |\lambda_i| , \tag{1.70}$$

where λ_i is the i-th eigenvalue of the approximation operator $\mathbf{A}$. The stability is unconditional if equation (1.69) is satisfied for any step length Δt. Otherwise, the stability depends on the ratio $\Delta t/T$, where T is the eigenperiod of studied simulation. The Newmark generalized acceleration operator is unconditionally stable for $\beta \leqslant 1/4$. The stability of the Wilson modification of the linear acceleration methods depends on the value of the parameter Θ. The relation between $\Theta_{\min}$ and $\Delta t/T$ is shown in Figure 11. The lowest value for the parameter $\Theta_{\min}$ is obtained for the time step ratio $\sqrt{3}/\pi = 0.551$, corresponding to the stability limit for the linear acceleration method. With increasing time step ratio, the $\Theta_{\min}$ value increases asymptotically towards $\Theta_{\min} = 1.5$. Unconditional stability of the incremental formulation of the Wilson Θ-method is thus obtained for $\Theta = 1.5$. The corresponding value for the direct formulation is $\Theta = 1.37$ [6].

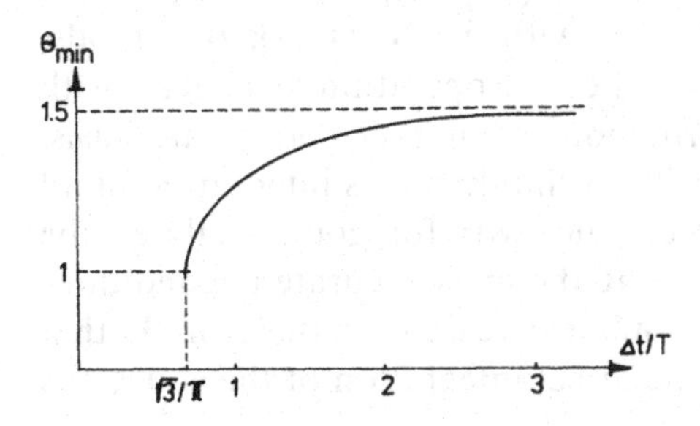

Fig. 11. Instability limit $\Theta_{\min}$ versus time step ratio $\Delta t/T$ for the incremental form of the Wilson method.

For the nonlinear analysis case the approximation operator $\mathbf{A}$ will be displacement-dependent. Statements on the boundedness of its eigenvalues will therefore be not applicable for proving stability. Since mathematical evidence of stability properties for temporal operators in the nonlinear range is lacking, various authors have made numerical studies on this subject [80, 116]. The results indicate that when the Newton—Raphson scheme is used to assure dynamic equilibrium at each step, the integration methods seem to carry their stability properties over to the nonlinear range. For the constant average acceleration method some analytical studies [7, 35, 36] also conclude that this operator will be unconditionally stable for many nonlinear problems. A numerical algorithm is defined to be stable if the sum of kinetic and internal energies is bounded within each time step relative to the external work and kinetic and external energies in the previous time step, viz.

$$(\boldsymbol{T}_{i+1} + \boldsymbol{U}_{i+1}) \leqslant (1 + \xi)(\boldsymbol{T}_i + \boldsymbol{U}_i + \Delta \boldsymbol{W}), \tag{1.71}$$

where ξ is an arbitrary small constant and discrete energies are defined by the relations

$$\boldsymbol{U}_{i+1} = \boldsymbol{U}_i + \frac{1}{2}\Delta \boldsymbol{r}_i^{\mathrm{T}}(\boldsymbol{V}_{i+1}^{\mathrm{S}} + \boldsymbol{V}_i^{\mathrm{S}}) \quad \text{(internal energy)}, \tag{1.72}$$

$$\boldsymbol{W}_{i+1} = \boldsymbol{W}_i + \frac{1}{2}\Delta \boldsymbol{r}_i^{\mathrm{T}}(\boldsymbol{R}_{i+1} + \boldsymbol{R}_i) \quad \text{(external work)}, \tag{1.73}$$

$$\boldsymbol{T}_i = \frac{1}{2}\dot{\boldsymbol{r}}_i^{\mathrm{T}}\mathbf{M}_i\dot{\boldsymbol{r}}_i \quad \text{(kinetic energy)}. \tag{1.74}$$

The convergence criterion

$$\Delta \boldsymbol{r}_i^{\mathrm{T}}(\Delta \boldsymbol{V}_{i+1} + \Delta \boldsymbol{V}_i) < \varepsilon(\boldsymbol{U}_i + \boldsymbol{T}_i), \tag{1.75}$$

will be sufficient to prove stability for the constant average acceleration integration operator also in the nonlinear range. The convergence limit ε will directly influence the energy bounds.

The computational procedure for direct step-by-step integration of the incremental dynamic equilibrium equations with the Newmark β-methods, including the Wilson Θ-method, is termed direct operation as it deals with equations of motion without transformation to another coordinate basis. This concept corresponds mathematically to simultaneous integration of all modes using the same time step. In accordance with foregoing analyses, the constant average acceleration method being the most accurate unconditionally stable method in linear as well as nonlinear regions, is therefore both a convenient and favourable choice for the direct integration of the equations of motion.

Chapter 2

Transfer Matrix Method — Linear Approach

2.1. Basic Concepts

The transfer matrix method is a variant of the generalized discrete concepts for solving linear and nonlinear problems of mechanics. In combination with algorithms of generalized analysis of motion outlined in preceding chapter, this method may be regarded as effective tool for complex numerical analysis of structural systems.

The development of the transfer matrix method was initiated first of all by the efforts to obtain vectorized line algorithms capable of dealing with the modern vector computer codes.

One way to model the complex structures is by the finite element method. Whereas the finite element method lends itself to modelling very complex structures, it has the major disadvantage that it often requires a very large number of degrees of freedom (sometimes tens of thousands) to obtain accurate estimates. Because it is impractical to work with such a large number of degrees of freedom, the methods reducing the number of degrees of freedom are desirable. Quite often the complex structure can be conveniently regarded as an assemblage of a few simpler substructures. For such cases it is possible to construct a mathematical model containing a substantially reduced number of degrees of freedom compared with the finite element method. The algorithm may be produced by the chaining of matrix operations and thus obtain the vectorized programmatory code in accordance with the requirements of optimal computer calculation. One of the methods utilizing this principle is termed the transfer matrix method.

The method is based on theoretical principles developed in references [19, 23, 42, 48, 122, 126]. Such principles are further developed, modified and adopted for discrete analyses of complex structural configurations in the next three chapters.

The transfer matrix method principally consists of four typical matrix expressions defined by:

1. transfer matrix,
2. nodal matrix,

3. matrix of initial parameters,
4. matrix of boundary conditions.

The method works with variable matrix blocks, allowing the transfer of state vectors over the multitude of elements or substructures in the assumed structural simulation. Such matrices are termed the transfer and nodal matrices. The transfer matrix performs the transfer of state vector from initial into the end points of an arbitrary element of the assumed simulation. The nodal matrix allows the transfer of a state vector over the nodal points coupling together adjacent elements or substructures to form the whole structure. The coupling process requires that certain geometric compatibility conditions are satisfied at each point of an internal boundary between two adjacent substructures. In nodal points may be further concentrated the physical discontinuities; i.e. the elastic or rigid supports, inertial mass and loading parameters, structural couplings, etc. The complete set of boundary conditions of a studied discrete simulation is defined by the matrices of initial parameters and boundary conditions. The matrix of initial parameters defines the boundary conditions in initial point of a structure, whereas the matrix of boundary conditions defines the conditions in the end point of an assumed line simulation.

The geometry of a structural configuration is modelled by matrix multiplications during motion of transfer or nodal matrices over the simulation. The stiffness and internal damping parameters which are present in corresponding generalized equation of motion (1.16) are incorporated in the transfer matrices of individual elements. The inertial mass and external damping parameters as well as the loading influences appearing in equation (1.16) are summed up in nodal matrices of the analysed simulation. The intermediate elastic or rigid supports are also embodied in the nodal matrices. Further, the transfer matrices may contain the transformation terms for modelling of real geometry of configuration. Elastic or rigid boundary conditions of a total structural simulation are modelled in the matrices of initial parameters and in the matrices of boundary conditions. Both of these matrices are interactively coupled.

This principle can be applied to linear as well as nonlinear incremental analyses of one-dimensional or multidimensional structural configurations, using the discretization in space and time. The principal advantages of this concept are:

1. simple modelling of updated complex simulations in arbitrary linear and nonlinear regions of their static and dynamic behaviour,
2. dynamic modelling of various intermediate and total boundary conditions which may be variable during deformation process,
3. statically redundant components are directly implemented which allows

the analyses of highly redundant systems using the same algorithm of line solution as for statically determinant structures,

4. chaining of the calculation operations that allows the total vectorization of programmatory code in accordance with the requirements of vector computer techniques,

5. small demands on the computer time and storage owing to the small systems of algebraic equations to be solved.

As disadvantage of this method are to be mentioned some problems with numerical instability of the computer calculation, especially in advanced nonlinear regions of complex simulations. Therefore some procedures for overcoming numerical difficulties are mentioned in this chapter which have proved to be effective for the solution of such problems.

In this chapter the principles of the transfer matrix method (TMM) are outlined when adopted for the solution of structural configurations in linear regions of their static behaviour. Attention is further paid to the solution of periodic and aperiodic responses. In the following chapter TMM is modified, resulting in the so-called FETM-variant, which is adopted for the solution of nonlinear problems. Both methods are presented in their discrete modifications which are suitable for optimal numerical solutions of advanced linear and nonlinear problems of complex structural configurations.

2.2. Algorithm of the Method

The algorithm of the transfer matrix method is defined in two principal steps. During the first step the unknown initial parameters defined by the matrix of initial parameters are transferred using the matrix multiplications of transfer and nodal matrices into the end point of the applied simulation. The boundary conditions in the end point, implemented in the matrix of boundary conditions, define the set of algebraic equations for determination of the unknown initial parameters. When solving the eigenvalue problem, the first step defines the corresponding eigenvalue matrix. In the second step the calculated initial parameters are put into the state vector in initial point of simulation and its repeated multiplications with transfer and nodal matrices determine the set of resulting state vectors, stress and strain components in nodal points of the used discrete model.

The described algorithm of the transfer matrix method will be explained in detail using the simple beam system subjected to the action of static and dynamic loads as an illustration. The elementary bending behaviour of the beam system is studied with elevation and discretization schemes as shown in Figure 12.

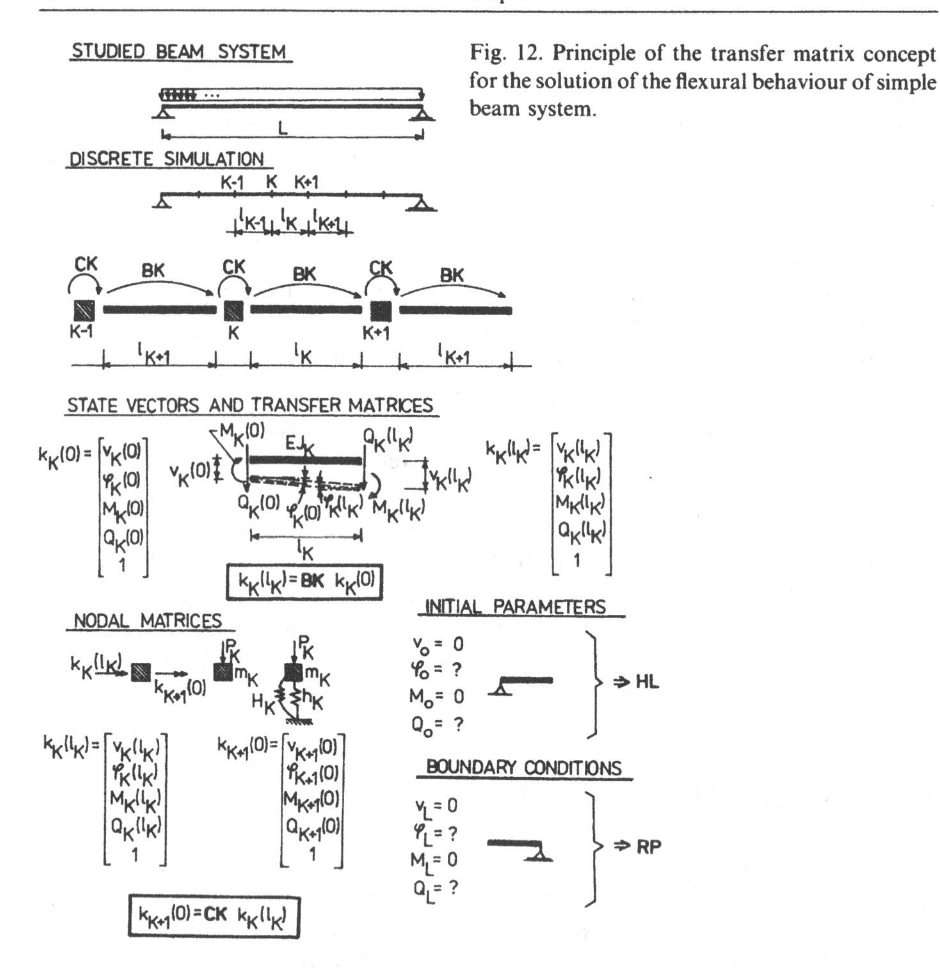

Fig. 12. Principle of the transfer matrix concept for the solution of the flexural behaviour of simple beam system.

Static Analysis

The state vector of bending components in the initial point of an arbitrary one-dimensional element of the discrete simulation assumed in Figure 12 is given by

$$\boldsymbol{k}_K(0) = (v_K(0),\ \varphi_K(0),\ M_K(0),\ Q_K(0),\ 1)^T, \tag{2.1}$$

where v_K, φ_K, M_K and Q_K are the displacement, slope, moment and lateral force in node K, respectively. The last term of the vector $\boldsymbol{k}$ is a preliminary unit and has importance for incorporation of external loading or deformation components into calculation process.

When deriving the corresponding transfer matrix **BK**, the known transformation equations for the couplings of components of the state vector $\boldsymbol{k}_K$ between initial and end points of the studied element can be written as

$$v_K(l_K) = v_K(0) + l_K\varphi_K(0) - \frac{l_K^2}{2EJ_K} M_K(0) - \frac{l_K^3}{6EJ_K} Q_K(0) \ , \tag{2.2}$$

$$\varphi_K(l_K) = \varphi_K(0) - \frac{l_K}{EJ_K} M_K(0) - \frac{l_K^2}{2EJ_K} Q_K(0) \ , \tag{2.3}$$

$$M_K(l_K) = M_K(0) + l_K Q_K(0) \ , \tag{2.4}$$

$$Q_K(l_K) = Q_K(0) \ , \tag{2.5}$$

$$1 = 1, \tag{2.6}$$

where l_K is the length of the element and EJ_K is its stiffness.

These equations may be written as the vector—matrix relation

$$\boldsymbol{k}_K(l_K) = \mathbf{BK}\ \boldsymbol{k}_K(0), \tag{2.7}$$

where **BK** is the transfer matrix of the studied problem, given by

$$\mathbf{BK} = \begin{bmatrix} 1 & l_K & -\dfrac{l_K^2}{2EJ_K} & -\dfrac{l_K^3}{6EJ_K} & 0 \\ & 1 & -\dfrac{l_K}{EJ_K} & -\dfrac{l_K^2}{2EJ_K} & 0 \\ & & 1 & l_K & 0 \\ & & & 1 & 0 \\ & & & & 1 \end{bmatrix}, \tag{2.8}$$

which produces the desired couplings between state vectors.

As has been stated earlier, in nodes of adjacent elements the loading or inertial parameters may be concentrated as well as various elastic or rigid intermediate supports. The transfer of the state vector over the nodal points of adjacent elements is performed using nodal matrices. The coupling equations in a single node of two elements of the simulation under investigation are

$$v_{K+1}(0) = v_K(l_K) \ , \tag{2.9}$$

$$\varphi_{K+1}(0) = \varphi_K(l_K) \ , \tag{2.10}$$

$$M_{K+1}(0) = M_K(l_K) \ , \tag{2.11}$$

$$Q_{K+1}(0) = Q_K(l_K) \ , \tag{2.12}$$

$$1 = 1, \tag{2.13}$$

which, written in the vector-matrix form used earlier

$$\begin{bmatrix} v_{K+1}(0) \\ \varphi_{K+1}(0) \\ M_{K+1}(0) \\ Q_{K+1}(0) \\ 1 \end{bmatrix} = \begin{bmatrix} 1 & & & & \\ & 1 & & & \\ & & 1 & & \\ & & & 1 & \\ & & & & 1 \end{bmatrix} \begin{bmatrix} v_K(l_K) \\ \varphi_K(l_K) \\ M_K(l_K) \\ Q_K(l_K) \\ 1 \end{bmatrix}, \tag{2.14}$$

determines the nodal matrix **CK** as

$$\mathbf{CK} = \begin{bmatrix} 1 & & & & \\ & 1 & & & \\ & & 1 & & \\ & & & 1 & \\ & & & & 1 \end{bmatrix}. \tag{2.15}$$

When in the node under investigation the external force $\pm P_K$ is present, equation (2.12) is modified via

$$Q_{K+1} = Q_K(l_K) \pm P_K, \tag{2.16}$$

and the nodal matrix assumes form

$$\mathbf{CK} = \begin{bmatrix} 1 & & & & \\ & 1 & & & \\ & & 1 & & \\ & & & 1 & \pm P_K \\ & & & & 1 \end{bmatrix}, \tag{2.17}$$

with the nonzero force term in the last column. Other external state components (initial deformation, slope or moment) concentrated in this node may similarly be incorporated into the last column of the nodal matrix. If in addition, the elastic support with stiffness characteristics h_K and H_K is located in this node, as shown in Figure 12, equations (2.11) and (2.12) become

$$M_{K+1} = \pm H_K \varphi_K(l_K) + M_K(l_K), \tag{2.18}$$

$$Q_{K+1} = \pm h_K v_K(l_K) + Q_K(l_K) \pm P_K, \tag{2.19}$$

and the nodal matrix is given by

$$\mathbf{CK} = \begin{bmatrix} 1 & & & & \\ & 1 & & & \\ & \pm H_K & 1 & & \\ \pm h_K & & & 1 & \pm P_K \\ & & & & 1 \end{bmatrix}. \tag{2.20}$$

For the case of rigid support the stiffness characteristics h_K or H_K tend to infinity (10^{20} and greater, depending on the type of computer). Hence, the degree of static redundancy of the structure does not influence the line algorithm of the solution. Generally, the significance of nodal matrix **CK** is defined as

$$\boldsymbol{k}_{K+1}(0) = \mathbf{CK}\ \boldsymbol{k}_K(l_K). \tag{2.21}$$

The nodal matrix, like the transfer matrix, allows the general variability of all transported physical parameters in individual nodal points.

Attention will be paid next to the implementation of boundary conditions into the algorithm of the TMM. The boundary conditions defined in the initial point of the studied discrete simulation are given by

$$v_0(0) = 0, \tag{2.22}$$

$$\varphi_0(0) = ?, \tag{2.23}$$

$$M_0(0) = 0, \tag{2.24}$$

$$Q_0(0) = ?, \tag{2.25}$$

where $\varphi_0(0)$ and $Q_0(0)$ are the unknown initial parameters.

The state vector $\boldsymbol{k}_0$ and the vector of initial parameters $\boldsymbol{u}_0$ in the initial point of the applied model are then given by

$$\boldsymbol{k}_0 = \begin{bmatrix} v_0(0) \\ \varphi_0(0) \\ M_0(0) \\ Q_0(0) \\ 1 \end{bmatrix}, \quad \boldsymbol{u}_0 = \begin{bmatrix} \varphi_0(0) \\ Q_0(0) \\ 1 \end{bmatrix}. \tag{2.26}$$

The relation between these two vectors is given by

$$\boldsymbol{k}_0 = \mathbf{HL}\ \boldsymbol{u}_0, \tag{2.27}$$

where **HL** is the matrix of initial parameters.

The matrix of initial parameters for the given boundary conditions is then written as

$$\mathbf{HL} = \begin{bmatrix} 0 & 0 & 0 \\ 1 & 0 & 0 \\ 0 & 0 & 0 \\ 0 & 1 & 0 \\ 0 & 0 & 1 \end{bmatrix}. \tag{2.28}$$

The boundary conditions assumed in the end point of the studied simulation

$$v_n(L) = 0, \tag{2.29}$$

$$\varphi_n(L) = ?, \tag{2.30}$$

$$M_n(L) = 0, \tag{2.31}$$

$$Q_n(L) = ?, \tag{2.32}$$

are implemented into algorithm of method over the matrix of boundary conditions **RP**, derived similarly as **HL**, and expressed as

$$\mathbf{RP} = \begin{bmatrix} 1 & 0 & 0 & 0 & 0 \\ 0 & 0 & 1 & 0 & 0 \end{bmatrix}. \tag{2.33}$$

Using the present principles, all other applicable boundary conditions may be defined including elastic supports, occurring in both end points of the studied simulation.

Owing to the aforementioned analysis, the four typical matrices utilized in the algorithm of the TMM are then defined as **BK**, **CK**, **HL** and **RP**. Such matrices allow the transfer operations amongst the state vectors $\boldsymbol{k}_i$.

The algorithm of the TMM in a typical two-step concept is defined by two following operations:

step 1 — determination of unknown initial parameters

$$\left(\mathbf{RP}\left(\left(\prod_{i=1}^{n}(\mathbf{CK}\,\mathbf{BK})\right)\mathbf{HL}\right)\right)u_0 \Rightarrow \begin{matrix} \varphi_0(0), \\ Q_0(0), \end{matrix} \tag{2.34}$$

step 2 — input of calculated initial parameters into the initial state vector $\boldsymbol{k}_0$ and determination of the state vectors in each node as

$$\boldsymbol{k}_i = \prod_{i=1}^{n}(\mathbf{CK}\,\mathbf{BK})\,\boldsymbol{k}_0, \tag{2.35}$$

i.e.

$$\boldsymbol{k}_1 = (\mathbf{CK}\,\mathbf{BK})\,\boldsymbol{k}_0, \tag{2.36}$$

$$\boldsymbol{k}_2 = (\mathbf{CK}\,\mathbf{BK})\,\boldsymbol{k}_1, \tag{2.37}$$

$$\boldsymbol{k}_3 = (\mathbf{CK}\,\mathbf{BK})\,\boldsymbol{k}_2, \tag{2.38}$$

$$\vdots$$

$$\boldsymbol{k}_n = (\mathbf{CK}\,\mathbf{BK})\,\boldsymbol{k}_{n-1}. \tag{2.39}$$

The present line concept may be adopted for the analysis of one-dimensional or multidimensional structural simulations as will be shown later.

Dynamic Analysis

The analysis of the dynamic behaviour of the present beam system is focused on three typical problems, viz.

a) natural vibration,
b) harmonically excited damped vibration,
c) periodic or aperiodic damped time response.

For the last item two concepts will be applied based on two different schemes of time discretization. The first concept incorporates the Fourier integral transformation into the algorithm of the TMM. The second concept is based on the combination of the TMM with Wilson's method of direct time integration. Both concepts are compared and numerically evaluated. In sequence, the algorithms of all of these dynamic analyses will be described.

a) Natural Vibration

For the solution of the eigenvalue problem of natural vibration the same algorithm of the TMM as described above is used with the following modifications:

— only the first step of analysis is performed,
— the inertial parameters, concentrated in nodal points of the assumed simulation, are implemented into equations (2.11) and (2.12) via

$$M_{K+1}(0) = -\tau\omega^2\varphi_K(l_K) + M_K(l_K), \tag{2.40}$$

$$Q_{K+1}(0) = -m\omega^2 v_K(l_K) + Q_K(l_K), \tag{2.41}$$

where τ is the mass moment of inertia, m is the inertial mass and ω is the frequency of natural vibration.

— the nodal matrix **CK** is then modified as

$$\mathbf{CK} = \begin{bmatrix} 1 & & & & \\ & 1 & & & \\ & -\tau\omega^2 & 1 & & \\ -m\omega^2 & & & 1 & \\ & & & & 1 \end{bmatrix}. \tag{2.42}$$

All other matrices of the algorithm of the transfer matrix method, derived for the present simple bending analysis and termed as **BK**, **RP** and **HL**, remain unchanged.

The modified algorithm of the first step concept of the TMM, in accordance with equation (2.34) can be written as

$$\mathbf{UL} = \mathbf{RP}\left(\left(\prod_{i=1}^{n}(\mathbf{CK}\,\mathbf{BK})\right)\mathbf{HL}\right). \tag{2.43}$$

The matrix **UL** may be split into the matrix of coefficients ε_i and the vector γ_i (last column of matrix **UL**)

$$\mathbf{UL} = (\varepsilon_i, \gamma_i). \tag{2.44}$$

When calculating the natural frequencies of an unloaded structure, the vector γ_i equals zero and the problem leads to the determination of eigenvalues and eigenvectors following from the singularity condition of the matrix $\boldsymbol{\varepsilon}_i$.

b) Harmonically Induced Damped Vibration

The modifications of the algorithm of the transfer matrix method when directly solving the harmonically excited, damped vibration of the studied beam simulation with regard to the internal or external (structural) damping, may be defined as follows:

— all calculations are performed in complex arithmetics,

— the complex moduli of elasticity containing the influences of internal damping, in accordance with equation (1.30) are incorporated into complex stiffness terms of the complex transfer matrix **BK** (see equation (2.8)),

— the complex stiffness characteristics K_B, containing the influences of external or structural damping in accordance with equation (1.32), are incorporated into the complex nodal matrix **CK**,

— when assuming the action of a harmonically variable external force $P(t)$, given by expression

$$P(t) = P_0 e^{i\Omega t}, \tag{2.45}$$

where P_0 is the amplitude and Ω is the frequency of forced vibration, in accordance with equation (2.42) the nodal matrix in such node is defined by

$$\mathbf{CK} = \begin{bmatrix} 1 & & & & \\ & 1 & & & \\ & -\tau\Omega^2 & 1 & & \\ -m\Omega^2 & & & 1 & -P_0 \\ & & & & 1 \end{bmatrix}, \tag{2.46}$$

or with consideration of external complex stiffness characteristics K_B by

$$\mathbf{CK} = \begin{bmatrix} 1 & & & & \\ & 1 & & & \\ & -\tau\Omega^2 & 1 & & \\ -m\Omega^2 - K_B & & & 1 & -P_0 \\ & & & & 1 \end{bmatrix}. \tag{2.47}$$

Each component of the state vectors in individual nodes is complex and phase shifted against exciting dynamic forces. Generally, the state vectors may be expressed in form

$$\mathbf{k}_i(z,t) = \mathbf{k}_i(z) e^{i(\Omega t + \alpha)} = \bar{\mathbf{k}}_i(z) e^{i\Omega t}, \tag{2.48}$$

where $\bar{k}_i$ is the complex value, z is the direction of longitudinal axis of the studied beam simulation, and α denotes the angle of phase shift.

All external dynamic loads, all components of the state vectors and all inertial factors can be expressed by the formula (2.48) and thus the problem of harmonically excited, damped vibrations is similar to the problem of static behaviour and may be solved using the previously described algorithm of transfer matrix method in accordance with equations (2.34) and (2.35). However, the algorithm is to be performed in complex arithmetics with consideration of the aforementioned modifications. The first step of the algorithm leads to the solution of the set of complex equations for the determination of unknown initial parameters in complex form. The calculation of the dynamic state vectors and resonance frequencies are determined directly. The preceding determination of natural frequencies and modes of vibration is not necessary. All calculated dynamic state components are determined by absolute values and by angles of phase shift in relation to the exciting forces.

EXAMPLE 1. The explained method for solving harmonically excited vibrations of structures with consideration of internal and external (structural) damping is numerically verified on the example of resonance analysis of illustrative beam system with elevation and physical parameters listed in Figure 13. The system is assumed to be subjected to the action of a harmonically variable midspan force and is provided with elastic supports as shown in Figure 13. The internal damping is considered on the basis of the complex modulus of elasticity as stated earlier. The external damping due to the friction in elastic supports is assumed by the complex spring characteristics $K_B = h_K(1 + i\eta_B)$, where η_B is the factor of external damping.

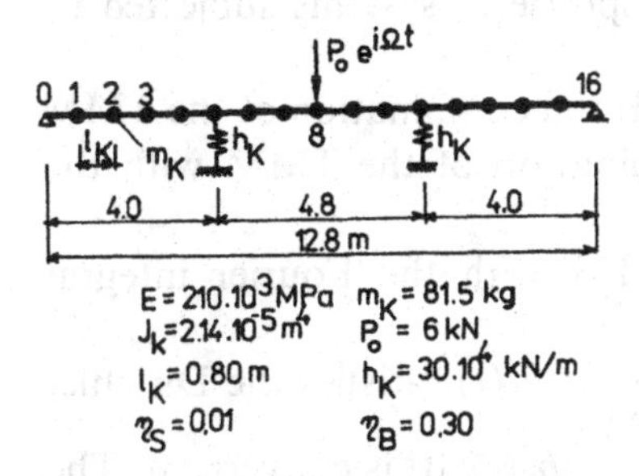

Fig. 13. Example 1 — Scheme of analysed beam system.

The stationary state of excited damped vibrations was investigated in a frequency region $\Omega \in (0, 2000)$ Hz, in which seven resonance frequencies of bending vibration were registered. Owing to the symmetrically positioned exciting force any antisymmetric resonance frequencies and antimetric forms of vibration were registered. The resonance curve with corresponding diagram of the phase shift for the midspan bending deflection in nodal point No. 8, in accordance with the scheme in Figure 13, are shown in Figure 14.

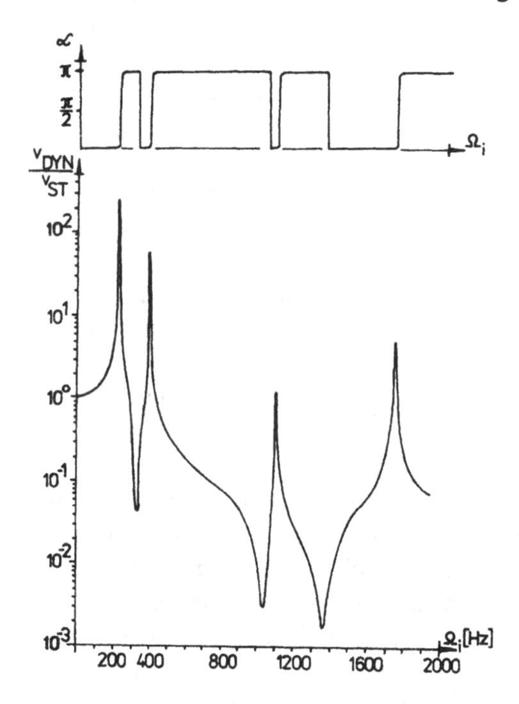

Fig. 14. Example 1 — Diagram of the angle of phase shift and resonance curve of flexural deflection midspan of studied system.

c) Periodic and Aperiodic Damped Time Response

The foregoing analysis was based on the assumption of the stationary harmonic vibrations of the studied beam system. In this section the application of the TMM for the solution of the time response of systems subjected to general time variable loads is described.

As stated earlier, two concepts are applied: the combination of the TMM with integral transformations and the combination of the TMM with the methods of direct time integration.

First of all, the combination of the TMM with the Fourier integral transformation will be presented.

Assumed is that an arbitrary function of time $b_F(t)$ satisfies the Dirichlet conditions together with the condition that $\int_{-\infty}^{\infty} |b_F(t)|\,dt$ is convergent. The following transformation equations are used

$$H_F(\omega) = \int_0^{\infty} b_F(t)\,e^{-i\omega t}\,dt\,, \tag{2.49}$$

$$b_{\mathrm{F}}(t)=\frac{1}{2\pi}\int_{-\infty}^{\infty}H_{\mathrm{F}}(\omega)\,\mathrm{e}^{\mathrm{i}\omega t}\,\mathrm{d}\omega=\frac{1}{\pi}\int_{0}^{\infty}\mathrm{Re}[H_{\mathrm{F}}(\omega)]\,\mathrm{e}^{\mathrm{i}\omega t}\,\mathrm{d}\omega\,. \tag{2.50}$$

Assumed is the excitation of the structure by the impulse shown in Figure 15a. The transform $H_{\mathrm{F}_1}(\omega)$ is given by

$$H_{\mathrm{F}_1}(\omega)=\int_{0}^{\infty}\mathrm{e}^{-at}\,\mathrm{e}^{-\mathrm{i}\omega t}\,\mathrm{d}t=\frac{1}{a+\mathrm{i}\omega}\,. \tag{2.51}$$

For the impulse in Figure 15b the transform $H_{\mathrm{F}_2}(\omega)$ is given by

$$H_{\mathrm{F}_2}(\omega)=\int_{0}^{T}\mathrm{e}^{-\mathrm{i}\omega t}\,\mathrm{d}t=\frac{1-\mathrm{e}^{-\mathrm{i}\omega t}}{\mathrm{i}\omega}\,. \tag{2.52}$$

The transforms (2.51) or (2.52) are embodied into the loading columns of nodal matrices in the nodal points in which the impulse loads are assumed to be acting.

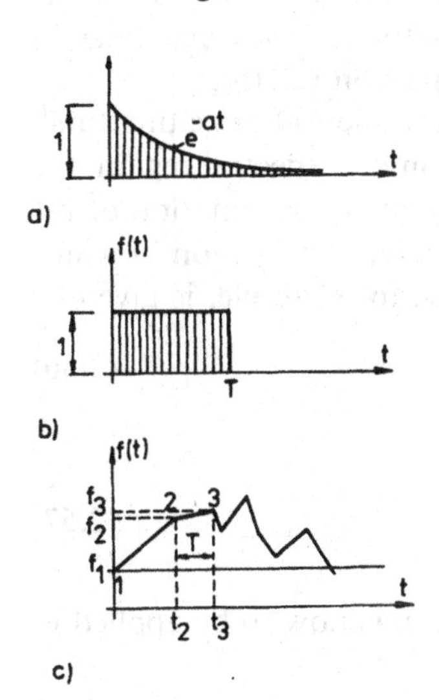

Fig. 15. Analysed impulses.

When assuming the operator

$$A=\frac{\partial}{\partial t}, \tag{2.53}$$

the transforms of inertial forces $-m(z)A^2(v(z,t))$ for both cases are given by

$$\int_{0}^{\infty}(-m(z)A^2(v(z,t)))\,\mathrm{e}^{\mathrm{i}\omega t}\,\mathrm{d}t=$$

$$=m(z)\,\omega^2 v(z,\omega)+\mathrm{i}\omega m(z)v(z,0)+m(z)A(v(z,0))\,, \tag{2.54}$$

with the assumption that $v(z, t)$ and $A(v(z, t))$, are zero in the limit $t \to \infty$. The transform of inertial forces is then given by

$$\int_0^\infty (-m(z)A^2(v(z,t)))\,e^{i\omega t}\,dt = m(z)\omega^2 v(z,\omega). \tag{2.55}$$

Equation (2.55) is included into the nodal matrix as an inertial component. The transfer matrices in their complex forms remain unchanged.

Using the TMM the transforms of functions of each of the state components within the interval $\omega \in (0, \infty)$ will be determined for loads expressed by equations (2.51), (2.52) and (2.55), when considering the inertial forces. The inertial influences are concentrated in each of the discrete points of the structure. The transforms of functions of state components are complex. Applying the formula (2.50) and performing the numerical integration continued for fixed time t, the originals of the corresponding state components are determined. The numerical integration can be carried out in the interval $\omega \in (-\infty, \infty)$ or in the interval $\omega \in (0, \infty)$ with double real values of transforms, whereas the originals of individual state components are expressed as the real parts of the resulting complex values (equation (2.50)).

In the case of the transformation of an arbitrary aperiodic exciting function in accordance with Figure 15c from time region into spectral region, the numerical solution is to be used. A solution using an approximation of the impulse function in the form of a polygon is effective, each section of which is to be transformed separately. The Section 2—3, for example, is given by

$$f_{2-3} = f_2 + f_K t - f_K t_2, \tag{2.56}$$

where

$$f_K = \frac{f_3 - f_2}{T}. \tag{2.57}$$

For the analysis the Fourier integral transformation is now to be applied in the form

$$H_{FF}(\omega) = \frac{1}{2\pi}\int_0^\infty b_{FF}(t)\,e^{-i\omega t}\,dt, \tag{2.58}$$

$$b_{FF}(t) = \int_{-\infty}^\infty H_{FF}(\omega)\,e^{i\omega t}\,d\omega = 2\int_0^\infty \mathrm{Re}[H_{FF}(\omega)]\,e^{i\omega t}\,d\omega, \tag{2.59}$$

the use of which yields

$$H_{FF}(\omega) = \frac{1}{2\pi}\int_{t_2}^{t_3} (f_2 + f_K t - f_K t_2)\,e^{-i\omega t}\,dt. \tag{2.60}$$

The calculation of the integral determines the expressions for real and imaginary parts of the transform

$$\mathrm{Re}\,(H_{FF}) = \frac{1}{2\pi}\left(\frac{f_K}{\omega^2}(\cos\omega t_3 - \cos\omega t_2) + (\omega t_3 - \omega t_2)\sin\omega t_3 + \right.$$

$$\left. + \frac{f_2}{\omega}(\sin\omega t_3 - \sin\omega t_2)\right), \tag{2.61}$$

$$\mathrm{Im}\,(H_{FF}) = \frac{1}{2\pi}\left(\frac{f_K}{\omega^2}(\sin\omega t_2 - \sin\omega t_3) + (\omega t_3 - \omega t_2)\cos\omega t_3 + \right.$$

$$\left. + \frac{f_2}{\omega}(\cos\omega t_3 - \cos\omega t_2)\right). \tag{2.62}$$

The transform of an aperiodic exciting function is determined by the superposition of transforms of partial sections of the polygon. The inverse transformation into the time region is expressed in the form

$$b_{FF}(t) = 2\left\{\int_0^\infty [\mathrm{Re}\,(H_{FF})\cos\omega t]\,\mathrm{d}\omega - \int_0^\infty [\mathrm{Im}\,(H_{FF})\sin\omega t]\,\mathrm{d}\omega\right\}. \tag{2.63}$$

For the sake of numerical integration, the curves $\mathrm{Re}\,(H_{FF})$ and $\mathrm{Im}\,(H_{FF})$ are again approximated by polygons. The time history of the investigated state components in the studied time interval is determined by the numerical integration over the whole of the analysed spectral region.

The solution of the aperiodic excited vibrations of structures over the frequency spectrum of the exciting signal, with application of the transfer function of the system (see equations (1.13) and (1.15)), is applicable also for the solution of cases in which the exciting function has a stochastic character [104]. The analogous analysis can also be performed for the solution of the problems of vibrations of structures subjected to initial conditions [104].

A further extension of periodic and aperiodic dynamic analysis represents an application of the transfer matrix method to the solution of structures subjected to the action of moving constant and harmonically variable forces. For this purpose again the combination of the transfer matrix method with integral transformations will be used.

For the case of a constant force P_0, moving with a constant speed c, the transform $H_{M_c}(\omega)$ will be determined by the equation

$$H_{M_c}(\omega) = \frac{P_0}{c}\int_0^\infty \delta_D(z-u)\,\mathrm{e}^{-\mathrm{i}\omega u/c}\,\mathrm{d}u = \frac{P_0}{c}\mathrm{e}^{-\mathrm{i}\omega z/c}, \tag{2.64}$$

with Dirac delta function δ_D and with substitution $u = ct$. The expression

(2.64) as loading term, is included into loading columns of nodal matrices in each discrete point of the studied structural simulation.

For the case of a harmonically variable force P_M, moving over the structure with constant speed c, the transform H_{M_r} is given by

$$H_{M_r}(\omega) = \frac{P_M}{c}\int_0^{\infty} \delta_D(z-u)\sin\left(\frac{\Omega}{c}u\right)e^{-i\omega u/c}\,du =$$

$$= \frac{P_M}{c}\sin\left(\frac{\Omega}{c}\right)z\,e^{-i\omega z/c}, \tag{2.65}$$

where Ω is the inducing frequency of the moving force.

The transforms of inertial forces for both cases are given by expression (2.55). The transforms of state components are again complex. The originals of the individual state components in the studied time interval are determined by the numerical integration of transforms in the interval $\omega\in(-\infty, \infty)$ or $\omega\in(0, \infty)$ (2.50), with a convenient choice of the integration step. In the local peaks of transforms (corresponding to resonance regions) and for small parameters of damping refined integration steps are to be used. For the investigation of the low speeds of moving forces or loads it is inevitable to apply the distinct refinement of discretization meshes in space and in time.

Attention is further paid to the application of the present concept for the analysis of structures subjected to the action of moving loads.

The moving concentrated load G_L can be written as

$$G_L = \delta_D(z-u)(m_L g - m_L A^2(v(u,t)))\,. \tag{2.66}$$

Its transform is given by

$$H_{M_L}(\omega) = \frac{m_L g}{c}\int_0^{\infty} \delta_D(z-u)\,e^{-i\omega u/c}\,du -$$

$$- \frac{m_L}{c}\int_0^{\infty} \delta_D(z-u)B^2\left(v\left(u,\frac{u}{c}\right)\right)e^{-i\omega u/c}\,du =$$

$$= \frac{m_L g}{c}e^{-i\omega z/c} - \frac{m_L}{c}D^2(v(z,\omega))\,e^{-i\omega z/c} =$$

$$= \frac{m_L}{c}(g - D^2(v(z,\omega)))\,e^{-i\omega z/c}, \tag{2.67}$$

where B and D are the operators

$$B = \frac{\partial}{\partial u}, \quad D = \frac{\partial}{\partial z}, \tag{2.68}$$

m_L is the mass of the moving load, c is its speed and g is the acceleration due to gravity. The second term of the last expression of equation (2.67), determining the inertial components of moving load, can be conveniently expressed in form of a difference formula

$$D^2(v(z, \omega)) = \frac{v_k - 2v_{k-1} + v_{k-2}}{l_k^2}. \tag{2.69}$$

which is implemented into the corresponding terms of nodal matrices of the studied simulation.

EXAMPLE 2. The explained concepts of application of the TMM to the analysis of structures subjected to moving or impulsive loads are numerically illustrated in Example 2. The simple beam system subjected first of all to an impulsive loading function, as shown in Figure 15a, is studied. Assumed impulse is positioned midspan of studied beam regarding the parameter $a = 10$. Later is realized the solution of the dynamic time response of the studied beam when subjected to influences of constant and harmonically variable forces as well as to the action of a constant load, which move with high speeds over the structure. The physical parameters of the studied structure are shown in Figure 16.

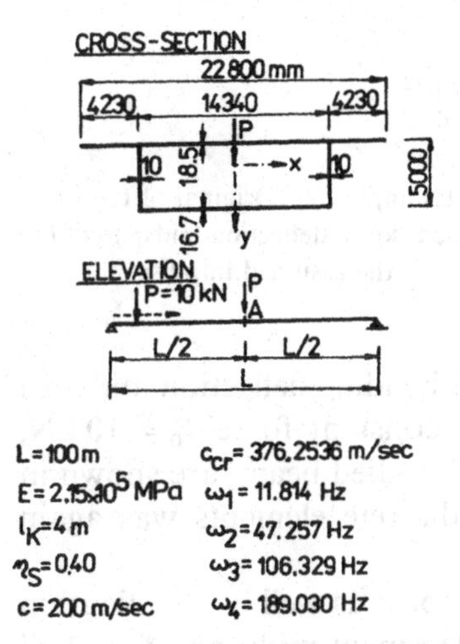

Fig. 16. Example 2 — Scheme of the analysed box beam.

The studied example must be accepted as an illustrative solution. This was the reason why numerical computations were performed on the example of the homogeneous beam with constant cross-sectional parameters. In order to reduce the amount of numerical operations, the choice of some parameters of the solved beam is not whole realistic. The analogous algorithm of the

TMM but may be adopted for the analysis of structures having general variability of all physical parameters and various boundary conditions.

The transform and original of the function of bending deflection midspan of the studied beam due to the assumed impulse are illustrated in Figures 17 and 18. In the resonance regions the step-size for numerical integration of the obtained transform was refined. The number of assumed elements of space discretization was $n = 25$.

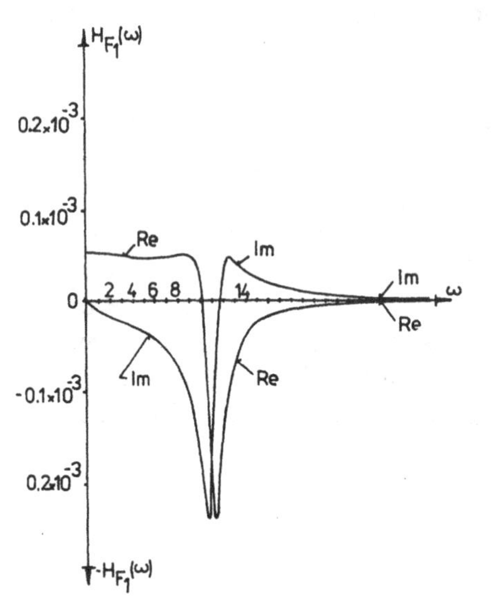

Fig. 17. Example 2 — Transform of the function of the flexural deflection midspan of the beam due to the assumed impulse.

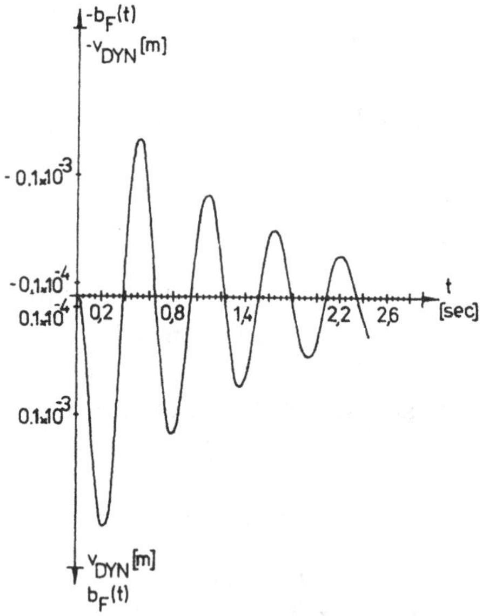

Fig. 18. Example 2 — Original of the function of the flexural deflection midspan of the beam due to the assumed impulse.

The transform and original of the function of bending deflection midspan of the studied beam due to the action of the constant force $P_0 = 10\,\text{kN}$, moving with the velocity $c = 200\,\text{m s}^{-1}$ over the studied beam, are shown in Figures 19 and 20. The number of assumed discrete elements was again $n = 25$.

The real and imaginary patterns of the transform as well as the diagram of the original of the function of bending displacement midspan of studied beam due to the action of harmonically variable force $P_M = P_0 e^{i\Omega t}$, which is moving with the velocity $c = 200\,\text{m s}^{-1}$ over the studied beam, are shown in Figures 21 and 22. In accordance with the foregoing example, the amplitude of the moving force is assumed to be $P_0 = 10\,\text{kN}$. In order to be able to distinguish the difference in displacements when comparing the action of constant and harmonically variable moving forces, the forcing frequency of the latter force is assumed to be corresponding with the first natural frequency of the studied structure, i.e. $\omega_{\mathrm{I}} = \Omega = 11.8$ Hz. The comparison of displacements for both these cases is also illustrated in Figure 22. The number of assumed elements of space discretization was once again $n = 25$.

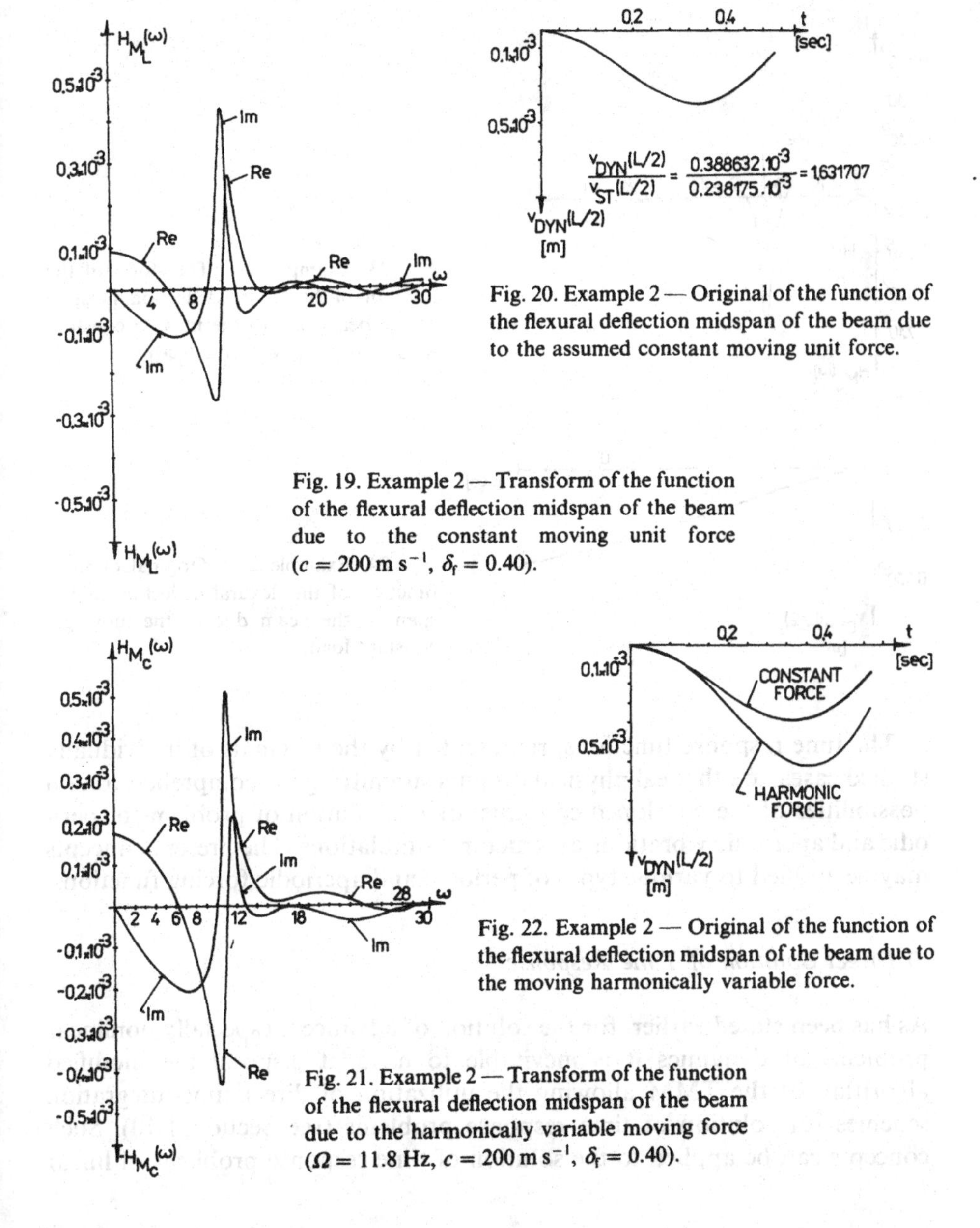

Fig. 19. Example 2 — Transform of the function of the flexural deflection midspan of the beam due to the constant moving unit force ($c = 200$ m s^{-1}, $\delta_f = 0.40$).

Fig. 20. Example 2 — Original of the function of the flexural deflection midspan of the beam due to the assumed constant moving unit force.

Fig. 21. Example 2 — Transform of the function of the flexural deflection midspan of the beam due to the harmonically variable moving force ($\Omega = 11.8$ Hz, $c = 200$ m s^{-1}, $\delta_f = 0.40$).

Fig. 22. Example 2 — Original of the function of the flexural deflection midspan of the beam due to the moving harmonically variable force.

The real and imaginary patterns of the transform as well as the diagram of the original of the function of bending displacement midspan of the studied beam due to the action of a moving constant load $m_L = m_0 l_k$ ($c = 400\,\mathrm{m\,s^{-1}}$, $n = 40$, $l_k = 2.5\,\mathrm{m}$ and m_0 is the mass per unit length of the beam in accordance with Figure 16), are illustrated in Figures 23 and 24.

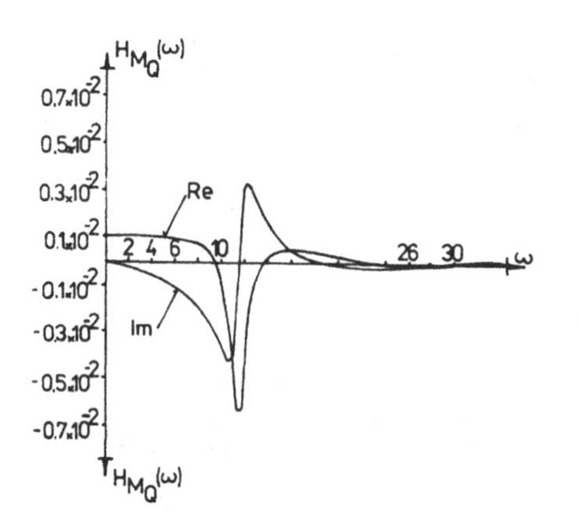

Fig. 23. Example 2 — Transform of the function of the flexural deflection midspan of the beam due to the moving constant load ($c = 400\,\mathrm{m\,s^{-1}}$, $\delta_f = 0.40$).

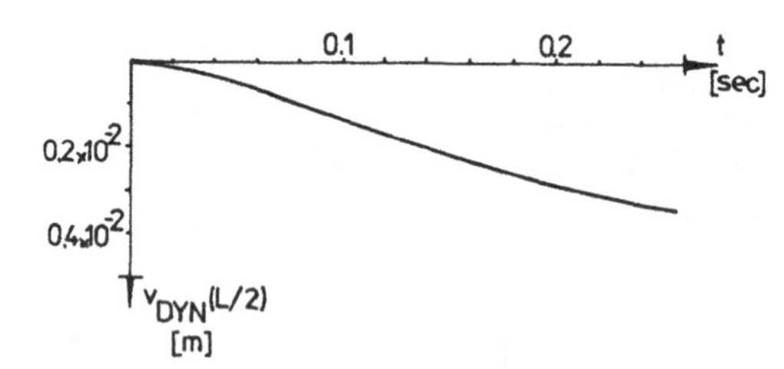

Fig. 24. Example 2 — Original of the function of the flexural deflection midspan of the beam due to the moving constant load.

The time response functions, represented by the originals of individually studied cases, are the real physical outputs submitting the comprehension on possibilities of the developed concepts for the solution of problems of periodic and aperiodic vibrations of structural simulations. The present concepts may be applied to various types of periodic and aperiodic forcing functions.

d) Direct Solution of Time Response

As has been stated earlier, for the solution of advanced, especially nonlinear problems of dynamics it is inevitable to have at disposal the modified algorithm of the TMM allowing the utilization of direct time-integration schemes for solution of time response problems (see Section 1.10). Such concepts can be applied to the solution of time response problems of linear or nonlinear structural systems subjected to the arbitrary periodic or aperiodic forcing signals. Methods of direct time integration, when compared with aforementioned integral transformation techniques, allow a more generalized approach for solving such problems.

The incorporation of the Wilson method of direct time integration into the algorithm of the TMM will be described next ([105, 106]). The flow-chart of the present concept is given by the following operations:

A. Initial stage of solution

1. Using the known algorithm of the TMM the calculation of displacements $v_0(t)$ in the set of all nodes of simulation is performed. In accordance with the aforementioned, the loading and discretization schemes, as shown in Figure 25, are assumed. The transfer and nodal matrices of bending vibration, derived in the foregoing analysis, are also utilized for the present case.

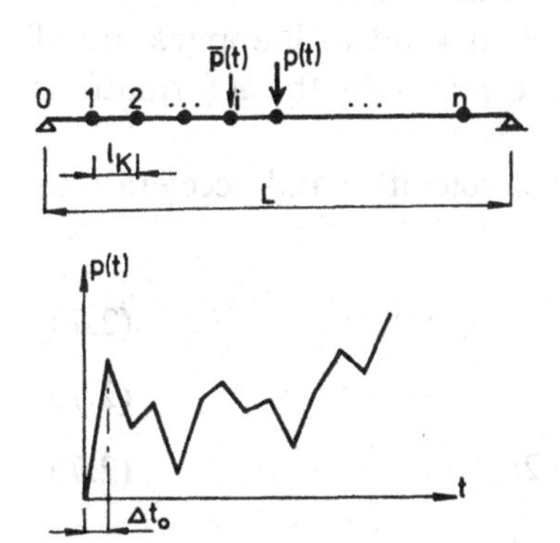

Fig. 25. Loading scheme and discretization mesh of the studied case.

2. The calculation of initial velocities $\dot{v}_0(t)$ and accelerations $\ddot{v}_0(t)$ at the start of time step Δt_0 (a dot indicates the differentiation with respect to time)

$$\dot{v}_0(t) = v_0(t)/\Delta t_0, \quad \ddot{v}_0(t) = \dot{v}_0(t)/\Delta t_0 = v_0(t)/\Delta t_0^2. \tag{2.70}$$

3. The choice of the time step Δt and of the parameter Θ of the Wilson method ($\Theta = 1.37$).

4. The calculation of integration constants for a chosen time step Δt and for the parameter $\tau = \Theta \Delta t$

$$\begin{aligned} &A_0 = 6/\tau^2, && A_1 = 3/\tau, && A_2 = 2A_1, \\ &A_3 = \tau/2, && A_4 = A_0/\Theta, && A_5 = -A_2/\Theta, \\ &A_6 = 1 - 3/\Theta, && A_7 = \Delta t/2, && A_8 = \Delta t^2/6. \end{aligned} \tag{2.71}$$

5. The formulation of the effective stiffness $\bar{K}(t)$

$$\bar{K}(t) = K(t) + A_0 m + A_1 \delta_f. \tag{2.72}$$

Implementation into the algorithm of the transfer matrix method: In nodal matrices the inertial mass m for each element is multiplied with the parameter A_0. The parameters of damping δ_f in complex moduli of elasticity located in the transfer matrices of each element, are multiplied by the integration parameter A_1.

B. For each time-integration step

6. The calculation of the modified load

$$\bar{p}(t + \Theta\Delta t) = p(t) + \Theta(p(t + \Delta t) - p(t)) + m(A_0 v(t) + A_2 \dot{v}(t) + + 2\ddot{v}(t)) + \delta_f(t)(A_1 v(t) + 2\dot{v}(t) + A_3 \ddot{v}(t)). \tag{2.73}$$

Implementation into the algorithm of the transfer matrix method: the loads $\bar{p}(t)$ are put into corresponding terms of the loading column of nodal matrices in each point of applied discrete simulation. For the case of analysed bending response it is the term (4, 5) of nodal matrix **CK**.

7. Using the algorithm of the transfer matrix method with application of stiffness and loading parameters $\bar{K}(t)$ and $\bar{p}(t)$, respectively, the deformations $v(t + \Theta\Delta t)$ are calculated.

8. The calculation of resulting deformations, velocities and accelerations is as follows:

$$v(t + \Delta t) = A_4(v(t + \Theta\Delta t) - v(t)) + A_5 \dot{v}(t) + A_6 \ddot{v}(t), \tag{2.74}$$

$$\dot{v}(t + \Delta t) = \dot{v}(t) + A_7(\ddot{v}(t + \Delta t) + \ddot{v}(t)), \tag{2.75}$$

$$\ddot{v}(t + \Delta t) = v(t) + \Delta t \dot{v}(t) + A_8(\ddot{v}(t + \Delta t) + 2\ddot{v}(t)). \tag{2.76}$$

C. Repetition of calculation from the point B

In the CTM the present concept is listed in the illustrative computer program A-2, using the set of matrices A-1 for the solution of the studied dynamic flexural time response of one-dimensional structural simulations in their linear approach.

At this point it is useful to consider the general performance and efficiency of this unconditionally stable Wilson method of time integration when combined with the TMM [105].

The size of the errors which may be introduced by any numerical integration scheme will depend on the structural dynamic parameters and on the size of the time step. The general nature of the computational errors may be expressed in terms of an artificial change of the period and the reduction of amplitude of vibration. To control the stability, convergence and precision of computer calculations, the size of the time step Δt, the value of the ratio $\Delta t/T$ where T is the period of vibration, and finally the period elongation (PE) and amplitude decay (AD) may be used. Into the set of these parameters is to be further arrayed the numerical damping expressed as a decrease of the amplitude of vibration during computer calculation independently on the parameters of internal or external damping of the analysed simulation.

The ratio of calculated (T_V) and actual (T) periods of vibration as the function of time step versus period is illustrated in Figure 26. The dependence of time step, period and amplitude decay on various parameters Θ when solving the undamped vibration, is shown in Figure 27. The equivalent dependence of period elongation and of time step on period ratio for various Θ is illustrated in Figure 28. These relations were obtained using the combined concept of the Wilson method with the TMM approach [105]. Further, it has been found that the degree of spatial discretization does not influence the validity of the present conclusions.

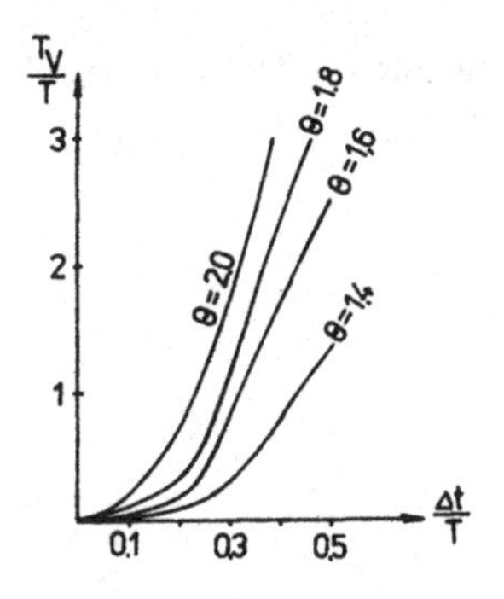

Fig. 26. Period elongation as a function of the time step versus period when the Wilson method is adopted.

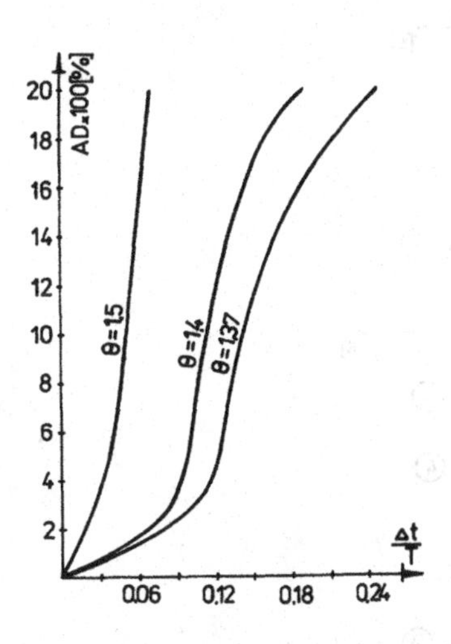

Fig. 27. Dependence of the time step, period and amplitude decay (AD) for various parameters Θ when assuming the undamped vibration.

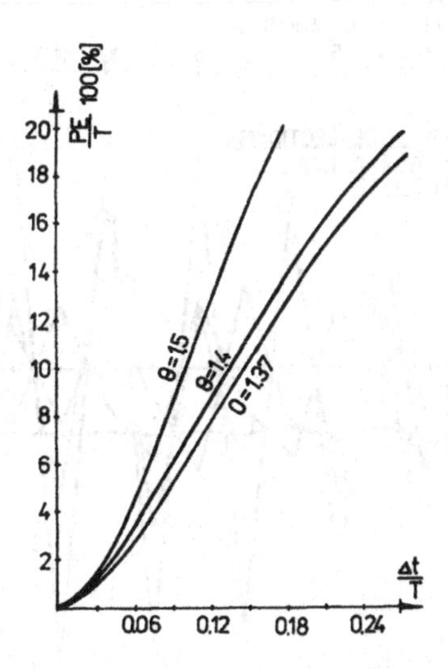

Fig. 28. Dependence of the period elongation (PE) and of the ratio time step to period for various parameters Θ when solving the undamped vibration.

EXAMPLE 3. The time response is studied of the one-dimensional approach of thin pipe system subjected to the action of a time variable midspan force $P_D = 10$ kN as shown in Figure 29. The solution method used is the aforementioned Wilson method combined with the TMM approach which is outlined in the paragraphs A-1 and A-2 of the Catalogue of Transfer Matrices.

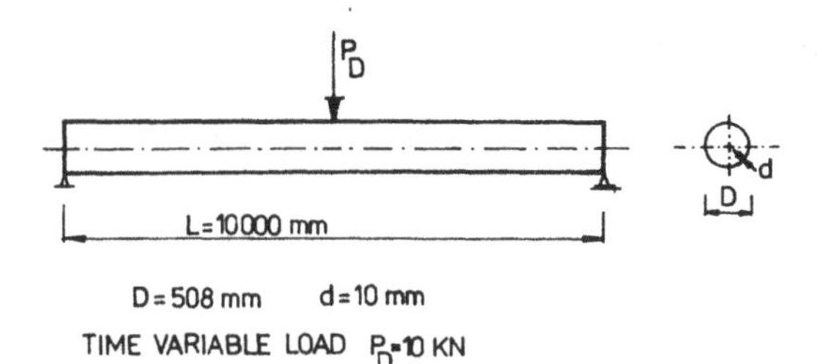

Fig. 29. Example 3 — The studied steel pipe system.

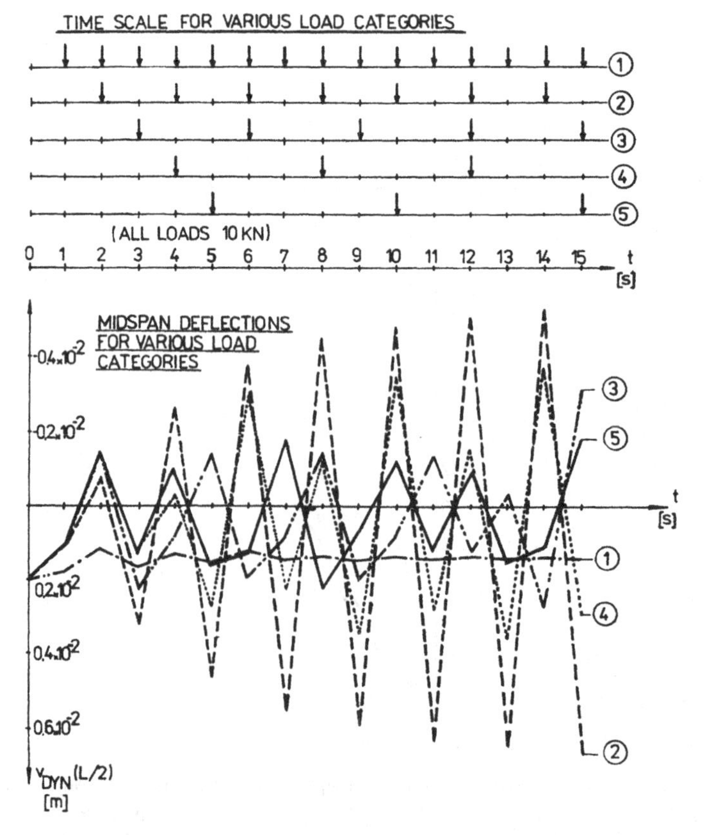

Fig. 30. Example 3 — Midspan deflection time response of the studied pipe when subjected to various load categories.

Five different patterns of the time history are assumed for the midspan force, which are illustrated in Figure 30. The response of the midspan bending deflection due to each of the assumed forcing signals is shown in Figure 30. It is evident, owing to the point of view of maximum amplitude, that the load time history of the second type represents the most important case for this example.

The requirements on the amount of machine calculation are low compared with other applicable concepts of present dynamic analyses, and may be performed using standard packages of common personal computers.

2.3. Techniques for Deriving Transfer Matrices

Several basic techniques for deriving transfer matrices are presented. The most important and verified transfer matrices are readily available in the Catalogue of Transfer Matrices at the end of this book. When a problem is to be solved, the individual matrices can be linked together as a set of building blocks. Examples of applications of these transfer matrices are found in Chapter 4.

Sometimes, however, the required transfer matrices are not catalogued, and the various methods for their derivation must be adopted. Some of such methods are now described. It is not necessary to understand this material in order to apply the transfer matrices mentioned in this book. This section is intended for readers wishing to verify given transfer matrices, develop matrices for special cases or learn more about the underlying theory of the transfer matrix method. Only methods capable of general application will be considered in sequence.

a) *Derivation of Transfer Matrices from Differential Equations*

The common method for solving problems of one independent variable is to eliminate $n - 1$ of the dependent variables to obtain an n-th order ordinary differential equation. If it is possible to find n-closed form solutions of such equation, it is then a straightforward matter to develop the corresponding transfer matrix. The n constants associated with the n-th order differential equation are determined by the boundary conditions of the problem [63].

An ordinary linear homogeneous differential equation of the n-th order, with variable coefficients is analysed, expressed as

$$D_n(y(z)) = a_n(z)y^{(n)}(z) + \cdots + a_1(z)\,y'(z) + a_0 y(z)\,, \tag{2.77}$$

with differential operator D_n defined in the intervals $z \in (a, b)$ and $y \in (-\infty, \infty)$. Owing to the presence of variable coefficients, the exact solution may be obtained for special cases only.

The solution is discretized into the set of N subintervals in which the coefficients $a_i(z)$ are assumed to be constant. The solution in the subinterval $(z_j, z_{j+1}) \in (a, b)$ is assumed to be

$$y(z) = \sum_{i=1}^{n} f_r(z_j) C_i, \tag{2.78}$$

where C_i are the integration constants. In the vector-matrix form there may be written

$$\begin{bmatrix} y_f(z_j) \\ y_f'(z_j) \\ \vdots \\ y_f^{(n-1)}(z_j) \end{bmatrix} = \begin{bmatrix} f_1(z_j) & f_2(z_j) & \cdots f_m(z_j) \\ f_1'(z_j) & f_2'(z_j) & \cdots f_m'(z_j) \\ \vdots & & \\ f_1^{(n-1)}(z_j) & f_2^{(n-1)}(z_j) & \cdots f_m^{(n-1)}(z_j) \end{bmatrix} \begin{bmatrix} C_1 \\ C_2 \\ \vdots \\ C_m \end{bmatrix} \tag{2.79}$$

or in general

$$\boldsymbol{y}(z) = \mathbf{F}(z)\boldsymbol{C}. \tag{2.80}$$

For the solution in the studied subinterval is required

$$\det \mathbf{F}(z_j) = \mathbf{W}(z_j) \neq 0, \tag{2.81}$$

with $\mathbf{W}(z_j)$ as Wronskian of fundamental solutions $f_i(z_j)$. The condition (2.81) yields

$$\boldsymbol{C} = \mathbf{F}^{-1}(z_j)\boldsymbol{y}(z_j) \tag{2.82}$$

and equation (2.80) may then be modified like

$$\boldsymbol{y}(z) = \mathbf{F}(z)\mathbf{F}^{-1}(z_j)\boldsymbol{y}(z_j). \tag{2.83}$$

Introducing the matrix substitution

$$\boldsymbol{\Phi}(z) = \mathbf{F}(z)\mathbf{F}^{-1}(z_j), \tag{2.84}$$

the modified equation (2.83) obtains the form

$$\boldsymbol{y}(z) = \boldsymbol{\Phi}(z)\boldsymbol{y}(z_j). \tag{2.85}$$

Further arrangement into vector-matrix equation

$$\begin{bmatrix} \boldsymbol{y}(z) \\ 1 \end{bmatrix} = \begin{bmatrix} \boldsymbol{\Phi}(z) & 0 \\ 0 & 1 \end{bmatrix} \begin{bmatrix} \boldsymbol{y}(z_j) \\ 1 \end{bmatrix} \tag{2.86}$$

or

$$\bar{\mathbf{y}}(z) = \mathbf{\Phi}(z)\, \bar{\mathbf{y}}(z_j) \tag{2.87}$$

defines the generalized couplings of solutions $\bar{\mathbf{y}}(z)$ and $\bar{\mathbf{y}}(z_j)$. The vector $\bar{\mathbf{y}}$ in the end point of the subinterval (z_j, z_{j+1}) is given by equation (2.86) for $z = z_{j+1}$. Introducing the substitution

$$\mathbf{B}_{j+1,j} = \mathbf{\Phi}(z_{j+1}) \tag{2.88}$$

may be defined the relation

$$\bar{\mathbf{y}}_{j+1} = \bar{\mathbf{B}}_{j+1,j}\, \bar{\mathbf{y}}_j, \tag{2.89}$$

which represents the basic equation of the transfer matrix method. Symbol $\bar{\mathbf{B}}_{j+1,j}$ denotes the transfer matrix which performs the couplings of vectors $\bar{\mathbf{y}}_{j+1}$ and $\bar{\mathbf{y}}_j$ in the subinterval (z_j, z_{j+1}).

The present flow-chart may be increased over the whole interval (a, b). This gives

$$\bar{\mathbf{y}}_1 = \bar{\mathbf{B}}_{1,0}\, \bar{\mathbf{y}}_0, \tag{2.90}$$

$$\bar{\mathbf{y}}_2 = \bar{\mathbf{B}}_{2,1}\, \bar{\mathbf{y}}_1 = \bar{\mathbf{B}}_{2,1}\bar{\mathbf{B}}_{1,0}\, \bar{\mathbf{y}}_0, \tag{2.91}$$

$$\bar{\mathbf{y}}_3 = \bar{\mathbf{B}}_{3,2}\, \mathbf{y}_2 = \bar{\mathbf{B}}_{3,2}\bar{\mathbf{B}}_{2,1}\bar{\mathbf{B}}_{1,0}\, \bar{\mathbf{y}}_0, \tag{2.92}$$

$\vdots$

etc.

The above equations may be written as

$$\bar{\mathbf{y}}_N = \bar{\mathbf{y}}(z_n) = \prod_{i=N}^{1} \bar{\mathbf{B}}_{i,i-1}\bar{\mathbf{y}}(z_0) = \bar{\mathbf{B}}_{N,0}\bar{\mathbf{y}}(z_0), \tag{2.93}$$

with the generalized transfer matrix in the interval (a, b)

$$\bar{\mathbf{B}}_{N,0} = \prod_{i=N}^{1} \bar{\mathbf{B}}_{i,i-1} = \bar{\mathbf{B}}_{N,N-1}\bar{\mathbf{B}}_{N-1,N-2} \cdots \bar{\mathbf{B}}_{2,1}\bar{\mathbf{B}}_{1,0}. \tag{2.94}$$

The present concept is applicable in cases where the closed form solutions of differential equations of higher order can be derived. However, when determining the transfer matrices it is necessary to take into account the fact that closed form solutions of differential equations of higher order with constant or variable coefficients can be obtained only for some special cases. Generally, the situation may be simplified when introducing the approximate system of n linear differential equations of the first order. The application of the matrix calculus for such concept allows the utilization of the theory of differential equations of the first order which is the best known for deriving of transfer matrices.

Assume a set of n first-order differential equations which in the homogeneous matrix modification can be written as

$$D(\boldsymbol{y}(z)) = \mathbf{M}(z)\boldsymbol{y}(z), \tag{2.95}$$

with differential operator of the first order D and the matrix of coefficients $\mathbf{M}(z)$.

The fundamental system of n independent solutions is given by

$$\begin{aligned} y_1(z) &= c_1 y_{11}(z) + c_2 y_{12}(z) + \cdots + c_n y_{1n}(z), \\ y_2(z) &= c_1 y_{21}(z) + c_2 y_{22}(z) + \cdots + c_n y_{2n}(z), \\ &\vdots \\ y_n(z) &= c_1 y_{n1}(z) + c_2 y_{n2}(z) + \cdots + c_n y_{nn}(z), \end{aligned} \tag{2.96}$$

or in vector form

$$\begin{aligned} \boldsymbol{y}(z) &= c_1 y_1(z) + c_2 y_2(z) + \cdots + c_n y_n(z) = \\ &= (y_1(z), y_2(z), \cdots, y_n(z))\boldsymbol{c}_j. \end{aligned} \tag{2.97}$$

Arbitrary solutions $\boldsymbol{y}_j$ have to satisfy the initial conditions in initial point z_0, given by

$$\boldsymbol{y}_j(z_0) = \begin{matrix} y_{1j} \\ y_{2j} \\ y_{3j} \\ \vdots \\ y_{nj} \end{matrix} \begin{bmatrix} 1 & 0 & 0 & \cdots & 0 \\ 0 & 1 & 0 & & 0 \\ 0 & 0 & 1 & & 0 \\ \vdots & & & & \\ 0 & 0 & 0 & & 1 \end{bmatrix} \tag{2.98}$$

Using these initial conditions for the evaluation of the matrix equation (2.95) then yields

$$\boldsymbol{y}(z) = (y_1(z), y_2(z), y_3(z), \cdots, y_n(z))\boldsymbol{y}_j(z_0) = \mathbf{B}_N(z)\boldsymbol{y}_j(z), \tag{2.99}$$

with the vector of solutions in the initial point $\boldsymbol{y}_j(z_0)$.

The transfer matrix $\mathbf{B}_N(z)$ is constructed from the partial solutions of the studied problem, which are obtained analytically or by the numerical integration when considering the corresponding initial conditions.

For the general derivation of the transfer matrices, several methods for solving differential equations of the first order may be applied.

When considering the set of n first-order differential equations with constant coefficients, given by equation (2.95), the system can be expanded into an infinite series via

$$\mathbf{M}(z) = \mathbf{M}_0 + \mathbf{M}_1 z + \mathbf{M}_2 z^2 + \mathbf{M}_3 z^3 + \dots, \tag{2.100}$$

$$\boldsymbol{y}(z) = \boldsymbol{y}_0 + \boldsymbol{y}_1 z + \boldsymbol{y}_2 z^2 + \boldsymbol{y}_3 z^3 + \cdots, \tag{2.101}$$

where $\mathbf{M}_i$ are the values of the coefficient matrices in the discrete point z. Substituting these expressions for $\mathbf{M}(z)$ and $\boldsymbol{y}(z)$ into the system (2.95) and comparison of coefficients gives

$$\begin{aligned} \boldsymbol{y}_1 &= \mathbf{M}_0 \boldsymbol{y}_0, \\ 2\boldsymbol{y}_2 &= \mathbf{M}_0 \boldsymbol{y}_1 + \mathbf{M}_1 \boldsymbol{y}_0, \\ 3\boldsymbol{y}_3 &= \mathbf{M}_0 \boldsymbol{y}_2 + \mathbf{M}_1 \boldsymbol{y}_1 + \mathbf{M}_2 \boldsymbol{y}_0, \\ 4\boldsymbol{y}_4 &= \mathbf{M}_0 \boldsymbol{y}_3 + \mathbf{M}_1 \boldsymbol{y}_2 + \mathbf{M}_2 \boldsymbol{y}_1 + \mathbf{M}_3 \boldsymbol{y}_0, \\ &\vdots \\ \text{etc.} \end{aligned} \tag{2.102}$$

The transfer matrix of the studied problem is then given by

$$\begin{aligned} \mathbf{B}_N(z) = \mathbf{I} + \mathbf{M}_0 z + \frac{1}{2}(\mathbf{M}_0^2 + \mathbf{M}_1) z^2 + \frac{1}{3}\left(\frac{1}{2}(\mathbf{M}_0^3 + \mathbf{M}_0\mathbf{M}_1) + \right. \\ \left. + \mathbf{M}_1\mathbf{M}_0 + \mathbf{M}_2\right) z^3 + \frac{1}{4}\left(\frac{1}{6}(\mathbf{M}_0^4 + \mathbf{M}_0^2\mathbf{M}_1)) + \frac{1}{3}(\mathbf{M}_0\mathbf{M}_1\mathbf{M}_0 + \right. \\ \left. + \mathbf{M}_0\mathbf{M}_2\right) z^4, \cdots, \text{etc.}, \end{aligned} \tag{2.103}$$

with unit diagonal matrix $\mathbf{I}$.

In addition to the many other analytical methods which are applicable for solving differential equations of first order (*Laplace transformation* [63], *exponential expansion* [48, 63, 122], *matrizant* [48, 63, 122], *product integral* [48, 63, 122], etc.), there are various numerical methods for integrating equation (2.95) to provide the transfer matrix. The cases in which analytical solutions of equation (2.95) are difficult or even impossible to obtain, should be treated numerically. Typical methods suitable for numerical integrating of systems of the first-order equations are the *Runge—Kutta*, *predictor-corrector* [63], *Romberg* or *Picard* approaches [48, 122], etc. The first of these when adopted for the derivation of transfer matrices is discussed next.

When deriving the transfer matrix which is coupling the vectors $\boldsymbol{y}(z_{n+1})$ and $\boldsymbol{y}(z_n)$, the discretization of the analysed integration region into the set of subregions

$$h = (z_{n+1} - z_n)/n, \tag{2.104}$$

is assumed. The application of the Runge—Kutta technique for solving equation (2.95) gives

$$\mathbf{y}(z_{n+1}) = \mathbf{y}(z_n) + \frac{1}{6}(k_0 + 2k_1 + 2k_2 + k_3), \tag{2.105}$$

where

$$\begin{aligned} k_0 &= h\mathbf{M}(z_n)\mathbf{y}(z_n), \\ k_1 &= h\mathbf{M}(z_n + h/2)\mathbf{y}(z_n + k_0/2), \\ k_2 &= h\mathbf{M}(z_n + h/2)\mathbf{y}(z_n + k_1/2), \\ k_3 &= h\mathbf{M}(z_n + h/2)\mathbf{y}(z_n + k_2), \end{aligned} \tag{2.106}$$

$\mathbf{M}(z_n)$, $\mathbf{M}(z_n + h)$ and $\mathbf{M}(z_n + h/2)$ are, respectively, the values of the matrix $\mathbf{M}(z)$ at the nodes n and $n + 1$ and at the section halfway between these points.

Substituting expressions (2.106) into equation (2.105) determines the relation between $\mathbf{y}(z_{n+1})$ and $\mathbf{y}(z_n)$, viz.

$$\begin{aligned} \mathbf{y}(z_{n+1}) = \Bigg[\mathbf{I} &+ \frac{h}{6}\left(\mathbf{M}(z_n) + 4\mathbf{M}\left(z_n + \frac{h}{2}\right) + \mathbf{M}(z_n + h)\right) + \\ &+ \frac{h^2}{6}\left(\mathbf{M}\left(z_n + \frac{h}{2}\right)\mathbf{M}(z_n) + \mathbf{M}(z_n + h)\mathbf{M}\left(z_n + \frac{h}{2}\right) + \mathbf{M}^2\left(z_n + \frac{h}{2}\right)\right) + \\ &+ \frac{h^3}{12}\left(\mathbf{M}^2\left(z_n + \frac{h}{2}\right)\mathbf{M}(z_n) + \mathbf{M}(z_n + h)\,\mathbf{M}^2\left(z_n + \frac{h}{2}\right)\right) + \\ &+ \frac{h^4}{24}\mathbf{M}(z_n + h)\,\mathbf{M}^2\left(z_n + \frac{h}{2}\right)\mathbf{M}(z_n)\Bigg]\mathbf{y}(z_n), \end{aligned} \tag{2.107}$$

where the expression between the outer braces represents the transfer matrix linking the state vectors $\mathbf{y}(z_n)$ and $\mathbf{y}(z_{n+1})$. If the matrix $\mathbf{M}(z_n)$ contains only constant elements, equation (2.107) reduces to

$$\mathbf{y}(z_{n+1}) = \left(\mathbf{I} + \frac{\mathbf{M}(z_n)h}{1!} + \frac{(\mathbf{M}(z_n)h)^2}{2!} + \frac{(\mathbf{M}(z_n)h)^3}{3!} + \frac{(\mathbf{M}(z_n)h)^4}{4!}\right)\mathbf{y}(z_n), \tag{2.108}$$

in which the outer brackets contain the first five terms of the infinite expansion of the transfer matrix generally given by

$$\mathbf{B}_N(z) = \mathbf{I} + \mathbf{M}z + \mathbf{M}^2\frac{z^2}{2!} + \mathbf{M}^3\frac{z^3}{3!} + \mathbf{M}^4\frac{z^4}{4!}\cdots. \tag{2.109}$$

A disadvantage of the Runge—Kutta method is that it is impossible to estimate the error associated with a given incremental step of size h. The repeated calculation using the halfstep $h/2$ may be applied, which improves the precision according to h^5.

For the analysis of complicated structures, such as grid-works, plate or shell systems, orthotropic thinwalled structures or three-dimensional simulations, for which is possible to find the end forces in terms of the end displacements, is especially useful to derive the transfer matrices from corresponding stiffness matrices. Such concept is considered as the most effective method for the derivation of generalized transfer matrices for linear and nonlinear members or substructures. The principles of derivation of transfer matrices from stiffness matrices and the implementation into the algorithm of the FETM-method are described in Chapter 3 of this book.

The present concepts are applicable for deriving transfer matrices for various elements (rods, beams, plates, shells, three-dimensional solids, etc.) of structural configurations. When using such elements the substructures or macroelements (thinwalled beams, grillages, orthotropic walls, folded plates, corrugated sheets, etc.) may be constructed in accordance with the principles of *hybrid substructuring* in space and time adopted for advanced analyses in Chapter 3. Such substructuring allows the modelling of complicated structural simulations when considering the general variability of physical parameters in space and time. Subsequently the transfer and nodal matrices of typical substructures wil be derived for the discrete simulation of modern structural configurations in space and time.

2.4. Thinwalled Beam Substructure

First of all the theoretical analysis leading to the determination of a simultaneous system of partial differential equations expressing the spatial behaviour of thinwalled box substructure will be outlined.

In the following analysis the generalized Bornscheuer system [10] of characteristic cross-sectional parameters will be dealt with which is generally used in theoretical investigations of thinwalled structures. Wagner's hypothesis [115] is assumed to be valid. The axial, bending, torsion-bending and distortion-bending kinematic movability of the cross-section is considered. These primary influences are intercoupled with the secondary degrees of freedom of supplementary deformations of thin walls of the analysed substructure.

In the case of the arbitrary virtual variation of the energetic level of a segment with a single biaxially antisymmetric box cross-section with variable longitudinal parameters as shown in Figure 31, the general energetic equivalence of external and internal works of the investigated substructure remains constant. On the basis of energetic balance of the substructure in time t, in accordance with equation (1.4), the virtual work of internal forces due to the coupled biaxial bending, torsion-bending, distortion-bending and axial behaviour of the analysed box substructure with a length l_K, positioned in a global coordinate system $\bar{x} - \bar{y} - \bar{z}$, is expressed by a matrix equation in form

$$\delta A_i = \int_0^{l_K} \Big\{ \mathbf{E}(\bar{z}, t)\, D^2(\boldsymbol{u}_i(\bar{z}, t)) \Big[\oint_{\mathbf{F}_c(s(\bar{z}))} (\gamma(s(\bar{z}))) \circ$$

$$\circ (\gamma(s(\bar{z})))\, \mathrm{d}\mathbf{F}_c(s(\bar{z})) \Big] \delta(D^2(\boldsymbol{u}_i(\bar{z}, t))) -$$

$$- \mathbf{G}(\bar{z}, t)\, D(\boldsymbol{u}_i(\bar{z}, t)) \Big[\oint_{\mathbf{F}_c(s(\bar{z}))} (\boldsymbol{\Psi}_{\mathrm{I}}(s(\bar{z}))\, t_{\mathrm{w}}^{-1}(s(\bar{z}))) \circ$$

$$\circ (\boldsymbol{\Psi}_{\mathrm{I}}(s(\bar{z}))\, t_{\mathrm{w}}^{-1}(s(\bar{z})))\, \mathrm{d}\mathbf{F}_c(s(\bar{z})) + \mathbf{J}_{\mathrm{I}}^{\mathrm{T}}(s(\bar{z}))\, \mathbf{K}_{\mathrm{S_I}} +$$

$$+ \mathbf{J}_{\mathrm{T}}(s(\bar{z})) \Big] \delta(D(\boldsymbol{u}_i(\bar{z}, t))) -$$

$$- \mathbf{G}(\bar{z}, t)\, D(\boldsymbol{u}_i(\bar{z}, t)) \Big[\oint_{\mathbf{F}_c(s(\bar{z}))} (\boldsymbol{\Psi}_{\mathrm{II}}(s(\bar{z}))\, t_{\mathrm{w}}^{-1}(s(\bar{z}))) \circ$$

$$\circ (\boldsymbol{\Psi}_{\mathrm{II}}(s(\bar{z}))\, t_{\mathrm{w}}^{-1}(s(\bar{z})))\, \mathrm{d}\mathbf{F}_c(s(\bar{z})) + \mathbf{J}_{\mathrm{II}}^{\mathrm{T}}(s(\bar{z}))\, \mathbf{K}_{\mathrm{S_{II}}} +$$

$$+ \mathbf{J}_{\mathrm{D}}(s(\bar{z})) \Big] \delta(D(\boldsymbol{u}_i(\bar{z}, t))) + [\mathbf{G}(\bar{z}, t)\, \mathbf{J}_{\mathrm{D}}(s(\bar{z}))]\, \delta(\boldsymbol{u}_i(\bar{z}, t)) \Big\} \mathrm{d}\bar{z}\,. \quad (2.110)$$

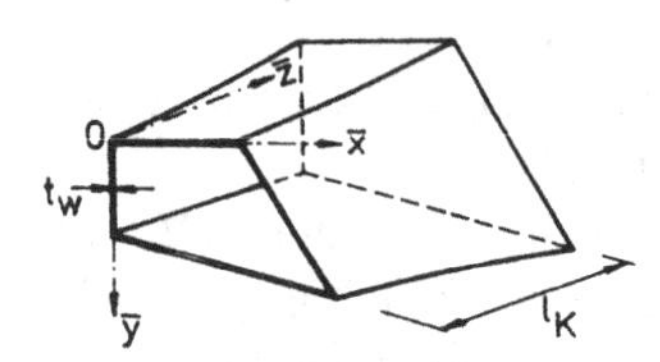

Fig. 31. Thinwalled single box substructure with variable cross-sectional parameters.

$\mathbf{E}(\bar{z}, t)$ and $\mathbf{G}(\bar{z}, t)$ are matrices of elasticity moduli in space and time. Symbol $\circ$ denotes the dyadic product of two vectors, D is the partial differential operator

$$D = \frac{\partial}{\partial \bar{z}} \quad (2.111)$$

and $\boldsymbol{u}_i(\bar{z}, t)$ is the vector of amplitudes of the elastic deformation state components

$$\boldsymbol{u}_i(\bar{z}, t) = \begin{bmatrix} v_z(\bar{z}, t) \\ v_x(\bar{z}, t) \\ v_y(\bar{z}, t) \\ \vartheta_T(\bar{z}, t) \\ \varrho_V(\bar{z}, t) \end{bmatrix}. \tag{2.112}$$

All deformation state components also functionally depend on the middle-line s and on the thickness of the wall t_w. In the vector $\boldsymbol{u}_i(\bar{z}, t)$ besides displacements $v_i(\bar{z}, t)$ occur $\vartheta_T(\bar{z}, t)$ and $\varrho_V(\bar{z}, t)$ as angles of torsion and distortion, respectively. $\gamma(s(\bar{z}))$ is the vector of warping functions with consideration of shear deformations of symmetric and antisymmetric kinds of deformations. $\boldsymbol{\Psi}_I s(\bar{z}))$ and $\boldsymbol{\Psi}_{II}(s(\bar{z}))$ are the vectors of shear functions due to torsion-bending and distortion-bending behaviour of a single box structure. Matrix $\mathbf{J}_I(s(\bar{z}))$ determines the matrix product

$$\mathbf{J}_I(s(\bar{z})) = \mathbf{J}_p(s(\bar{z}))\, \mathbf{K}_{S_I}, \tag{2.113}$$

where $\mathbf{J}_p(s(\bar{z}))$ is the matrix of Saint-Venant torsional rigidities $J_i(s(\bar{z}))$ of the individual walls of the box cross-section of the studied substructure

$$\mathbf{J}_p(s(\bar{z})) = \begin{bmatrix} J_1(s(\bar{z})) & 0 & 0 & 0 & 0 \\ 0 & J_2(s(\bar{z})) & 0 & 0 & 0 \\ 0 & 0 & J_3(s(\bar{z})) & 0 & 0 \\ 0 & 0 & 0 & J_4(s(\bar{z})) & 0 \\ 0 & 0 & 0 & 0 & 0 \end{bmatrix} \tag{2.114}$$

and the transformation matrix $\mathbf{K}_{S_I}$ is given by

$$\mathbf{K}_{S_I} = \begin{bmatrix} 0 & 0 & 0 & 1 & 0 \\ 0 & 0 & 0 & 1 & 0 \\ 0 & 0 & 0 & 1 & 0 \\ 0 & 0 & 0 & 1 & 0 \\ 0 & 0 & 0 & 1 & 0 \end{bmatrix}. \tag{2.115}$$

Superscript T is the symbol for transposition. $\mathbf{J}_T(s(\bar{z}))$ is the matrix of Bredt torsional resistance of the cross-section of the substructure

$$\mathbf{J}_T(s(\bar{z})) = \begin{bmatrix} 0 & 0 & 0 & 0 & 0 \\ 0 & 0 & 0 & 0 & 0 \\ 0 & 0 & 0 & 0 & 0 \\ 0 & 0 & 0 & J_{TB}(s(\bar{z})) & 0 \\ 0 & 0 & 0 & 0 & 0 \end{bmatrix}. \tag{2.116}$$

The vector of warping functions is determined by the expression

$$\gamma(s(\bar{z})) = \oint_{s(\bar{z})} [\boldsymbol{r}_i(s(\bar{z})) - \boldsymbol{\Psi}_{\mathrm{I}}(s(\bar{z}))\, t_{\mathrm{w}}^{-1}(s(\bar{z})) -$$

$$- \boldsymbol{\Psi}_{\mathrm{II}}(s(\bar{z}))\, t_{\mathrm{w}}^{-1}(s(\bar{z}))]\, \mathrm{d}(s(\bar{z})) =$$

$$= \oint_{s(\bar{z})} \left\{ \begin{bmatrix} 1 \\ \cos\beta(s(\bar{z})) \\ \sin\beta(s(\bar{z})) \\ r_{\mathrm{T}}(s(\bar{z})) \\ r_{\mathrm{D}}(s(\bar{z})) \end{bmatrix} - \begin{bmatrix} 0 \\ 0 \\ 0 \\ \Psi_{\mathrm{I}}(s(\bar{z})) \\ 0 \end{bmatrix} t_{\mathrm{w}}^{-1}(s(\bar{z})) - \right.$$

$$\left. - \begin{bmatrix} 0 \\ 0 \\ 0 \\ 0 \\ \Psi_{\mathrm{II}}(s(\bar{z})) \end{bmatrix} t_{\mathrm{w}}^{-1}(s(\bar{z})) \right\} \mathrm{d}(s(\bar{z})) = \begin{bmatrix} 1 \\ w_x \\ w_y \\ w_{\mathrm{I}} \\ w_{\mathrm{II}} \end{bmatrix}, \tag{2.117}$$

where $\boldsymbol{r}_i(s(\bar{z}))$ is the symbol for the vector of radius vectors as a function of the coordinate $\bar{z}$; $\beta(s(\bar{z}))$ is the angle between the tangent to the centre-line and the basic coordinate system in the investigated point of the box cross-section.

The angle of relative distortion ε_i due to the first antisymmetric distortional form of the box cross-section (see Figure 32) is given by

$$\varepsilon_i(s(\bar{z}), t) = \varrho_{m+1}(s(\bar{z}), t) + \varrho_m(s(\bar{z}), t)\,. \tag{2.118}$$

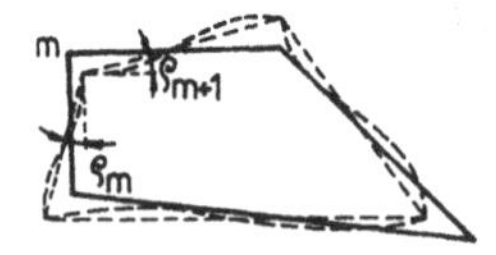

Fig. 32. Angles of relative distortion of the box cross-section.

The transformation matrix $\mathbf{K}_{\mathrm{S_{II}}}$ has the form

$$\mathbf{K}_{\mathrm{S_{II}}} = \begin{bmatrix} 0 & 0 & 0 & 0 & 1 \\ 0 & 0 & 0 & 0 & 1 \\ 0 & 0 & 0 & 0 & 1 \\ 0 & 0 & 0 & 0 & 1 \\ 0 & 0 & 0 & 0 & 1 \end{bmatrix} \tag{2.119}$$

and the matrix $\mathbf{J}_{\mathrm{II}}(s(\bar{z}))$ is defined by the matrix product

$$\mathbf{J}_{\mathrm{II}}(s(\bar{z})) = \mathbf{H}_{\mathrm{B}}(s(\bar{z}))\, \mathbf{K}_{\mathrm{S_{II}}}, \tag{2.120}$$

where $\mathbf{H}_{\mathrm{B}}(s(\bar{z}))$ is the matrix of bending rigidities $J_{\mathrm{B}_i}(s(\bar{z}))$ of each wall of the cross-section, which is simulated as a transverse frame of unit width in the direction of the $\bar{z}$-axis

$$\mathbf{H}_{\mathrm{B}}(s(\bar{z})) = \begin{bmatrix} J_{\mathrm{B}_1}(s(\bar{z})) & 0 & 0 & 0 & 0 \\ 0 & J_{\mathrm{B}_2}s((\bar{z})) & 0 & 0 & 0 \\ 0 & 0 & J_{\mathrm{B}_3}(s(\bar{z})) & 0 & 0 \\ 0 & 0 & 0 & J_{\mathrm{B}_4}(s(\bar{z})) & 0 \\ 0 & 0 & 0 & 0 & 0 \end{bmatrix} . \tag{2.121}$$

$\mathbf{J}_{\mathrm{D}}(s(\bar{z}))$ is the matrix of frame rigidity $J_{\mathrm{D}_1}(s(\bar{z}))$ of a unit frame of the studied segment

$$\mathbf{J}_{\mathrm{D}}(s(\bar{z})) = \begin{bmatrix} 0 & 0 & 0 & 0 & 0 \\ 0 & 0 & 0 & 0 & 0 \\ 0 & 0 & 0 & 0 & 0 \\ 0 & 0 & 0 & 0 & 0 \\ 0 & 0 & 0 & 0 & J_{\mathrm{D}_1}(s(\bar{z})) \end{bmatrix} . \tag{2.122}$$

The virtual work of loading and inertial factors may be written as

$$A_e = \int_0^{l_K} [\boldsymbol{p}_i(\bar{z}, t)\, \boldsymbol{r}_i(s(\bar{z}))\, \delta(\boldsymbol{u}_i(\bar{z}, t))]\, \mathrm{d}\bar{z} +$$

$$+ \int_0^{l_K} \{[\bar{\boldsymbol{\varepsilon}}(s(\bar{z}))\, A^2(\boldsymbol{u}_i(\bar{z}, t)) - \bar{\boldsymbol{\varphi}}(s(\bar{z}))\, C^2(\boldsymbol{u}_i(\bar{z}, t))]\, \delta(\boldsymbol{u}_i(\bar{z}, t))\}\, \mathrm{d}z\,. \tag{2.123}$$

$\boldsymbol{p}_i(\bar{z}, t)$ is the vector of exciting components acting on the structure, $\bar{\boldsymbol{\varepsilon}}(s(\bar{z}))$ and $\bar{\boldsymbol{\varphi}}(s(\bar{z}))$ are the matrices of inertial influences, A and C are the partial differential operators

$$A = \frac{\partial}{\partial t}, \quad C = \frac{\partial^2}{\partial z\, \partial t}. \tag{2.124}$$

Introduce the matrix substitutions

$$\bar{\mathbf{F}}_1(s(\bar{z})) = \oint_{\mathbf{F}_c(s(\bar{z}))} [\boldsymbol{\gamma}(s(\bar{z})) \circ \boldsymbol{\gamma}(s(\bar{z}))]\, \mathrm{d}\mathbf{F}_c(s(\bar{z}))\,, \tag{2.125}$$

$$\bar{\mathbf{F}}_2(s(\bar{z})) = \oint_{\mathbf{F}_c(s(\bar{z}))} [(\boldsymbol{\Psi}_{\mathrm{I}}(s(\bar{z}))\, t_{\mathrm{w}}^{-1}(s(\bar{z}))) \circ (\boldsymbol{\Psi}_{\mathrm{I}}(s(\bar{z}))\, t_{\mathrm{w}}^{-1}(s(\bar{z})))]\, \mathrm{d}\mathbf{F}_c(s(\bar{z})) +$$

$$+ \mathbf{J}_{\mathrm{I}}^{\mathrm{T}}(s(\bar{z}))\, \mathbf{K}_{\mathrm{S}_{\mathrm{I}}} + \mathbf{J}_{\mathrm{T}}(s(\bar{z}))\,, \tag{2.126}$$

$$\bar{\mathbf{F}}_3(s(\bar{z})) = \oint_{\mathbf{F}_c(s(\bar{z}))} [(\boldsymbol{\Psi}_{\mathrm{II}}(s(\bar{z}))t_{\mathrm{w}}^{-1}(s(\bar{z}))) \circ (\boldsymbol{\Psi}_{\mathrm{II}}(s(\bar{z}))\, t_{\mathrm{w}}^{-1}(s(\bar{z})))]\, \mathrm{d}\mathbf{F}_c(s(\bar{z})) +$$

$$+ \mathbf{J}_{\mathrm{II}}^{\mathrm{T}}(s(\bar{z}))\, \mathbf{K}_{\mathrm{S}_{\mathrm{II}}} + \mathbf{J}_{\mathrm{D}}(s(\bar{z}))\,. \tag{2.127}$$

On the basis of the modified energetic declaration (1.4)

$$\delta\pi_V = \delta A_i - \delta A_e = 0\,, \tag{2.128}$$

the virtual variation of the energetic potential can be expressed in form

$$\begin{aligned}\delta\pi_V = \int_0^{l_K} \{&[\mathbf{E}\bar{\mathbf{F}}_1(s)](\bar{z})\, D^2(\boldsymbol{u}_i(\bar{z},t))\, \delta(D^2(\boldsymbol{u}_i(\bar{z},t))) - \\ &- [\mathbf{G}\bar{\mathbf{F}}_2(s)](\bar{z})\, D(\boldsymbol{u}_i(\bar{z},t))\, \delta(D(\boldsymbol{u}_i(\bar{z},t))) - \\ &- [\mathbf{G}\bar{\mathbf{F}}_3(s)](\bar{z})\, D(\boldsymbol{u}_i(\bar{z},t))\, \delta(D(\boldsymbol{u}_i(\bar{z},t))) + \\ &+ [\mathbf{G}(\bar{z})\, \mathbf{J}_D(s(\bar{z}))]\, \delta(\boldsymbol{u}_i(\bar{z},t)\boldsymbol{p}_i(\bar{z},t)\, \boldsymbol{r}_i(s(\bar{z}))\, \delta(\boldsymbol{u}_i(\bar{z},t)) - \\ &- \bar{\varepsilon}(s(\bar{z}))\, A^2(\boldsymbol{u}_i(\bar{z},t)) - \bar{\varphi}(s(\bar{z}))\, C^2(\boldsymbol{u}_i(\bar{z},t))\, \delta(\boldsymbol{u}_i(\bar{z},t))\}\, \mathrm{d}\bar{z} = 0\,. \end{aligned} \tag{2.129}$$

For a general set of variations $\delta(\boldsymbol{u}_i(\bar{z},t))$ the condition for satisfying equation (2.129) is characterized by the elimination of expression

$$\begin{aligned}&[\mathbf{E}\bar{\mathbf{F}}_1(s)](\bar{z})\, D^4(\boldsymbol{u}_i(\bar{z},t)) - [\mathbf{G}\,\bar{\mathbf{F}}_2(s)](\bar{z})\, D^2(\boldsymbol{u}_i(\bar{z},t)) - \\ &- [\mathbf{G}\bar{\mathbf{F}}_3(s)](\bar{z})\, D^2(\boldsymbol{u}_i(\bar{z},t)) + [\mathbf{G}(\bar{z})\, \mathbf{J}_D(s(\bar{z}))](\boldsymbol{u}_i(\bar{z},t)) - \\ &- \boldsymbol{p}_i(s(\bar{z}))\boldsymbol{u}_i(\bar{z},t)\, \boldsymbol{r}_i(\bar{z},t) - \bar{\varepsilon}(s(\bar{z}))\, A^2(\boldsymbol{u}_i(\bar{z},t)) - \\ &- \bar{\varphi}(s(\bar{z}))\, C^2(\boldsymbol{u}_i(\bar{z},t))]\, \boldsymbol{u}_i(\bar{z},t) = 0\,. \end{aligned} \tag{2.130}$$

Equation (2.130) gives the matrix description of a simultaneous system of partial differential equations determining the space behaviour of a generally antisymmetric single box substructure by arbitrary choice of the corresponding global coordinate system $\bar{x} - \bar{y} - \bar{z}$. Equation (2.130) must be satisfied at each point of the longitudinal axis $\bar{z}$.

For deriving of generalized transfer matrix of the studied substructure, the homogeneous form of the matrix equation (2.130) with eliminated loading factors is used. This equation represents a simultaneous system of $n + 1$ differential equations given in Table I (n is a number of edges of the cross-section of substructure). The first three equations express the symmetric kinds of behaviour, e.g., the axial and bending action of substructure; the following two equations express the asymmetric kinds of behaviour due to torsion-bending and distortion-bending deformations, coupled in the global coordinate system $\bar{x} - \bar{y} - \bar{z}$ (Figure 31). The modified equation (2.130) becomes

$$\begin{aligned}&[\mathbf{E}\bar{\mathbf{F}}_1(s)](\bar{z})\, D^4(\boldsymbol{u}_i(\bar{z},t)) - [\mathbf{G}\bar{\mathbf{F}}_2(s)](\bar{z})\, D^2(\boldsymbol{u}_i(\bar{z},t)) - \\ &- [\mathbf{G}\bar{\mathbf{F}}_3(s)](\bar{z})\, D^2(\boldsymbol{u}_i(\bar{z},t)) + [\mathbf{G}(\bar{z})\, \mathbf{J}_D(s(\bar{z}))]\, \boldsymbol{u}_i(\bar{z},t) = \\ &= [\bar{\varepsilon}(\bar{z})\, A^2(\boldsymbol{u}_i(\bar{z},t)) - \bar{\varphi}(\bar{z})\, C^2(\boldsymbol{u}_i(\bar{z},t))]\, \boldsymbol{u}_i(\bar{z},t)\,. \end{aligned} \tag{2.131}$$

The given system of differential equations (see Table I) holds in the case of an arbitrary choice of the global coordinate system $\bar{x} - \bar{y} - \bar{z}$. The sub-

Table I. Simultaneous system of differential equations of space vibration for a thinwalled single-box substructure

$$
\begin{aligned}
&- E(\bar z)\, D^2(v_{\bar z}(\bar z, t))\, F_c(s(\bar z)) + E(\bar z)\, D^3(v_{\bar x}(\bar z, t))\, F_{\bar x}(s(\bar z)) + E(\bar z)\, D^3(v_{\bar y}(\bar z, t))\, F_{\bar y}(s(z)) + E(\bar z)\, D^4(\vartheta_T(\bar z, t))\, F_{\bar w_I}(s(\bar z)) + \\
&\qquad + E(\bar z)\, D^4(\varrho_V(\bar z, t))\, F_{\bar w_{II}}(s(\bar z)) = \\
&= \quad m(\bar z)\, A^2(v_{\bar z}(\bar z, t)
\end{aligned}
$$

$$
\begin{aligned}
&- E(\bar z)\, D^3(v_{\bar z}(\bar z, t))\, F_{\bar x}(s(\bar z)) + E(\bar z) D^4(v_{\bar x}(\bar z, t))\, F_{\bar x\bar x}(s(\bar z)) + E(\bar z)\, D^4(v_{\bar y}(\bar z, t))\, F_{\bar x\bar y}(s(\bar z)) + E(\bar z)\, D^4(\vartheta_T(\bar z, t))\, F_{\bar x\bar w_I}(s(\bar z)) + \\
&\qquad + E(\bar z)\, D^4(\varrho_V(\bar z, t))\, F_{\bar x\bar w_{II}}(s(\bar z)) = \\
&= -m(\bar z)\, A^2(v_{\bar x}(\bar z, t)) - (me_{TK_y})(\bar z)\, A^2(\vartheta_T(\bar z, t)) - (me_{TP_y})(\bar z)\, A^2(\varrho_V(\bar z, t)) + m(\bar z) \frac{F_{\bar x\bar x}(s(\bar z))}{F_c(s(\bar z))} C^2(v_{\bar x}(\bar z, t))
\end{aligned}
$$

$$
\begin{aligned}
&- E(\bar z)\, D^3(v_{\bar z}(\bar z, t))\, F_{\bar y}(s(\bar z)) + E(\bar z)\, D^4(v_{\bar x}(\bar z, t))\, F_{\bar y\bar x}(s(\bar z)) + E(\bar z)\, D^4(v_{\bar y}(\bar z, t))\, F_{\bar y\bar y}(s(\bar z)) + E(\bar z)\, D^4(\vartheta_T(\bar z, t))\, F_{\bar y\bar w_I}(s(\bar z)) + \\
&\qquad + E(\bar z)\, D^4(\varrho_V(\bar z, t))\, F_{\bar y\bar w_{II}}(s(\bar z)) = \\
&= -m(\bar z)\, A^2(v_{\bar y}(\bar z, t)) - (me_{TK_x})(\bar z)\, A^2(\vartheta_T(\bar z, t)) - (me_{TP_x})(\bar z) A^2(\varrho_V(\bar z, t)) + m(\bar z) \frac{F_{\bar y\bar y}(s(\bar z))}{F_c(s(\bar z))} C^2(v_{\bar y}(\bar z, t))
\end{aligned}
$$

$$
\begin{aligned}
&- E(\bar z)\, D^3(v_{\bar z}(\bar z, t))\, F_{\bar w_I}(s(\bar z)) + E(\bar z)\, D^4(v_{\bar x}(\bar z, t))\, F_{\bar w_I\bar x}(s(\bar z)) + E(\bar z)\, D^4 v_{\bar y}(\bar z, t))\, F_{\bar w_I\bar y}(s(\bar z)) + E(\bar z)\, D^4(\vartheta_T(z, t)\, F_{\bar w_I\bar w_I}(s(\bar z)) + \\
&\qquad + E(\bar z)\, D^4(\varrho_V(\bar z, t))\, F_{\bar w_I\bar w_{II}}(s(\bar z)) - G(\bar z)\, D^2(\vartheta_T(\bar z, t))\, J_{TB}(\bar z) = \\
&= \quad m(e_{TK_y} + e_{TP_y})(\bar z)\, A^2(v_{\bar x}(s(\bar z)) - m(e_{TK_x} + e_{TP_x})(\bar z)\, A^2(v_{\bar y}(s(\bar z)) - \chi_V(\bar z)\, A^2(\vartheta_T(s(\bar z)) + m(\bar z) \frac{F_{\bar w_I\bar w_I}(s(\bar z))}{F_c(s(\bar z))} C^2(\vartheta_T(\bar z, t))
\end{aligned}
$$

$$
\begin{aligned}
&- E(\bar z)\, D^3(v_{\bar z}(\bar z, t))\, F_{\bar w_{II}}(s(\bar z)) + E(\bar z)\, D^4(v_{\bar x}(\bar z, t))\, F_{\bar w_{II}\bar x}(s(\bar z)) + E(\bar z)\, D^4(v_{\bar y}(\bar z, t))\, F_{\bar w_{II}\bar y}(s(\bar z)) + E(\bar z)\, D^4(\vartheta_T(\bar z, t))\, F_{\bar w_{II}\bar w_I}(s(\bar z)) + \\
&\qquad + E(\bar z)\, D^4(\varrho_V(\bar z, t))\, F_{\bar w_{II}\bar w_{II}}(s(\bar z)) - G(\bar z)\, D^2(\varrho_V(\bar z, t))\, J_{D_I}(\bar z) + G(\bar z)\, \varrho_V(\bar z, t))\, J_{D_I}(\bar z) = \\
&= \quad m(e_{TK_y} + e_{TP_y})(\bar z)\, A^2(v_{\bar x}(s(\bar z)) - m(e_{TK_x} + e_{TP_x})(\bar z)\, A^2(v_{\bar y}(s(\bar z)) - \tau_V(\bar z)\, A^2(\varrho_V(s(\bar z)) + m(\bar z) \frac{F_{\bar w_{II}\bar w_{II}}(s(\bar z))}{F_c(s(\bar z))} C^2(\varrho_V(\bar z, t))
\end{aligned}
$$

stitutions of quadratic integrals are summed up in Table II. The generalized matrix of the used cross-sectional quadratic integrals $\bar{\mathbf{F}}_1(s(\bar{z}))$, the matrices of the torsion-bending and distortion-bending rigidities, together with the matrices of inertial influences are listed in Table III.

For the determination of the generalized transfer matrix of the studied box substructure, the separation of equation system (2.131) must be realized on the basis of the orthogonalization of the stiffness matrices of the studied problem. The transformation matrix $\mathbf{K}_R$ allows the diagonalization of at most two of the three analysed stiffness matrices. The analysis of investigated box cross-section is based on the norm-conditions in field ξ, given by

$$\begin{aligned} \mathbf{F}_{V_1}(s(\xi)) &= \mathrm{Diag}\,(\mathbf{K}_R \bar{\mathbf{F}}_1(s(\xi))\,\mathbf{K}_R^T)\,, \\ \mathbf{F}_{V_2}(s(\xi)) &= \mathrm{Diag}\,(\mathbf{K}_R \bar{\mathbf{F}}_2(s(\xi))\,\mathbf{K}_R^T)\,, \\ \mathbf{F}_{V_3}(s(\xi)) &= \mathrm{Diag}\,(\mathbf{K}_R \bar{\mathbf{F}}_3(s(\xi))\,\mathbf{K}_R^T)\,. \end{aligned} \tag{2.132}$$

The conditions (2.132) are satisfied if there pays the equation of the eigenvalue problem

$$[\mathbf{F}_{V_2}(s(\xi)) - \lambda_i \mathbf{F}_{V_1}(s(\xi))]\, \boldsymbol{q}_{ev} = 0\,, \tag{2.133}$$

where λ_i are the eigenvalues resulting from the condition for nontrivial solutions

$$\mathrm{Det}\,[\mathbf{F}_{V_2}(s(\xi)) - \lambda_i \mathbf{F}_{V_1}(s(\xi))] = 0 \tag{2.134}$$

and $\boldsymbol{q}_{ev}$ are the corresponding eigenvectors. From the norm-condition of eigenvectors follows

$$\mathbf{F}_{V_1}(s(\xi)) = \mathrm{Diag}\,(1)\,, \tag{2.135}$$

$$\mathbf{F}_{V_2}(s(\xi)) = \mathrm{Diag}\,(\lambda_i)\,. \tag{2.136}$$

The basic assumption of the equation (2.133) is the condition of a nonsimultaneous singularity of matrices $\mathbf{F}_{V_1}(s)\xi))$ and $\mathbf{F}_{V_2}(s(\xi))$. It means that the warping functions $\gamma(s(\xi))$ cannot be affine to two basic deflection functions.

First of all the diagonalization of the stiffness matrix $\bar{\mathbf{F}}_1(s(\xi))$ must be realized, consisting of five degrees of orthogonalization, in which the nondiagonal quadratic integrals of matrix $\bar{\mathbf{F}}_1(s(\xi))$ are eliminated.

The first three degrees of orthogonalization on the basis of coupling of corresponding state components on the centre of gravity T, on the principal axes of the cross-section and on the centre of torsion K, with appertaining local coordinate systems (Figure 33) are generally known and are not dealt with in this section. After orthogonalization of the third degree the stiffness matrix $\mathbf{F}_1(s(\xi))$ is given by

Table II. Substitutions of quadratic integrals of matrix $\mathbf{F}_{\mathrm{I}}(s(\bar{z}))$

$F_c(s(\bar{z})) = \left(\oint_s 1^2 t_w \, ds\right)(\bar{z})$	$F_{\bar{x}}(s(\bar{z})) = \left(\oint_s 1\,\bar{x} t_w \, ds\right)(\bar{z})$	$F_{\bar{y}}(s(\bar{z})) = \left(\oint_s 1\,\bar{y} t_w \, ds\right)(\bar{z})$	$F_{\bar{w}_{\mathrm{I}}}(s(\bar{z})) = \left(\oint_s 1\,\bar{w}_{\mathrm{I}} t_w \, ds\right)(\bar{z})$	$F_{\bar{w}_{\mathrm{II}}}(s(\bar{z})) = \left(\oint_s 1\,\bar{w}_{\mathrm{II}} t_w \, ds\right)(\bar{z})$
	$F_{\bar{x}\bar{x}}(s(\bar{z})) = \left(\oint_s \bar{x}^2 t_w \, ds\right)(\bar{z})$	$F_{\bar{x}\bar{y}}(s(\bar{z})) = \left(\oint_s \bar{x}\bar{y} t_w \, ds\right)(\bar{z})$	$F_{\bar{x}\bar{w}_{\mathrm{I}}}(s(\bar{z})) = \left(\oint_s \bar{x}\bar{w}_{\mathrm{I}} t_w \, ds\right)(\bar{z})$	$F_{\bar{x}\bar{w}_{\mathrm{II}}}(s(\bar{z})) = \left(\oint_s \bar{x}\bar{w}_{\mathrm{II}} t_w \, ds\right)(\bar{z})$
	$F_{\bar{y}\bar{x}}(s(\bar{z})) = \left(\oint_s \bar{y}\bar{x} t_w \, ds\right)(\bar{z})$	$F_{\bar{y}\bar{y}}(s(\bar{z})) = \left(\oint_s \bar{y}^2 t_w \, ds\right)(\bar{z})$	$F_{\bar{y}\bar{w}_{\mathrm{I}}}(s(\bar{z})) = \left(\oint_s \bar{y}\bar{w}_{\mathrm{I}} t_w \, ds\right)(\bar{z})$	$F_{\bar{y}\bar{w}_{\mathrm{II}}}(s(\bar{z})) = \left(\oint_s \bar{y}\bar{w}_{\mathrm{II}} t_w \, ds\right)(\bar{z})$
	$F_{\bar{w}_{\mathrm{I}}\bar{x}}(s(\bar{z})) = \left(\oint_s \bar{w}_{\mathrm{I}}\bar{x} t_w \, ds\right)(\bar{z})$	$F_{\bar{w}_{\mathrm{I}}\bar{y}}(s(\bar{z})) = \left(\oint_s \bar{w}_{\mathrm{I}}\bar{y} t_w \, ds\right)(\bar{z})$	$F_{\bar{w}_{\mathrm{I}}\bar{w}_{\mathrm{I}}}(s(\bar{z})) = \left(\oint_s \bar{w}_{\mathrm{I}}^2 t_w \, ds\right)(\bar{z})$	$F_{\bar{w}_{\mathrm{I}}\bar{w}_{\mathrm{II}}}(s(\bar{z})) = \left(\oint_s \bar{w}_{\mathrm{I}}\bar{w}_{\mathrm{II}} t_w \, ds\right)(\bar{z})$
	$F_{\bar{w}_{\mathrm{II}}\bar{x}}(s(\bar{z})) = \left(\oint_s \bar{w}_{\mathrm{II}}\bar{x} t_w \, ds\right)(\bar{z})$	$F_{\bar{w}_{\mathrm{II}}\bar{y}}(s(\bar{z})) = \left(\oint_s \bar{w}_{\mathrm{II}}\bar{y} t_w \, ds\right)(\bar{z})$	$F_{\bar{w}_{\mathrm{II}}\bar{w}_{\mathrm{I}}}(s(\bar{z})) = \left(\oint_s \bar{w}_{\mathrm{II}}\bar{w}_I t_w \, ds\right)(\bar{z})$	$F_{\bar{w}_{II}\bar{w}_{\mathrm{II}}}(s(\bar{z})) = \left(\oint_s \bar{w}_{II}^2 t_w \, ds\right)(\bar{z})$

Table III. The matrix of cross-sectional quadratic integrals, the matrices of torsional and distortional rigidities and the matrices of inertial components of space vibration of the thinwalled single-box substructure

$$\bar{\mathbf{F}}_1(s(\bar{z})) = \begin{bmatrix} F_c(s(\bar{z})) & F_{\bar{x}}(s(\bar{z})) & F_{\bar{y}}(s(\bar{z})) & F_{\bar{w}_{\mathrm{I}}}(s(\bar{z})) & F_{\bar{w}_{\mathrm{II}}}(s(\bar{z})) \\ F_{\bar{x}}(s(\bar{z})) & F_{\bar{x}\bar{x}}(s(\bar{z})) & F_{\bar{x}\bar{y}}(s(\bar{z})) & F_{\bar{x}\bar{w}_{\mathrm{I}}}(s(\bar{z})) & F_{\bar{x}\bar{w}_{\mathrm{II}}}(s(\bar{z})) \\ F_{\bar{y}}(s(\bar{z})) & F_{\bar{y}\bar{x}}(s(\bar{z})) & F_{\bar{y}\bar{y}}(s(\bar{z})) & F_{\bar{y}\bar{w}_{\mathrm{I}}}(s(\bar{z})) & F_{\bar{y}\bar{w}_{\mathrm{II}}}(s(\bar{z})) \\ F_{\bar{w}_{\mathrm{I}}}(s(\bar{z})) & F_{\bar{w}_{\mathrm{I}}\bar{x}}(s(\bar{z})) & F_{\bar{w}_{\mathrm{I}}\bar{y}}(s(\bar{z})) & F_{\bar{w}_{\mathrm{I}}\bar{w}_{\mathrm{I}}}(s(\bar{z})) & F_{\bar{w}_{\mathrm{I}}\bar{w}_{\mathrm{II}}}(s(\bar{z})) \\ F_{\bar{w}_{\mathrm{II}}}(s(\bar{z})) & F_{\bar{w}_{\mathrm{II}}\bar{x}}(s(\bar{z})) & F_{\bar{w}_{\mathrm{II}}\bar{y}}(s(\bar{z})) & F_{\bar{w}_{\mathrm{II}}\bar{w}_{\mathrm{I}}}(s(\bar{z})) & F_{\bar{w}_{\mathrm{II}}\bar{w}_{\mathrm{II}}}(s(\bar{z})) \end{bmatrix},$$

$$\bar{\mathbf{F}}_2(s(\bar{z})) = \begin{bmatrix} 0 & 0 & 0 & 0 & 0 \\ 0 & 0 & 0 & 0 & 0 \\ 0 & 0 & 0 & 0 & 0 \\ 0 & 0 & 0 & J_{\mathrm{I}}(s(\bar{z})) & 0 \\ 0 & 0 & 0 & 0 & 0 \end{bmatrix}, \quad \bar{\mathbf{F}}_3(s(\bar{z})) = \begin{bmatrix} 0 & 0 & 0 & 0 & 0 \\ 0 & 0 & 0 & 0 & 0 \\ 0 & 0 & 0 & 0 & 0 \\ 0 & 0 & 0 & 0 & 0 \\ 0 & 0 & 0 & 0 & J_{\mathrm{II}}(s(\bar{z})) \end{bmatrix},$$

$$\bar{\boldsymbol{\varepsilon}}(s(\bar{z})) = \begin{bmatrix} m & 0 & 0 & 0 & 0 \\ 0 & -m & 0 & -me_{\mathrm{TK}_y}(s(\bar{z})) & -me_{\mathrm{TP}_y}(s(\bar{z})) \\ 0 & 0 & -m & me_{\mathrm{TK}_x}(s(\bar{z})) & me_{\mathrm{TP}_y}(s(\bar{z})) \\ 0 & m(e_{\mathrm{TK}_y}+e_{\mathrm{TP}_y})(s(\bar{z})) & -m(e_{\mathrm{TK}_x}+e_{\mathrm{TP}_x})(s(\bar{z})) & -\chi_{\mathrm{V}} & 0 \\ 0 & m(e_{\mathrm{TK}_y}+e_{\mathrm{TP}_y})(s(\bar{z})) & -m(e_{\mathrm{TK}_x}+e_{\mathrm{TP}_x})(s(\bar{z})) & 0 & -\tau_{\mathrm{V}} \end{bmatrix},$$

$$\chi_{\mathrm{V}} = \chi + [m(e^2_{\mathrm{TK}_x} + e^2_{\mathrm{TK}_y})](s(\bar{z})), \quad \tau_{\mathrm{V}} = \tau + [m(e^2_{\mathrm{TP}_x} + e^2_{\mathrm{TP}_x})](s(\bar{z})),$$

$$\bar{\boldsymbol{\varphi}}(s(\bar{z})) = \begin{bmatrix} 0 & 0 & 0 & 0 & 0 \\ 0 & -m\left(\frac{F_{\bar{x}\bar{x}}}{F_c}\right)(s(\bar{z})) & 0 & 0 & 0 \\ 0 & 0 & -m\left(\frac{F_{\bar{y}\bar{y}}}{F_c}\right)(s(\bar{z})) & 0 & 0 \\ 0 & 0 & 0 & -m\left(\frac{F_{\bar{w}_{\mathrm{I}}\bar{w}_{\mathrm{I}}}}{F_c}\right)(s(\bar{z})) & 0 \\ 0 & 0 & 0 & 0 & -m\left(\frac{F_{\bar{w}_{\mathrm{II}}\bar{w}_{\mathrm{II}}}}{F_c}\right)(s(\bar{z})) \end{bmatrix}.$$

$$\tilde{\mathbf{F}}_1(s(\xi)) = \begin{bmatrix} F_c & 0 & 0 & 0 & 0 \\ 0 & F_{xx} & 0 & 0 & F_{x\mathring{w}_{\mathrm{II}}} \\ 0 & 0 & F_{yy} & 0 & F_{y\mathring{w}_{\mathrm{II}}} \\ 0 & 0 & 0 & F_{w_{\mathrm{I}}w_{\mathrm{I}}} & F_{\mathring{w}_{\mathrm{I}}\mathring{w}_{\mathrm{II}}} \\ 0 & F_{\mathring{w}_{\mathrm{II}}x} & F_{\mathring{w}_{\mathrm{II}}y} & F_{\mathring{w}_{\mathrm{II}}w_{\mathrm{I}}} & F_{w_{\mathrm{II}}\mathring{w}_{\mathrm{II}}} \end{bmatrix}, \tag{2.137}$$

with integrals F_{ij} as functions of s and ξ in accordance with Table II.

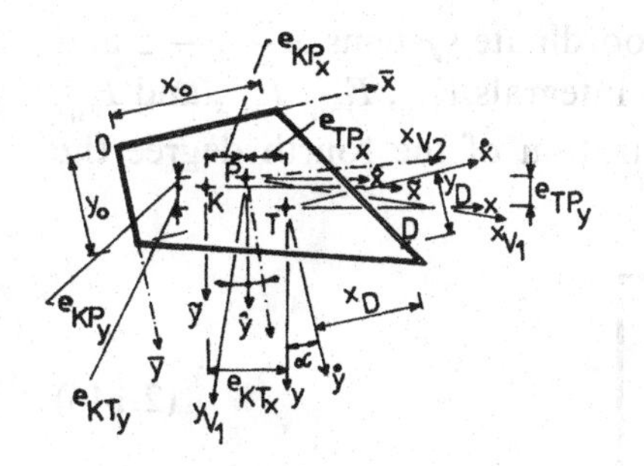

Fig. 33. Reference of the coordinate systems to three characteristic points of the cross-section denoted by the centre of gravity T, the centre of torsion K and the centre of distortion P.

In the fourth degree of orthogonalization the combined bending-distortional terms $F_{x\mathring{w}_{II}}$, $F_{y\mathring{w}_{II}}$, $F_{\mathring{w}_{II}x}$ and $F_{\mathring{w}_{II}y}$ of the stiffness matrix $\tilde{\mathbf{F}}_1(s(\xi))$ are eliminated. The coordinate system $\tilde{x}-\tilde{y}-\tilde{z}$, bounded on the centre of torsion K, is transformed into the new system $\hat{x}-\hat{y}-\hat{z}$ with the initial point in the centre of distortion P (Figure 33). By coupling of the distortional components at the centre of gravity, for the case of the basic distortional warping deflection of arbitrary point of the centre-line D (see Figure 33) there pays

$$w_{II}(s(\xi)) = \mathring{w}_{II}(s(\xi)) + e_{TP_y}(s(\xi))\,x_D(s(\xi)) - e_{TP_x}(s(\xi))\,y_D(s(\xi))\,. \tag{2.138}$$

Substituting into the equivalency conditions of the orthogonalization of the first degree

$$F_{\tilde{x}\mathring{w}_{II}} = 0\,,\quad F_{\tilde{y}\mathring{w}_{II}} = 0\,, \tag{2.139}$$

derived implying the condition of zero bending moments of the distortional behaviour of the cross-section, there yields the following set of equations

$$F_{\tilde{x}\mathring{w}_{II}} + e_{TP_y}F_{\tilde{x}\tilde{x}} - e_{TP_x}F_{\tilde{x}\tilde{y}} = 0\,, \tag{1.140}$$

$$F_{\tilde{y}\mathring{w}_{II}} + e_{TP_y}F_{\tilde{y}\tilde{x}} - e_{TP_x}F_{\tilde{y}\tilde{y}} = 0\,, \tag{2.141}$$

from which follows the position of the centre of distortion P in the coordinate system $\bar{x}-\bar{y}-\bar{z}$

$$e_{TP_x} = \frac{F_{\tilde{y}\mathring{w}_{II}}F_{\tilde{x}\tilde{x}} - F_{\tilde{x}\mathring{w}_{II}}F_{\tilde{x}\tilde{y}}}{F_{\tilde{x}\tilde{x}}F_{\tilde{y}\tilde{y}} - F^2_{\tilde{x}\tilde{y}}}\,, \tag{2.142}$$

$$e_{TP_y} = \frac{F_{\tilde{y}\mathring{w}_{II}}F_{\tilde{x}\tilde{y}} - F_{\tilde{x}\mathring{w}_{II}}F_{\tilde{y}\tilde{y}}}{F_{\tilde{x}\tilde{x}}F_{\tilde{y}\tilde{y}} - F^2_{\tilde{x}\tilde{y}}}\,. \tag{2.143}$$

The position of the centre of distortion in the coordinate system $x-y-z$ ($F_{xy}=0$) is given by

$$e_{TP_x} = \frac{F_{y\mathring{w}_{II}}}{F_{yy}}\,,\quad e_{TP_y} = -\frac{F_{x\mathring{w}_{II}}}{F_{xx}}\,. \tag{2.144, 2.145}$$

In relation to the centre of distortion, in coordinate systems $\mathring{x} - \mathring{y} - \mathring{z}$ and $x - y - z$, the condition that the quadratic integrals $F_{x\mathring{w}_{II}}$, $F_{y\mathring{w}_{II}}$, $F_{\mathring{w}_{II}x}$ and $F_{\mathring{w}_{II}y}$ are zero is satisfied. After the orthogonalization of the fourth degree the stiffness matrix $\mathbf{\hat{F}}_I(s(\xi))$ has form

$$\mathbf{\hat{F}}_I(s(\xi)) = \begin{bmatrix} F_c & 0 & 0 & 0 & 0 \\ 0 & F_{xx} & 0 & 0 & 0 \\ 0 & 0 & F_{yy} & 0 & 0 \\ 0 & 0 & 0 & F_{w_I w_I} & F_{w_I \mathring{w}_{II}} \\ 0 & 0 & 0 & F_{\mathring{w}_{II} w_I} & F_{\mathring{w}_{II} \mathring{w}_{II}} \end{bmatrix}. \tag{2.146}$$

In the final, the fifth orthogonalization degree of stiffness matrix $\hat{F}_I(s(\xi))$, the residual combined torsional and distortional terms $F_{w_I \mathring{w}_{II}}$ and $F_{\mathring{w}_{II} w_I}$ are eliminated. The coordinate system $\hat{x} - \hat{y} - \hat{z}$ bounded on the centre of distortion P, is transformed into new coordinate system $x_{v_1} - y_{v_1} - z_{v_1}$ or $x_{v_2} - y_{v_2} - z_{v_2}$ which form the angle $\pm\beta$ with the coordinate system $\hat{x} - \hat{y} - \hat{z}$. For an arbitrary point D placed on the centre-line of the cross-section of the studied substructure, the following coordinate substitutions can be introduced

$$x_D = \mathring{x} \cos \beta_V + \mathring{y} \sin \beta_V\,, \tag{2.147}$$

$$y_D = \mathring{y} \cos \beta_V + \mathring{x} \sin \beta_V\,. \tag{2.148}$$

Satisfying the equivalency condition

$$\oint_{s(\xi)} w_I \mathring{w}_{II} \, \mathrm{d}(s(\xi)) = 0\,, \tag{2.149}$$

determines the equation

$$\begin{aligned} & F_{\mathring{w}_I \mathring{w}_{II}} + e_{TK_x} e_{TP_y} F_{\mathring{x}\mathring{x}} \cos^2 \beta_V + e_{TK_y} e_{TP_y} F_{\mathring{y}\mathring{y}} \sin^2 \beta_V + \\ & + e_{TP_x} e_{TK_y} F_{\mathring{x}\mathring{x}} \cos \beta_V \sin \beta_V - e_{TK_y} e_{TP_x} F_{\mathring{y}\mathring{y}} \sin \beta_V \cos \beta_V + \\ & + e_{TK_y} e_{TP_x} F_{\mathring{y}\mathring{y}} \cos^2 \beta_V + e_{TK_x} e_{TP_x} F_{\mathring{x}\mathring{x}} \sin^2 \beta_V + \\ & + e_{TK_x} e_{TP_y} F_{\mathring{x}\mathring{x}} \cos \beta_V \sin \beta_V - e_{TK_x} e_{TP_y} F_{\mathring{y}\mathring{y}} \cos \beta_V = 0\,. \end{aligned} \tag{2.150}$$

After some arrangements there a quadratic equation is obtained the solution of which gives the sought angles $\pm\beta_V$ determining the location of the coordinate systems $x_{v_1} - y_{v_1} - z_{v_1}$ and $x_{v_2} - y_{v_2} - z_{v_2}$ in relation to the coordinate system $\hat{x} - \hat{y} - \hat{z}$. In such coordinate systems the quadratic integrals $F_{w_I \mathring{w}_{II}}$ and $F_{\mathring{w}_{II} w_I}$ have zero values. The resulting distortional components due to the torsion-bending behaviour act on the eccentricities e_{KP_x} and e_{KP_y} and cause the mutual coupling of torsion-bending and distortion-bending effects in accordance with reference [75].

After the fifth degree of orthogonalization the matrix $\mathbf{F}_V(s(\xi))$ has the form

$$\mathbf{F}_V(s(\xi)) = \begin{bmatrix} F_c & & & & \\ & F_{xx} & & & \\ & & F_{yy} & & \\ & & & F_{w_I w_I} & \\ & & & & F_{w_{II} w_{II}} \end{bmatrix}. \tag{2.151}$$

The orthogonalized stiffness matrix $F_V(s(\xi))$ does until now not involve the shear strains due to individual separate kinds of spatial behaviour of the studied thinwalled box substructure. Especially the shear strains occurring due to the action of antisymmetric kinds of behaviour (torsion-bending and distortion-bending) are of great importance in the analysis of thinwalled box systems [14, 28, 76]. The shear stresses, variable over the cross-section, cause the nonlinear warping deflections of a thinwalled box substructure and an exact analysis of this problem requires to give up from the Wagner hypothesis, which would lead to theoretical discrepancies when compared with other theoretical analyses in this section. This was the reason for the application of the following refined approximate method for the solution of shear deformations on the studied box substructure.

The angles of shear deformations, variable at the centre-line of the cross-section and corresponding to symmetric and antisymmetric kinds of behaviour, are replaced by the average constant angles in field ξ

$$\eta_x(s(\xi), t) = \frac{v_x(\xi, t)}{v_x(s(\xi))(GF_x)(s(\xi))}, \tag{2.152}$$

$$\eta_y(s(\xi), t) = \frac{v_y(\xi, t)}{v_x(s(\xi))(GF_y)(s(\xi))}, \tag{2.153}$$

$$\eta_{w_I}(s(\xi), t) = \frac{\vartheta_T(\xi, t)}{v_{w_I}(s(\xi))(GJ_{TB})(s(\xi))}, \tag{2.154}$$

$$\eta_{w_{II}}(s(\xi), t) = \frac{\varrho_V(\xi, t)}{v_{w_{II}}(s(\xi))(GJ_{D_I})(s(\xi))}, \tag{2.155}$$

which after substitution into

$$\begin{aligned} v_z(\xi, t) = v_z(\xi, t) - [D(v_x(\xi, t)) + \eta_x(s(\xi), t)]\, x_D(s(\xi)) - \\ - [D(v_y(\xi, t)) + \eta_y(s(\xi), t)]\, y_D(s(\xi)) - \end{aligned}$$

$$- [D(\vartheta_{\mathrm{T}}(\xi, t)) + \eta_{w_{\mathrm{I}}}(s(\xi), t)]\, w_{\mathrm{I_D}}(s(\xi)) -$$

$$- [D(\varrho_{\mathrm{V}}(\xi, t)) + \eta_{w_{\mathrm{II}}}(s(\xi), t)]\, w_{\mathrm{II_D}}(s(\xi)), \tag{2.156}$$

determine the instantaneous location of the point D, with the coordinates $x_{\mathrm{D}}(s(\xi))$, $y_{\mathrm{D}}(s(\xi))$, $w_{\mathrm{I_D}}(s(\xi))$, $w_{\mathrm{II_D}}(s(\xi))$, which is positioned on the centre-line of the cross-section. The vector of coefficients

$$\boldsymbol{b}^{-1}(s(\xi)) = \begin{bmatrix} v_x(s(\xi)) \\ v_y(s(\xi)) \\ v_{w_{\mathrm{I}}}(s(\xi)) \\ v_{w_{\mathrm{II}}}(s(\xi)) \end{bmatrix} \tag{2.157}$$

determines the energetic average level of the relation of internal and external works of the primary shear flows of space behaviour of substructure. The individual coefficients are given by the relations

$$\frac{1}{v_x} = \frac{F_x}{F_{xx}} \int_{F_c} \left[\frac{\oint (F_x t_{\mathrm{w}}^{-1})\, \mathrm{d}(s(\xi))}{\oint t_{\mathrm{w}}^{-1}\, \mathrm{d}(s(\xi))} - F_x \right]^2 \frac{\mathrm{d}F_c}{t_{\mathrm{w}}}, \tag{2.158}$$

$$\frac{1}{v_y} = \frac{F_y}{F_{yy}} \int_{F_c} \left[\frac{\oint (F_y t_{\mathrm{w}}^{-1})\, \mathrm{d}(s(\xi))}{\oint t_{\mathrm{w}}^{-1}\mathrm{d}(s(\xi))} - F_y \right]^2 \frac{\mathrm{d}F_c}{t_{\mathrm{w}}}, \tag{2.159}$$

$$\frac{1}{v_{w_{\mathrm{I}}}} = \frac{J_{\mathrm{TB}}}{F_{w_{\mathrm{I}}w_{\mathrm{I}}}} \int_{F_c} \left[\frac{\oint (F_{w_{\mathrm{I}}} t_{\mathrm{w}}^{-1})\, \mathrm{d}(s(\xi))}{\oint t_{\mathrm{w}}^{-1}\, \mathrm{d}(s(\xi))} - F_{w_{\mathrm{I}}} \right]^2 \frac{\mathrm{d}F_c}{t_{\mathrm{w}}}, \tag{2.160}$$

$$\frac{1}{v_{w_{\mathrm{II}}}} = \frac{J_{\mathrm{D_I}}}{F_{w_{\mathrm{II}}w_{\mathrm{II}}}} \int_{F_c} \left[\frac{\oint (F_{w_{\mathrm{II}}} t_{\mathrm{w}}^{-1})\, \mathrm{d}(s(\xi))}{\oint t_{\mathrm{w}}^{-1}\, \mathrm{d}(s(\xi))} - F_{w_{\mathrm{II}}} \right]^2 \frac{\mathrm{d}F_c}{t_{\mathrm{w}}}. \tag{2.161}$$

The corrected terms of the vector of amplitudes of deformation state components, taking into consideration the shear strains, are determined by the formulae

$$D(v_x^{\mathrm{s}}(\xi, t)) = D(v_x(\xi, t)) + \eta_x(s(\xi), t), \tag{2.162}$$

$$D(v_y^{\mathrm{s}}(\xi, t)) = D(v_y(\xi, t)) + \eta_y(s(\xi), t)), \tag{2.163}$$

$$D(\vartheta_{\mathrm{T}}^{\mathrm{s}}(\xi, t)) = D(\vartheta_{\mathrm{T}}(\xi, t)) + \eta_{w_{\mathrm{I}}}(s(\xi), t), \tag{2.164}$$

$$D(\varrho_{\mathrm{V}}^{\mathrm{s}}(\xi, t)) = D(\varrho_{\mathrm{V}}(\xi, t)) + \eta_{w_{\mathrm{II}}}(s(\xi), t). \tag{2.165}$$

When using the substitutions

$$F_{xx}^{\mathrm{s}} = \frac{F_{xx}}{\varkappa_x}, \quad F_{yy}^{\mathrm{s}} = \frac{F_{yy}}{\varkappa_y}, \quad F_{w_{\mathrm{I}}w_{\mathrm{I}}}^{\mathrm{s}} = \frac{F_{w_{\mathrm{I}}w_{\mathrm{I}}}}{\varkappa_{w_{\mathrm{I}}}}, \quad F_{w_{\mathrm{II}}w_{\mathrm{II}}}^{\mathrm{s}} = \frac{F_{w_{\mathrm{II}}w_{\mathrm{II}}}}{\varkappa_{w_{\mathrm{II}}}}, \tag{2.166}$$

with coefficients

$$\varkappa_x(s(\xi)) = \left(1 + \frac{F_x}{v_x GF_c}\right)^{-1}, \quad \varkappa_y(s(\xi)) = \left(1 + \frac{F_y}{v_y GF_c}\right)^{-1},$$

$$\varkappa_{w_I}(s(\xi)) = \left(1 + \frac{1}{v_{w_I}}\right)^{-1}, \quad \varkappa_{w_{II}}(s(\xi)) = \left(1 + \frac{1}{v_{w_{II}}}\right)^{-1}, \tag{2.167}$$

the orthogonalized matrix equation (2.131) becomes

$$\begin{aligned}&[\mathbf{EF}^s_V(s)](\xi)D^4(u^s_i(\xi,t)) - [\mathbf{GF}^s_2(s)](\xi)D^2(u^s_i(\xi,t)) - \\ &- [\mathbf{GF}^s_3(s)](\xi)D^2(u^s_i(\xi,t)) + [\mathbf{GJ}_D(s)](\xi)u^s_i(\xi,t) = \\ &= [\bar{\varepsilon}(\xi)A^2(u^s_i(\xi,t)) - \bar{\varphi}(\xi)C^2(u^s_i(\xi,t))]\,u^s_i(\xi,t)\,.\end{aligned} \tag{2.168}$$

Superscript s denotes the state components with consideration of the shear deformations. The moduli of elasticity are assumed in the aforementioned complex formulation.

For the thinwalled box substructure K, positioned in the global coordinate system $\bar{x} - \bar{y} - \bar{z}$ (Figure 31), the couplings of the state components are related to the three aforementioned characteristic points of the cross-section and to the local coordinate systems corresponding to the individual degrees of orthogonalization. The vectors of state components in the initial and end points of the investigated box substructure of length l_K are defined by

$$\begin{aligned}\boldsymbol{k}^s_{V,K}(0,t) = \; &[v^s_{z,K}(0,t),\; N^s_{z,K}(0,t),\\ &v^s_{x,K}(0,t),\; \varphi^s_{x,K}(0,t),\; M^s_{x,K}(0,t),\; Q^s_{x,K}(0,t),\\ &v^s_{y,K}(0,t),\; \varphi^s_{y,K}(0,t),\; M^s_{y,K}(0,t),\; Q^s_{y,K}(0,t),\\ &\vartheta^s_{T,K}(0,t),\; \zeta^s_{T,K}(0,t),\; W^s_{T,K}(0,t),\; T^s_{T,K}(0,t),\\ &\varrho^s_{V,K}(0,t),\; \beta^s_{V,K}(0,t),\; M^s_{W,K}(0,t),\; M^s_{V,K}(0,t),\; 1]^T\end{aligned} \tag{2.169}$$

and

$$\begin{aligned}\boldsymbol{k}^s_{V,K}(l_K,t) = \; &[v^s_{z,K}(l_K,t),\; N^s_{z,K}(l_K,t),\\ &v^s_{x,K}(l_K,t),\; \varphi^s_{x,K}(l_K,t),\; M^s_{x,K}(l_K,t),\; Q^s_{x,K}(l_K,t),\\ &v^s_{y,K}(l_K,t),\; \varphi^s_{y,K}(l_K,t),\; M^s_{y,K}(l_K,t),\; Q^s_{y,K}(l_K,t),\\ &\vartheta^s_{T,K}(l_K,t),\; \zeta^s_{T,K}(l_K,t),\; W^s_{T,K}(l_K,t),\; T^s_{T,K}(l_K,t),\\ &\varrho^s_{V,K}(l_K,t),\; \beta^s_{V,K}(l_K,t),\; M^s_{W,K}(l_K,t),\; M^s_{V,K}(l_K,t),\; 1]^T.\end{aligned} \tag{2.170}$$

where $v^s_{z,K}$ and $N^s_{z,K}$ are the deformation and force components of axial behaviour, $v^s_{x,K}$, $\varphi^s_{x,K}$, $M^s_{x,K}$ and $Q^s_{x,K}$ are the deformation and force components of bending behaviour in the x-direction, e.g., displacement, slope, bending moment and transverse force. A similar notation holds for the next four parameters corresponding to the bending action in direction y. Further, the state parameters $\vartheta^s_{T,K}$, $\zeta^s_{T,K}$, $W^s_{T,K}$ and $T^s_{T,K}$ determine the torsion-bending behaviour of the analysed substructure, e.g., the angle of torsion, corresponding slope, bimoment and torque. Finally, the last four parameters of both vectors represent the state components for distortion-bending behaviour defined as the angle of distortion $\varrho^{s}_{V,K}$, the first derivative of the angle of distortion $\beta^{s}_{V,K}$, distortional bimoment $M^{s}_{W,K}$ and distortional moment $M^{s}_{V,K}$. More information about the physical significance of the distortion-bending parameters is given, for example, in references [27, 83, 112].

Applying the techniques in Section 2.3 for solving of individual separated equations of the orthogonalized system (2.168) the transfer and nodal submatrices $\mathbf{BK}_i$ and $\mathbf{CK}_i$ were derived corresponding to various cases of symmetric and antisymmetric behaviour of the studied thinwalled box substructure [19, 23, 42, 48, 64, 81, 83, 112, 122, 126]. These matrices are given by A-3 — A-9 in the CTM. The scheme of a generalized transfer matrix of space behaviour $\mathbf{BK}_V$, constructed by the diagonal assembly of submatrices for axial-, biaxially bending-, torsion-bending and distortion-bending behaviour of studied substructure is given by A-10 in the CTM. When deriving the corresponding generalized nodal matrix $\mathbf{CK}_V$, the transformation of the system of three characteristic points of the cross-section, implied by the mutual couplings of symmetric and antisymmetric components of the space behaviour on the eccentricities between such points, is to be taken into account [83]. The coupling members of the mutual interdependence of individual separated kinds of behaviour on such eccentricities are listed in the generalized nodal matrix $\mathbf{CK}_V$ in the term A-11 of the CTM. In the generalized nodal matrix the mass and inertial terms are incorporated [83] as well as the stiffness characteristics of supports or stiffening elements when located in the nodal points of two- substructures.

The derived thinwalled box substructure represents the basic one-dimensional element for the construction of more complicated two- or three-dimensional substructures. The assemblage of the present box substructures allows the simulation of stiffening grids of orthotropic plates or shells as will be shown later. The present substructure may be further modified when solving thinwalled systems with multicell or open cross-sections or combinations thereof [75, 81, 86]. For nondeformable cross-sections the derivation of the generalized transfer and nodal matrices may be simplified by neglecting the distortion-bending behaviour of the substructure. Similarly, when is it possible to replace the torsion-bending behaviour by the pure torsion action (see term A-8 of the CTM), the simplified matrix expressions may be used as will be shown in the following section.

2.5. Grillages

Grillages are the typical substructures of two- or three-dimensional orthotropically stiffened structural configurations. The simulation of such substructures is performed over the structural set-up of one-dimensional members derived in the previous section. Applied is the simulation in accordance with Figure 34, which consists of primary load-bearing members in the longitudinal direction and of the secondary members in the transverse direction. The couplings of primary members are performed over the secondary members of the simulation. As mentioned earlier, for the analysis there will be assumed the simplified bending and pure torsion action of the present primary and secondary thinwalled members. The conjugate pairs of deformation and force components of the bending and pure torsion behaviour of the primary members in arbitrary discrete section K of the longitudinal axis of the applied simulation (see Figure 34) are put into the state vector as follows

$$\begin{aligned}\boldsymbol{k}_{\mathrm{K}}(z,t) = \ & [v_{x,1}(z,t),\ \varphi_{x,1}(z,t),\ M_{x,1}(z,t),\ Q_{x,1}(z,t),\\ & \vartheta_{\mathrm{T},1}(z,t),\ T_{\mathrm{T},1}(z,t),\\ & v_{x,2}(z,t),\ \varphi_{x,2}(z,t),\ M_{x,2}(z,t),\ Q_{x,2}(z,t),\\ & \vartheta_{\mathrm{T},2}(z,t),\ T_{\mathrm{T},2}(z,t),\\ & v_{x,3}(z,t),\ \varphi_{x,3}(z,t),\ M_{x,3}(z,t),\ Q_{x,3}(z,t),\\ & \vartheta_{\mathrm{T},3}(z,t),\ T_{\mathrm{T},3}(z,t),\\ & \vdots\\ & v_{x,n}(z,t),\ \varphi_{x,n}(z,t),\ M_{x,n}(z,t),\ Q_{x,n}(z,t),\\ & \vartheta_{\mathrm{T},n}(z,t),\ T_{\mathrm{T},n}(z,t)]^{\mathrm{T}},\end{aligned} \tag{2.171}$$

where the notation of the individual components is as defined earlier. The subscript n denotes the number of primary members in the assumed simulation. All state components are assumed with consideration of shear strains in accordance with foregoing analysis. In the notation the superscript s, for simplification, is omitted.

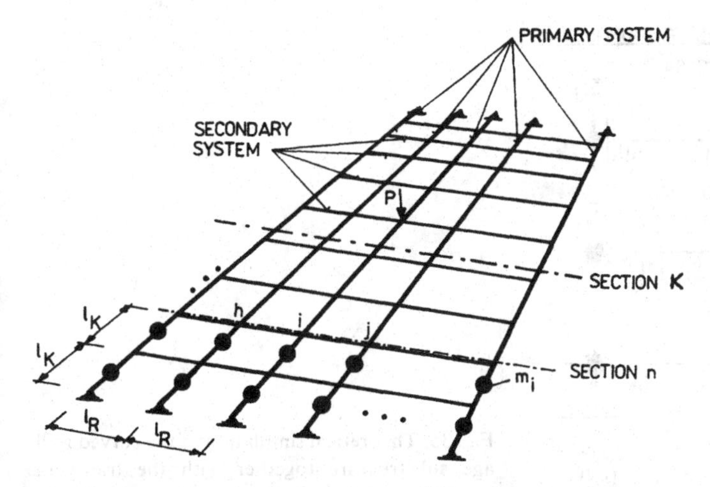

Fig. 34. Theoretical simulation of the grillage substructure.

The transfer submatrix $\mathbf{BK}_i$ for a single primary member is obtained via a diagonal assembly of transfer submatrices for bending and pure torsion action combining the corresponding transfer matrix expressions in terms A-4 and A-8 of the CTM. When the primary element is subjected to the simultaneous action of axial, bending and torsional influences, the corresponding transfer submatrix contains the terms A-5 or A-6 (depending on the tensile or compressive axial forces combined with the bending behaviour) as well as the term A-8 of the CTM. The scheme of such submatrices is illustrated in the terms B-1, B-2 and B-3 of the CTM. The resulting transfer matrices for all primary members of the studied nodal section are constructed by a diagonal assembly of such a submatrices as is shown schematically in the terms B-4, B-5 and B-6 of the CTM.

The information over the structural configuration of the generalized curved grid substructures is carried by the application of a similar curved substructuring simulation as shown in Figure 35. The primary load-bearing

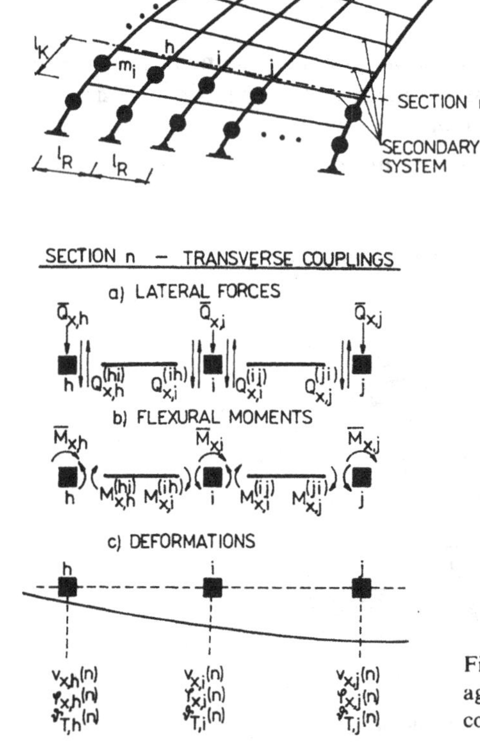

Fig. 35. Theoretical simulation of the curved grillage substructure together with the transverse couplings of primary elements.

system is created by longitudinally curved main members. The secondary load-bearing system is defined by the transversal couplings of primary elements. The parameters of the primary load-bearing system are summed up in the transfer matrices, whereas the parameters of the secondary load-bearing system are summed up in the corresponding nodal matrices [102].

The cross-sectional dimensions of the curved primary member are assumed to be small compared with their horizontal radii of curvature and their length, which consequently restricts the member to the range of slender, one-dimensional elements. The corresponding transfer submatrix and generalized transfer matrix of the curved grid substructure were derived, for example, in references [60] and [87] and are summed up in the terms B-7 and B-8 of the CTM.

Attention will be focused further on the determination of the influences of transverse couplings of the studied types of primary load-bearing members. Transverse couplings over the secondary members of the applied simulation will be analysed.

The state components of the secondary load-bearing members will be denoted further by the superscript (s). In the case of action of lateral forcing functions, the arbitrary nodal point will perform the vertical flexure $v_{x,i}$ together with longitudinal and transversal slopes and rotations $\varphi_{x,i}$ and $\vartheta_{T,i}$. The continuity conditions of displacements and force components in each direction of primary and secondary members of the load-bearing grid must be satisfied, given by

$$v_{x,i} = v^{(s)}_{x,i},\ \varphi_{x,i} = \vartheta^{(s)}_{T,i},\ \vartheta_{T,i} = -\ \varphi^{(s)}_{x,i}, \tag{2.172}$$

$$M_{x,i} = M^{(ih)}_{x,i} + M^{(ij)}_{x,i},\ Q_{x,i} = Q^{(ih)}_{x,i} - Q^{(ij)}_{x,i} \tag{2.173}$$

(see Figure 35). The components of the force state parameters of secondary load-bearing elements can be expressed as the functions of unknown nodal displacements. The individual terms of the nodal submatrix of transverse interaction were derived for the primary members elastically coupled on both edges and subjected to flexural behaviour. The functional couplings of displacement and force state components are expressed as

$$M^{(s)}_{x,i} = -\frac{6(EF_{xx})^{(s)}}{l^2_R} v^{(s)}_{x,j} + \frac{2(EF_{xx})^{(s)}}{l_R} \varphi^{(s)}_{x,j} + + \frac{6(EF_{xx})^{(s)}}{l^2_R} v^{(s)}_{x,i} + \frac{4(EF_{xx})^{(s)}}{l_R} \varphi^{(s)}_{x,i} + \bar{M}_{x,i}, \tag{2.174}$$

$$Q^{(s)}_{x,i} = \frac{12(EF_{xx})^{(s)}}{l^3_R} v^{(s)}_{x,j} - \frac{6(EF_{xx})^{(s)}}{l^2_R} \varphi^{(s)}_{x,j} -$$

$$- \frac{12(EF_{xx})^{(s)}}{l_R^3} v_{x,i}^{(s)} - \frac{6(EF_{xx})^{(s)}}{l_R^2} \varphi_{x,i}^{(s)} + \bar{Q}_{x,i}, \tag{2.175}$$

$$M_{x,j}^{(s)} = - \frac{6(EF_{xx})^{(s)}}{l_R^2} v_{x,j}^{(s)} + \frac{4(EF_{xx})^{(s)}}{l_R} \varphi_{x,j}^{(s)} +$$

$$+ \frac{6(EF_{xx})^{(s)}}{l_R^2} v_{x,i}^{(s)} + \frac{2(EF_{xx})^{(s)}}{l_R} \varphi_{x,i}^{(s)} + \bar{M}_{x,j}, \tag{2.176}$$

$$Q_{x,j}^{(s)} = \frac{12(EF_{xx})^{(s)}}{l_R^3} v_{x,j}^{(s)} - \frac{6(EF_{xx})^{(s)}}{l_R^2} \varphi_{x,j}^{(s)} -$$

$$- \frac{12(EF_{xx})^{(s)}}{l_R^3} v_{x,i}^{(s)} - \frac{6(EF_{xx})^{(s)}}{l_R^2} \varphi_{x,i}^{(s)} + \bar{Q}_{x,j}, \tag{2.177}$$

where $\bar{M}_{x,i}$, $\bar{Q}_{x,i}$, $\bar{M}_{x,j}$ and $\bar{Q}_{x,j}$ are the external moments and lateral forces, l_R is the distance between longitudinal members of the applied simulation and $(EF_{xx})^{(s)}$ is the corresponding flexural stiffness of the studied load-bearing members.

The analogous interaction of the force and deformation components of torsional behaviour of the secondary load-bearing members can be written as

$$M_{x,i}^{(s)} = \frac{(GJ_T)^{(s)}}{l_R} \vartheta_{T,i}^{(s)} - \frac{(GJ_T)^{(s)}}{l_R} \vartheta_{T,j}^{(s)}, \tag{2.178}$$

$$M_{x,j}^{(s)} = - \frac{(GJ_T)^{(s)}}{l_R} \vartheta_{T,i}^{(s)} + \frac{(GJ_T)^{(s)}}{l_R} \vartheta_{T,j}^{(s)}, \tag{2.179}$$

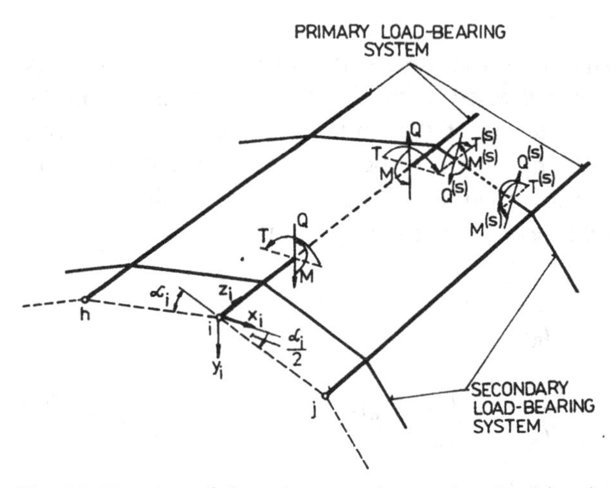

Fig. 36. Notation of the primary and secondary load-bearing systems of the studied spatial grillage substructure.

with continuity conditions

$$\vartheta_{T,i} = \varphi^{(s)}_{x,i},\ M_{x,i} = M^{(ih)}_{x,i} + M^{(ij)}_{x,i},\ T_{T,i} = M^{(s)}_{x,i}. \tag{2.180}$$

Regarding the aforementioned relations, the nodal submatrices as well as the generalized nodal matrix of transverse interaction of primary load-bearing members were derived, which are summed up in the terms B-9 and B-10 of the CTM.

The analogous generalized nodal matrix of the transverse interaction pays also for the analyses of curved grid simulations [60].

When studying the inplane action of the studied grid simulations, the inplane flexural behaviour combined with the axial action of individual members may be assumed. The corresponding transfer and nodal matrices are summed up in the terms B-11 to B-16 of the CTM.

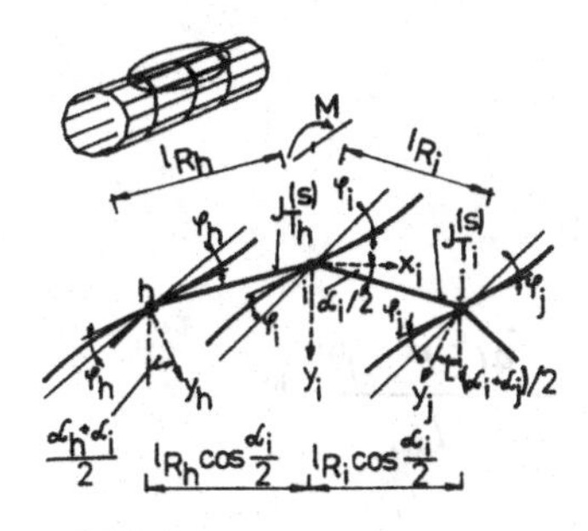

Fig. 37. Scheme of transverse interaction for the determination of relations bending moments versus displacements.

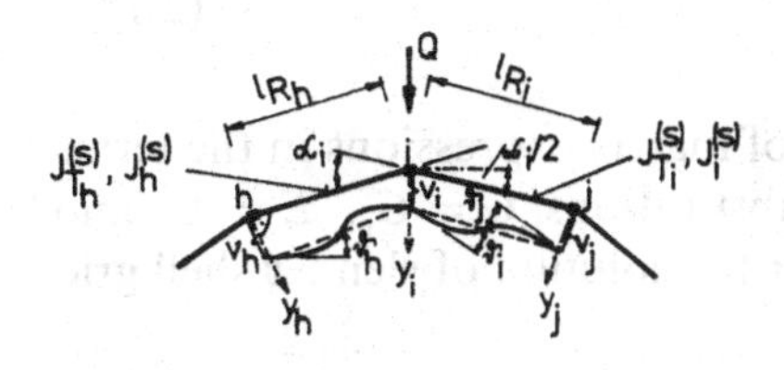

Fig. 38. Scheme of the transverse interaction for the determination of the relations between lateral forces and displacements.

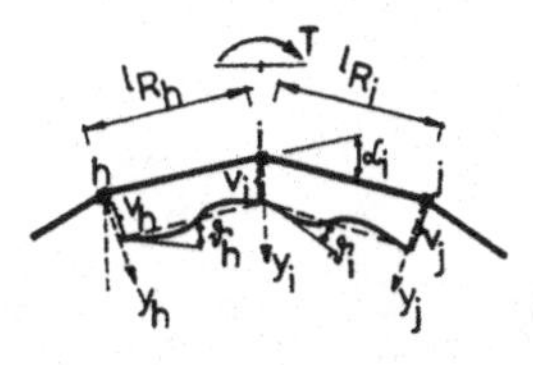

Fig. 39. Scheme of the transverse interaction for the determination of the relations between torques and displacements.

Further development of the present analysis is directed towards the solution of spatial shell grids as shown in Figure 36. In accordance with the Figures 37, 38 and 39, the transformation expressions for the bending moments, lateral forces and torsional moments may be derived, giving

$$M^{(s)}_{x,i} = \frac{(GJ_{T,h})^{(s)} \cos(\alpha_h + \alpha_i)/2}{l_{R,h} \cos(\alpha_i/2)}\, \varphi^{(s)}_{x,n} - \left(\frac{(GJ_{T,h})^{(s)}}{l_{R,h} \cos(\alpha_i/2)} + \frac{(GJ_{T,i})^{(s)}}{l_{R,i} \cos(\alpha_i/2)}\right) \varphi^{(s)}_{x,i} +$$

$$+\frac{(GJ_{T,i})^{(s)}\cos((\alpha_i+\alpha_j)/2)}{l_{R,i}\cos(\alpha_i/2)}\varphi^{(s)}_{x,j}, \tag{2.181}$$

$$\begin{aligned}Q^{(s)}_{x,i} &= \frac{-12(EF_{xx,h})^{(s)}\cos(\alpha_h/2)\cos(\alpha_i/2)}{l^3_{R,h}}v^{(s)}_{x,h}+\\ &+\left[\left(\frac{12(EF_{xx,h})^{(s)}}{l^3_{R,h}}+\frac{12(EF_{xx,i})^{(s)}}{l^3_{R,i}}\right)\cos^2(\alpha_i/2)\right]v^{(s)}_{x,i}-\\ &-\frac{12(EF_{xx,i})^{(s)}\cos(\alpha_i/2)\cos(\alpha_j/2)}{l^3_{R,i}}v^{(s)}_{x,j}-\\ &-\frac{6(EJ_{T,h})^{(s)}\cos(\alpha_i/2)}{l^2_{R,h}}\vartheta^{(s)}_{T,h}+\frac{6(EJ_{T,i})^{(s)}\cos(\alpha_i/2)}{l^2_{R,i}}\vartheta^{(s)}_{T,j},\end{aligned} \tag{2.182}$$

$$\begin{aligned}T^{(s)}_{T,i} &= -\frac{6(EF_{xx,h})^{(s)}\cos(\alpha_h/2)}{l^2_{R,h}}v^{(s)}_{x,h}+\\ &+\left(\frac{6(EF_{xx,h})^{(s)}}{l^2_{R,h}}-\frac{6(EF_{xx,i})^{(s)}}{l^2_{R,i}}\right)\cos(\alpha_i/2)\,v^{(s)}_{x,i}+\\ &+\frac{6(EF_{xx,i})^{(s)}\cos(\alpha_j/2)}{l^2_{R,i}}v^{(s)}_{x,j}-\frac{2(EF_{xx,h})^{(s)}}{l_{R,h}}\vartheta^{(s)}_{T,h}-\frac{4(EF_{xx,h})^{(s)}}{l_{R,h}}\vartheta^{(s)}_{T,i}+\\ &+\frac{4(EF_{xx,i})^{(s)}}{l_{R,i}}\vartheta^{(s)}_{T,i}-\frac{2(EF_{xx,j})^{(s)}}{l_{R,j}}\vartheta^{(s)}_{T,j}.\end{aligned} \tag{2.183}$$

These equations are summed up in the set of matrix expressions in the terms B-17 to B-19 of the CTM using the substitutions $G = G_p$, $E = E_p$ and $F^{(s)}_{xx} = J^s$. Such matrices may be adopted for the solution of slender shell grid configurations.

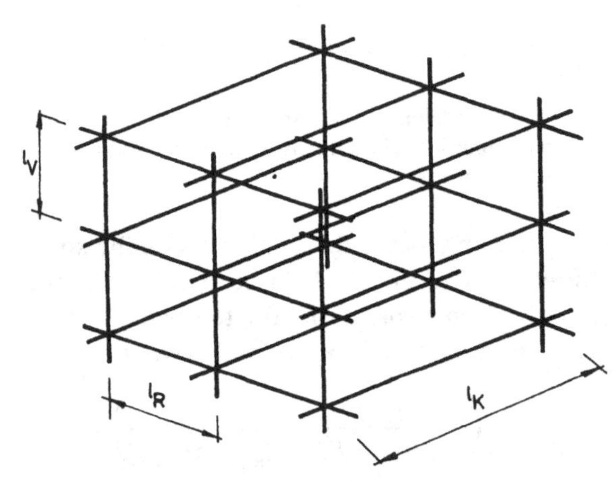

Fig. 40. Discrete simulation of the spatial grillage substructure.

When combining the aforementioned matrix terms for the transverse and in-plane behaviour of plane grid simulations, the transfer and nodal matrices for three-dimensional space grid substructure may be constructed as shown in Figure 40. Such a substructure may be adopted for the spatial simulation of solids or three-dimensional structural configurations. The corresponding matrix expressions are summed up in the terms B-20 to B-22 of the CTM.

When solving the present grillage simulations, the inertial, damping and forcing components as well as the initial parameters and boundary conditions are incorporated into the algorithm of the TMM in accordance with the aforementioned analysis. For example, the generalized matrices of initial parameters and of boundary conditions for the last case of the three-dimensional spatial grid simulation are illustrated in the terms B-23 and B-24 of the CTM.

2.6. Plane Substructure

Further will be derived the transfer and nodal matrices of plate substructures incorporated into primary and secondary members of load-bearing systems derived in the foregoing section.

First of all the determination of the transfer matrix for the plate substructure located between longitudinal and transversal members of grillage simulation, as shown in Figure 41, will be performed.

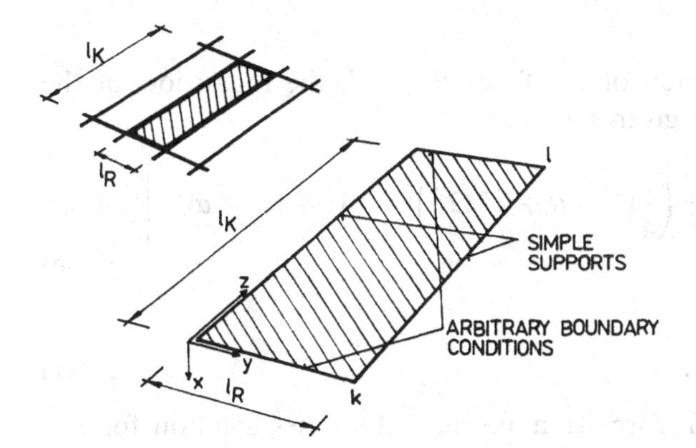

Fig. 41. Notation of the assumed plate substructure.

The differential equation of the plate element given by

$$AB^4(v_x^{(p)}) + \frac{2H}{z^2}F^2(v_x^{(p)}) + \frac{N}{z^4}D^4(v_x^{(p)}) + \frac{2A}{z}B^3(v_x^{(p)}) -$$

$$-\frac{N}{z^2}B^2(v_x^{(p)}) + \frac{2(N+H)}{z^4}D^2(v_x^{(p)}) + \frac{N}{z^3}B(v_x^{(p)}) + m\,C^2(v_x^{(p)}) = 0\,, \qquad (2.184)$$

in which the plate parameters, contrary to the previous analysis, are denoted with superscript (p), is dealt with. A, H and N are the plate stiffness parameters defined by

$$A = E^{(p)} J_x^{(p)}, \; N = E^{(p)} J_y^{(p)}, \; H = G^{(p)} J_T^{(p)}, \tag{2.185}$$

with corresponding plate rigidities in bending and torsion $J_i^{(p)}$. D, B, F and C are the additional differential operators given by

$$D = \partial/\partial y, \quad B = \partial/\partial x, \quad F = \partial^2/(\partial x \partial y), \quad C = \partial/\partial t. \tag{2.186}$$

A solution of equation (2.184) may be obtained assuming a series expansion of the form

$$v_x^{(p)} = \sum_{n=1}^{\infty} R_n \sin(\lambda y), \tag{2.187}$$

where R_n is a function of z and the parameter λ is given by

$$\lambda = n\pi/l_R. \tag{2.188}$$

Arbitrary transverse edge boundary conditions are specified by the function R_n. By introducing a new variable $v = \ln z$, equation (2.184) can be transformed into the following expression

$$A\,S^4(R_n) - 4A\,S^3(R_n) + (5A - 2H\lambda^2 - N)\,S^2(R_n) + \\ + (2A + 4H\lambda^2 + 2N)\,S(R_n) - (2H\lambda^2 + 2N\lambda^2 - N\lambda^4)\,R_n = 0 \tag{2.189}$$

with operator $S = \mathrm{d}/\mathrm{d}v$.

When assuming a solution of the form $R_n = \mathrm{e}^{s_i v}$, the four roots of the characteristic equation are given by

$$s_i = \pm\left[\frac{\alpha + 2\Omega\lambda^2 + 1}{2} \pm \left(\frac{1}{4}(1 + \alpha + 2\lambda\Omega)^2\right)^2 - (\lambda^2 - 1)^2 - \alpha)^{1/2}\right]^{1/2} + 1, \tag{2.190}$$

in which

$$\alpha = N/A \text{ and } \Omega = H/A. \tag{1.191}$$

Recalling the variable transformation $v = \ln z$, the series solution for displacements reduces to the expression

$$v_x^{(p)} = \sum_{n=1}^{\infty} (C_1 z^{s_1} + C_2 z^{s_2} + C_3 z^{s_3} + C_4 z^{s_4}) \sin(\lambda y). \tag{2.192}$$

This type of solution, however, will yield terms of high magnitude even for moderate values of the roots s_1, s_2, s_3 and s_4. To procure a more accurate computer solution, a new term Z is introduced such that

$$Z = z/z_0, \tag{2.193}$$

in which z_0 is the distance to the outside edge of the plate (l_K). Redefining the arbitrary constants of equation (2.192)

$$\begin{aligned} A1 &= C_1 z^{s_1}, \\ A2 &= C_2 z^{s_2}, \\ A3 &= C_3 z^{s_3}, \\ A4 &= C_4 z^{s_4}, \end{aligned} \tag{2.194}$$

the final form of the solution may be derived which is given by

$$v_x^{(p)} = \sum_{n=1}^{\infty} (A1\,Z^{s_1} + A2\,Z^{s_2} + A3\,Z^{s_3} + A4\,Z^{s_4}) \sin(\lambda y). \tag{2.195}$$

Taking account of substitutions

$$Y = \sin(\lambda y), \quad t_i = s_i - 1, \quad ZZ = -AY/z^2, \quad W = AY/z^3, \tag{2.196}$$

then the individual state components for the determination of the corresponding transfer matrix between the edges l and k ($Z = 1$), can be written as

$$\sum_{n=1}^{\infty} (A1 + A2 + A3 + A4)\,Y = v_k^{(p)}, \tag{2.197}$$

$$\sum_{n=1}^{\infty} (A1 s_1 + A2 s_2 + A3 s_3 + A4 s_4)\,Y = \varphi_k^{(p)}, \tag{2.198}$$

$$\sum_{n=1}^{\infty} AY(A1 s_1 t_1 + A2 s_2 t_2 + A3 s_3 t_3 + A4 s_4 t_4)/z^2 = M_k^{(p)}, \tag{2.199}$$

$$\begin{aligned} &\sum_{n=1}^{\infty} AY(A1(s_1 t_1 - 2\beta\lambda^2 t_1 - \alpha(s_1 - \lambda^2)) + \\ &+ A2(s_2 t_2^2 - 2\beta\lambda^2 t_2 - (s_2 - \lambda^2)) + A3(s_3 t_3 - 2\beta\lambda^2 t_3 - \alpha(s_3 - \lambda^2)) + \\ &+ A4(s_4 t_4^2 - 2\beta\lambda^2 t_4 - \alpha(s_4 - \lambda^2)))/z^3 = Q_k^{(p)}. \end{aligned} \tag{2.200}$$

Evaluation of the constants Ai there yields the relations

$$\begin{aligned} v_{x,1}^{(p)} &= v_{x,k}^{(p)} \left(\sum_{n=1}^{\infty} (DD1 + BB1 + CC1 + AA1) \right) + \\ &+ \varphi_{x,k}^{(p)} \left(\sum_{n=1}^{\infty} (DD2 + BB2 + CC2 + AA2) \right) + \\ &+ M_{x,k}^{(p)} \left(\sum_{n=1}^{\infty} \frac{Y}{ZZ} (DD3 + BB3 + CC3 + AA3) \right) + \end{aligned}$$

$$+ Q^{(p)}_{x,k}\left(\sum_{n=1}^{\infty}\frac{Y}{W}(DD4 + BB4 + CC4 + AA4)\right), \tag{2.201}$$

$$\begin{aligned}\varphi^{(p)}_{x,1} = {} & v^{(p)}_{x,k}\left(\sum_{n=1}^{\infty}(DD1\,s_1 + CC1\,s_2 + BB1\,s_3 + AA1\,s_4)\right) + \\ & + \varphi^{(p)}_{x,k}\left(\sum_{n=1}^{\infty}(DD2\,s_1 + CC2\,s_2 + BB2\,s_3 + AA2\,s_4)\right) + \\ & + M^{(p)}_{x,k}\left(\sum_{n=1}^{\infty}\frac{Y}{ZZ}(DD3\,s_1 + CC3\,s_2 + BB3\,s_3 + AA3\,s_4)\right) + \\ & + Q^{(p)}_{x,k}\left(\sum_{n=1}^{\infty}\frac{Y}{W}(DD4\,s_1 + CC4\,s_2 + BB4\,s_3 + AA4\,s_4)\right),\end{aligned} \tag{2.202}$$

$$\begin{aligned}M^{(p)}_{x,1} = {} & v^{(p)}_{x,k}\left(\sum_{n=1}^{\infty}\frac{ZZ}{Y}(DD1\,u_1 + CC1\,u_2 + BB1\,u_3 + AA1\,u_4)\right) + \\ & + \varphi^{(p)}_{x,k}\left(\sum_{n=1}^{\infty}\frac{ZZ}{Y}(DD2\,u_1 + CC2\,u_2 + BB2\,u_3 + AA2\,u_4)\right) + \\ & + M^{(p)}_{x,k}\left(\sum_{n=1}^{\infty}(DD3\,u_1 + CC3\,u_2 + BB3\,u_3 + AA3\,u_4)\right) + \\ & + Q^{(p)}_{x,k}\left(\sum_{n=1}^{\infty}\frac{ZZ}{W}(DD4\,u_1 + CC4\,u_2 + BB4\,u_3 + AA4\,u_4)\right),\end{aligned} \tag{2.203}$$

$$\begin{aligned}Q^{(p)}_{x,1} = {} & Q^{(p)}_{x,k}\left(\sum_{n=1}^{\infty}\frac{W}{Y}(DD1\,p_1 + CC1\,p_2 + BB1\,p_3 + AA1\,p_4)\right) + \\ & + \varphi^{(p)}_{x,k}\left(\sum_{n=1}^{\infty}\frac{W}{Y}(DD2\,p_1 + CC2\,p_2 + BB2\,p_3 + AA2\,p_4)\right) + \\ & + M^{(p)}_{x,k}\left(\sum_{n=1}^{\infty}\frac{W}{ZZ}(DD3\,p_1 + CC3\,p_2 + BB3\,p_3 + AA3\,p_4)\right) + \\ & + Q^{(p)}_{x,k}\left(\sum_{n=1}^{\infty}\frac{W}{ZZ}(DD4\,p_1 + CC4\,p_2 + BB4\,p_3 + AA4\,p_4)\right).\end{aligned} \tag{2.204}$$

The transfer matrix of the studied plate substructure together with the corresponding substitution parameters is shown in the term C-1 of the CTM.

Attention will be now directed to the determination of the nodal matrix of the transverse interaction for the studied plate substructure. Slope-deflection equations for the analysed plate substructure in the transverse direction will be derived. Restraining edge moments and reactions are proportional to the applied rotation and some as yet unknown plate constants.

Recalling that the sine series expansion was employed in the plate-deflection solution, the loads, plate-end slopes, end moments and reactions are also assumed to be given by a sine series expansion. Figure 42 shows these parameters as functions of $\sin(\gamma z)$. The consideration of the actual end slopes, and deflections specifies the four proposed conditions as shown in Figures 43, 44, 45 and 46. In Figures 43 and 44 an end rotation is applied to

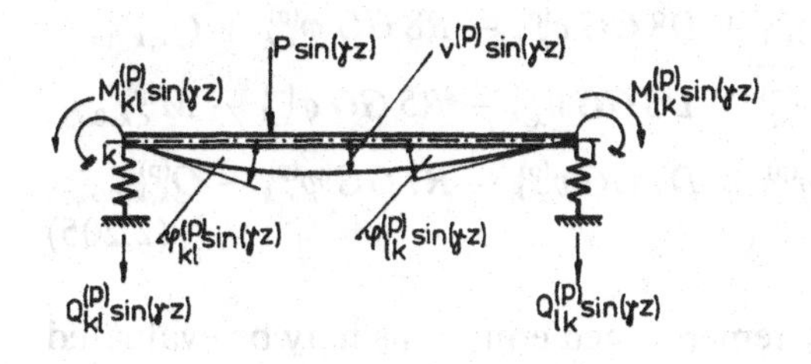

Fig. 42. State parameters of the plate element as functions of $\sin(\gamma z)$.

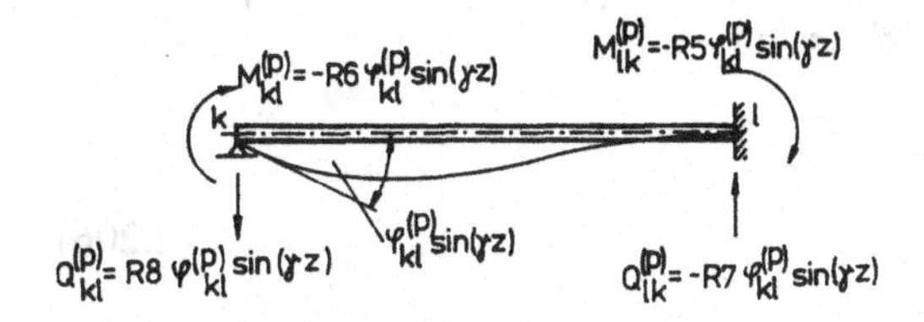

Fig. 43. The assumed boundary conditions I for the determination of plate constants.

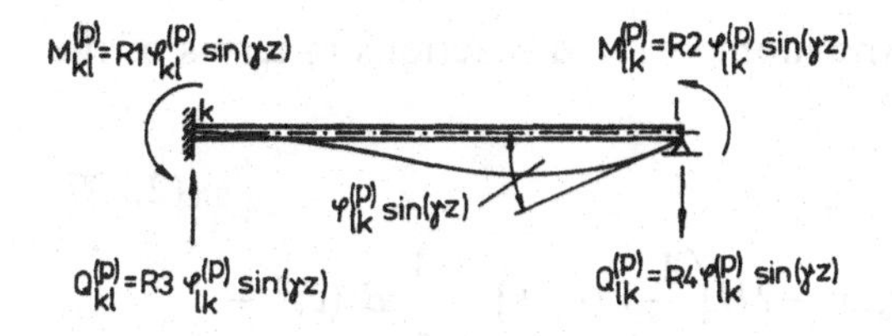

Fig. 44. The assumed boundary conditions II for the determination of plate constants.

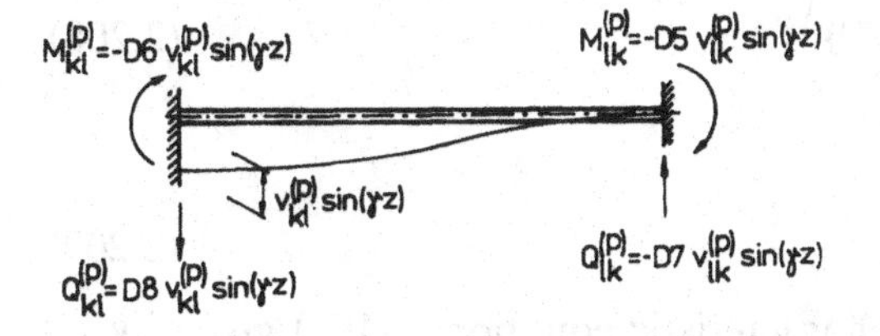

Fig. 45. The assumed boundary conditions III for the determination of plate constants.

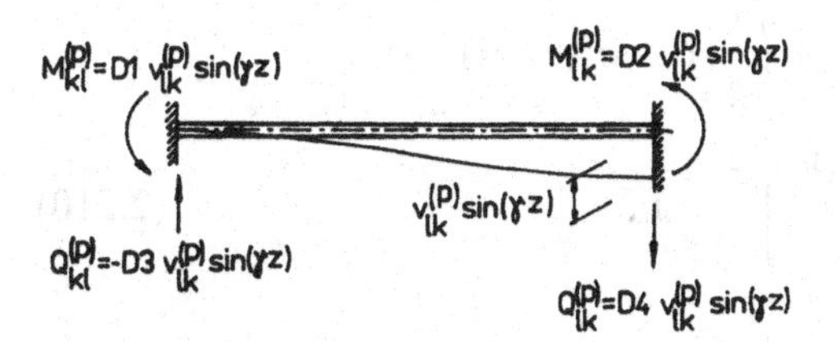

Fig. 46. The assumed boundary conditions IV for the determination of plate constants.

the plate at the ends k and l, with the opposite end fixed. Figures 45 and 46 represent the two conditions for deflection with resulting restraining forces proportional to the plate constants and the imposed deflection. Combining the effects of slope, deflection and external loads produces the following expressions for the plate edge forces

$$
\begin{aligned}
M_{x,k}^{(\mathrm{p})} &= D1\,GG\,v_{x,1}^{(\mathrm{p})} + R1\,GG\,\varphi_{x,1}^{(\mathrm{p})} - D6\,GG\,v_{x,k}^{(\mathrm{p})} + R6\,GG\,\varphi_{x,k}^{(\mathrm{p})} + M_{x,k,0}^{(\mathrm{p})}, \\
Q_{x,k}^{(\mathrm{p})} &= -\,D3\,GG\,v_{x,1}^{(\mathrm{p})} - R3\,GG\,\varphi_{x,1}^{(\mathrm{p})} + D8\,GG\,v_{x,k}^{(\mathrm{p})} - R8\,GG\,\varphi_{x,k}^{(\mathrm{p})} + Q_{x,k,0}^{(\mathrm{p})}, \\
M_{k,1}^{(\mathrm{p})} &= -D2\,GG\,v_{x,1}^{(\mathrm{p})} - R2\,GG\,\varphi_{x,1}^{(\mathrm{p})} + D5\,GG\,v_{x,k}^{(\mathrm{p})} - R5\,GG\,\varphi_{x,k}^{(\mathrm{p})} - M_{x,k,0}^{(\mathrm{p})}, \\
Q_{x,k}^{(\mathrm{p})} &= -\,D4\,GG\,v_{x,1}^{(\mathrm{p})} - R4\,GG\,\varphi_{x,1}^{(\mathrm{p})} + D7\,GG\,v_{x,k}^{(\mathrm{p})} - R7\,GG\,\varphi_{x,k}^{(\mathrm{p})} - Q_{x,k,0}^{(\mathrm{p})},
\end{aligned}
\tag{2.205}
$$

where $GG = \sin(\gamma z)$.

The 16 plate constants in the aforementioned equations may be evaluated by realizing that they are actually the moments and reactions at the plate edge for various boundary conditions given by

$$
\begin{array}{lcccc}
 & \mathrm{I} & \mathrm{II} & \mathrm{III} & \mathrm{IV} \\
\varphi_{x,l}^{(\mathrm{p})} = & 1 & 0 & 0 & 0\,, \\
\varphi_{x,k}^{(\mathrm{p})} = & 0 & 1 & 0 & 0\,, \\
v_{x,l}^{(\mathrm{p})} = & 0 & 0 & 1 & 0\,, \\
v_{x,k}^{(\mathrm{p})} = & 0 & 0 & 0 & 1\,.
\end{array}
\tag{2.206}
$$

The following equations express the moments and reactions in terms of the deflections

$$M_x^{(\mathrm{p})} = -\,AD^2(v_x^{(\mathrm{p})}), \tag{2.207}$$

$$
\begin{aligned}
Q_x^{(\mathrm{p})} = &-\left(A(D^3(v_x^{(\mathrm{p})}) + \frac{1}{z}\right)D^2(v_x^{(\mathrm{p})}) + 2H\left(\frac{1}{z^2}V(v_x^{(\mathrm{p})})\right) - \frac{1}{z^3}B^2(v_x^{(\mathrm{p})} - \\
&- N\left(\frac{1}{z^2}D(v_x^{(\mathrm{p})}) + \frac{1}{z^3}B^2(v_x^{(\mathrm{p})})\right),
\end{aligned}
\tag{2.208}
$$

with additional operator

$$V = \partial^3/(\partial y\,\partial z^2)\,. \tag{2.209}$$

Taking account of the modified characteristic equation (2.190) given by

$$
\begin{aligned}
a_1, a_2, a_3, a_4 = \pm\Bigg[&\frac{\varepsilon + 2\tau\gamma^2 + 1}{2} \pm \left(\frac{1}{4}(1 + \varepsilon + 2\gamma\tau)^2 - \right. \\
&\left. - (\gamma^2 - 1)^2 - \varepsilon\right)^{1/2}\Bigg]^{1/2} + 1\,,
\end{aligned}
\tag{2.210}
$$

where

$$\varepsilon = A/N, \quad \tau = H/N, \quad \gamma = n\pi/l_K, \tag{2.211]}$$

the following expressions for the evaluation of the plate stiffness constants can be derived

$$-\frac{M_x^{(p)}}{N} z^2 = \sum_{n=1}^{\infty} (A1\, a_1(a_1 - 1)^2 X^{a_1} + A2\, a_2(a_2 - 1)\, X^{a_2} + $$
$$+ A3\, a_3(a_3 - 1)\, X^{a_3} + A4\, a_4(a_4 - 1)\, X^{a_4})\, GG\,, \tag{2.212}$$

$$-\frac{Q_x^{(p)}}{N} z^3 = \sum_{n=1}^{\infty} (A1(a_1(a_1 - 1)^2 - 2\tau\gamma^2(a_1 - 1) - \varepsilon(a_1 - \gamma^2))\, X^{a_1} +$$
$$+ A2(a_2(a_2 - 1)^2 - 2\tau\gamma^2(a_2 - 1) - \varepsilon(a_2 - \gamma^2))\, X^{a_2} +$$
$$+ A3(a_3(a_3 - 1)^2 - 2\tau\gamma^2(a_3 - 1) - \varepsilon(a_3 - \gamma^2))\, X^{a_3} +$$
$$+ A4(a_4(a_4 - 1)^2 - 2\tau\gamma^2(a_4 - 1) - \varepsilon(a_4 - \gamma^2))\, X^{a_4})\, GG \tag{2.213}$$

where

$$X = y/y_0, \tag{2.214}$$

y_0 being the actual coordinate of the studied node.

The corresponding system of linear equations can then be written as

$$\begin{array}{lccccc} & & \text{I} & \text{II} & \text{III} & \text{IV} \\ A1\, a_1 + A2\, a_2 + A3\, a_3 + A4\, a_4 & = & 1 & 0 & 0 & 0, \\ A1\, a_1 X^{a_1} + A2\, a_2 X^{a_2} + A3\, a_3 X^{a_3} + A4\, a_4 X^{a_4} & = & 0 & 1 & 0 & 0, \\ A1 + A2 + A3 + A4 & = & 0 & 0 & 1 & 0, \\ A1\, X^{a_1} + A2\, X^{a_2} + A3\, X^{a_3} + A4\, X^{a_4} & = & 0 & 0 & 0 & 1. \end{array} \tag{2.215}$$

The plate constants are then given by

$$R1 = \sum_{n=1}^{\infty} \left(\frac{N}{z^2} A1(\text{I})\, a_1(a_1 - 1) X^{a_1} \right),$$

$$R2 = \sum_{n=1}^{\infty} \left(\frac{N}{z^2} A2(\text{I})\, a_2(a_2 - 1) X^{a_2} \right),$$

$$R3 = \sum_{n=1}^{\infty} \left(\frac{N}{z^2} A3(\text{I})\, a_3(a_3 - 1) X^{a_3} \right),$$

$$R4 = \sum_{n=1}^{\infty} \left(\frac{N}{z^2} A4(\text{I})\, a_4(a_4 - 1) X^{a_4} \right),$$

$$R5 = \sum_{n=1}^{\infty} \left(\frac{N}{z^3} A1(\text{II})\, a_1(a_1 - 1) X^{a_1} \right),$$

$$
\begin{aligned}
R6 &= \sum_{n=1}^{\infty}\left(\frac{N}{z^3}A2(\mathrm{II})a_2(a_2-1)X^{a_2}\right),\\
R7 &= \sum_{n=1}^{\infty}\left(\frac{N}{z^3}A3(\mathrm{II})a_3(a_3-1)X^{a_3}\right),\\
R8 &= \sum_{n=1}^{\infty}\left(\frac{N}{z^3}A4(\mathrm{II})\,a_4(a_4-1)X^{a_4}\right),\\
D1 &= \sum_{n=1}^{\infty}\left(\frac{N}{z^3}A1(\mathrm{III})a_1(a_1-1)^2-2\tau\gamma^2(a_1-1)-\varepsilon(a_1-\gamma^2)\right),\\
D2 &= \sum_{n=1}^{\infty}\left(\frac{N}{z^3}A2(\mathrm{III})a_2(a_2-1)^2-2\tau\gamma^2(a_2-1)-\varepsilon(a_1-\gamma^2)\right),\\
D3 &= \sum_{n=1}^{\infty}\left(\frac{N}{z^3}A3(\mathrm{III})a_3(a_3-1)^2-2\tau\gamma^2(a_3-1)-\varepsilon(a_3-\gamma^2)\right),\\
D4 &= \sum_{n=1}^{\infty}\left(\frac{N}{z^3}A4(\mathrm{III})a_4(a_4-1)^2-2\tau\gamma^2(a_4-1)-\varepsilon(a_4-\gamma^2)\right),\\
D5 &= \sum_{n=1}^{\infty}\left(\frac{N}{z^3}A1(\mathrm{IV})a_1(a_1-1)^2-2\tau\gamma^2(a_1-1)-\varepsilon(a_1-\gamma^2)\right),\\
D6 &= \sum_{n=1}^{\infty}\left(\frac{N}{z^3}A2(\mathrm{IV})a_2(a_2-1)^2-2\tau\gamma^2(a_2-1)-\varepsilon(a_2-\gamma^2)\right),\\
D7 &= \sum_{n=1}^{\infty}\left(\frac{N}{z^3}A3(\mathrm{IV})a_3(a_3-1)^2-2\tau\gamma^2(a_3-1)-\varepsilon(a_3-\gamma^2)\right),\\
D8 &= \sum_{n=1}^{\infty}\left(\frac{N}{z^3}A4(\mathrm{IV})a_4(a_4-1)^2-2\tau\gamma^2(a_4-1)-\varepsilon(a_4-\gamma^2)\right).
\end{aligned}
\qquad (2.216)
$$

The torsional influences are approximately considered regarding elastic couplings in accordance with the pure torsion analysis of primary and secondary grillage members in the foregoing section.

The generalized nodal matrix of the transverse interaction, corresponding to the generalized transfer matrix of studied problem depicted in the term C-2, is summed up in the term C-3 of the CTM. The corresponding partial matrices are summed up in the terms C-4, C-5 and C-6 of the CTM. The schemes of incorporation of derived matrix expressions into the generalized nodal matrices of the transverse interaction when solving orthotropic shell simulations with closed or open cross-sections are illustrated in the term C-7 of the CTM. The complex nodal matrices of inertial and forcing parameters concentrated in the nodal points of the theoretical solution of the studied spatial shell simulations with plate skin members are summed up in the terms C-8 and C-9 of the CTM. The matrices of initial parameters together with the matrices of boundary conditions are summed up in the terms C-10 and C-11 of the CTM. Analogous matrices of the present type may be used for all other possible boundary conditions of the studied simulation [103].

2.7. Shell Substructure

The actual geometry of the curved shell surface in the previous section was replaced by an assembly of flat elements. Plate and rib elements may be combined to simulate an orthotropically stiffened shell substructure. Provided the deformations of the shell are within the assumptions of the conventional small displacement theory, the membrane forces are separate from the plate bending behaviour in a flat element; thus the shell substructure can be constructed by superimposing a plane stress element and a plate bending element. The coupling between membrane and bending actions are obtained in the assemblage of elements to form the complete structure. For problems involving large displacements, the stretching and flexural behaviour becomes coupled and this increases the complexity of the formulation. Therefore, the analysis of shell substructures when incorporated into a developed grillage mesh is more involved than that for flat panels. The problem will be lumped using the discrete mass lines running parallel to the supporting stringers. The line masses are connected by massless segments of thin cylindrical shells. The discrete approximation of a shell panel is depicted in Figure 47. The panels are assumed to be simply supported along their curved edges.

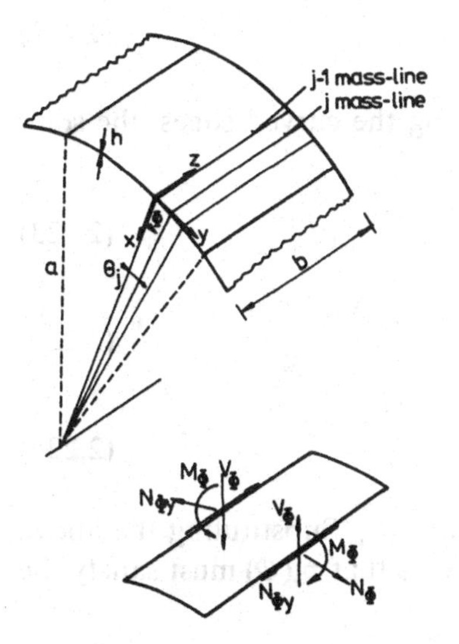

Fig. 47. Discrete approximation of the studied shell substructure.

First of all, the attention will be directed towards developing the transfer matrix across the massless shell segments between the mass lines. The derivation of the transfer matrix is based on the Donnell shell theory [18, 49]. For an unloaded and massless cylindrical shell element, the Donnell equations can be modified as

$$a^2B^2(v_z) + \frac{1}{2}(1 - \nu)P^2(v_z) + \frac{1}{2}a(1 + \nu)V(v_y) - \nu a\,B(v_x) = 0\,, \qquad (2.217)$$

$$\frac{1}{2}a(1 + \nu)V(v_z) + P^2(v_y) + \frac{1}{2}a^2(1 - \nu)B^2(v_y) - P(v_x) = 0\,, \qquad (2.218)$$

$$\nu a\,B(v_z) + P(v_y) - v_x - k[a^4B^4(v_x) + 2a^2V^2(v_x) + P^4(v_x)] = 0\,, \qquad (2.219)$$

with substitutions

$$B = \partial/\partial z\,, \quad P = \partial/\partial\Phi, \quad V = \partial^2/\partial z\partial\Phi, \qquad (2.220)$$

and where a is the radius of curvature that may change from segment to segment. Further, Φ is the angular coordinate of the shell segment, $k = h^2/(12a^2)$ and ν is Poisson's ratio.

Expressions (2.217), (2.218) and (2.219) are two second-order and one fourth-order partial differential equations in three unknowns, namely the displacements v_x, v_y and v_z. They can be rearranged giving a single eighth-order equation in one unknown, viz.

$$a^8k\nabla^8v_x + a^4(1 - \nu^2)B^4(v_x) = 0\,, \qquad (2.221)$$

where

$$\nabla^2 = \frac{\partial^2}{\partial z^2} + \frac{1}{a^2}\frac{\partial^2}{\partial\Phi^2} \quad \text{and} \quad \nabla^8 = (\nabla^2)^4\,. \qquad (2.222)$$

Regarding the simple support conditions along the curved edges, the solutions may be approximated by

$$v_z = e^{i\omega t}\,\alpha_n(\Phi)\cos\frac{n\pi z}{b}\,, \qquad (2.223)$$

$$v_y = e^{i\omega t}\,\beta_n(\Phi)\sin\frac{n\pi z}{b}\,, \qquad (2.224)$$

$$v_x = e^{i\omega t}\,\eta_n(\Phi)\sin\frac{n\pi z}{b}\,, \qquad (2.225)$$

with preliminary unknown functions α_n, β_n and η_n. Substituting the above expression for v_x into equation (2.221), it follows that $\eta_n(\Phi)$ must satisfy the following ordinary differential equation

$$\eta_n^{(8)} - 4q_n^2\eta_n^{(6)} + 6q_n^4\eta_n^{(4)} - 4q_n^6\eta_n^{(2)} + q_n^4(q_n^4 + p^4)\eta_n = 0\,, \tag{2.226}$$

where

$$q_n = \frac{n\pi a}{b} \quad \text{and} \quad p = 4\sqrt{\frac{12a^2(1-\nu^2)}{h^2}}\,. \tag{2.227}$$

The eight characteristic roots of this equation can be written as

$$\pm\sqrt{q_n + \frac{pq_n}{\sqrt{2}}(\pm 1 \pm \mathrm{i})}\,, \tag{2.228}$$

or in real and imaginary component form

$$\pm(\gamma_1 \pm \mathrm{i}\,\delta_1) \quad \text{and} \quad \pm(\gamma_2 \pm \mathrm{i}\,\delta_2)\,. \tag{2.229}$$

The values of the real and imaginary parts are

$$\gamma_1 = \sqrt{q_n}(q_n^2 + \sqrt{2}\,q_n p + p^2)^{1/4}\cos\left(\frac{1}{2}\arctan\frac{p}{\sqrt{2}\,q_n + p}\right),$$

$$\delta_1 = \sqrt{q_n}(q_n^2 + \sqrt{2}\,q_n p + p^2)^{1/4}\sin\left(\frac{1}{2}\arctan\frac{p}{\sqrt{2}\,q_n + p}\right),$$

$$\gamma_2 = \sqrt{q_n}(q_n^2 - \sqrt{2}\,q_n p + p^2)^{1/4}\cos\left[\frac{1}{2}\arctan p/(\sqrt{2}\,q_n - p)\right],$$

$$\delta_2 = \sqrt{q_n}(q_n^2 - \sqrt{2}\,q_n p + p^2)^{1/4}\sin\left[\frac{1}{2}\arctan p/(\sqrt{2}\,q_n - p)\right], \tag{2.230}$$

and, in all cases the principal values of the arctangents are to be used. When the characteristic roots are expressed in terms of their real and imaginary components, then equation (2.226) is satisfied by any of the following functions

$$\begin{array}{ll} \cosh(\gamma_1\Phi)\cos(\delta_1\Phi)\,, & \sinh(\gamma_1\Phi)\sin(\delta_1\Phi)\,, \\ \cosh(\gamma_2\Phi)\cos(\delta_2\Phi)\,, & \sinh(\gamma_2\Phi)\sin(\delta_2\Phi)\,, \\ \sinh(\gamma_1\Phi)\cos(\delta_1\Phi)\,, & \cosh(\gamma_1\Phi)\sin(\delta_1\Phi)\,, \\ \sinh(\gamma_2\Phi)\cos(\delta_2\Phi)\,, & \cosh(\gamma_2\Phi)\sin(\delta_2\Phi)\,. \end{array} \tag{2.231}$$

The eight solutions will be linear combinations of the above expressions and they facilitate the construction of the transfer matrix. The desired solution is then given by

$$\eta_n(\Phi) = \sum_{i=0}^{7} \varepsilon_i f_i(\Phi)\,. \tag{2.232}$$

For even subscripts of f_i even functions of Φ are implied and for $i = 0, 2, 4$ and 6 can be written

$$f_i = \alpha_{i1} \cosh(\gamma_1 \Phi) \cos(\delta_1 \Phi) + \alpha_{i2} \sinh(\gamma_1 \Phi) \sin(\delta_1 \Phi) + \\ + \alpha_{i3} \cosh(\gamma_2 \Phi) \cos(\delta_2 \Phi) + \alpha_{i4} \sinh(\gamma_2 \Phi) \sin(\delta_2 \Phi) . \tag{2.233}$$

The odd subscripts of f_i imply the odd functions of Φ and for $i = 1, 3, 5$ and 7 pays

$$f_i = \alpha_{i1} \sinh(\gamma_1 \Phi) \cos(\delta_1 \Phi) + \alpha_{i2} \cosh(\gamma_1 \Phi) \sin(\delta_1 \Phi) + \\ + \alpha_{i3} \sinh(\gamma_2 \Phi) \cos(\delta_2 \Phi) + \alpha_{i4} \cosh(\gamma_2 \Phi) \sin(\delta_2 \Phi) . \tag{2.234}$$

The functions $f_i(\Phi)$ in equation (2.232) are chosen such that

$$\begin{aligned} \varepsilon &= \eta_n(0) , & \varepsilon_1 &= P[\eta_n(0)] , \\ \varepsilon_2 &= P^2[\eta_n(0)] , & \varepsilon_3 &= P^3[\eta_n(0)] , \\ \varepsilon_4 &= P^4[\eta_n(0)] , & \varepsilon_5 &= \beta_n(0) , \\ \varepsilon_6 &= B[\beta_n(0)] , & \varepsilon_7 &= B^2[\beta_n(0)] . \end{aligned} \tag{2.235}$$

When introducing further

$$\beta_n(\Phi) = \sum_{i=0}^{7} \varepsilon_i g_i(\Phi) , \tag{2.236}$$

and analysing equations (2.218) and (2.219), the relation between the functions $g_i(\Phi)$ and $f_i(\Phi)$ for any particular i is given by

$$P^2(g_i) + \nu q_n^2 g_i = \tau_1(P(f_i) + \tau_2 P^3(f_i) + \tau_3 P^5(f_i) , \tag{2.237}$$

where

$$\tau_1 = 1 + \frac{1+\nu}{1-\nu} k\, q_n^4 , \tag{2.238}$$

$$\tau_2 = -2\left(\frac{1+\nu}{1-\nu}\right) k\, q_n^2 \tag{2.239}$$

and

$$\tau_3 = \frac{1+\nu}{1-\nu} k . \tag{2.240}$$

It is to be noted that $g_i(\Phi)$ is an odd function if $f_i(\Phi)$ is an even function and vice versa. The function $g_i(\Phi)$ may be written in form

$$g_i(\Phi) = \beta_{i1} \sinh(\gamma_1 \Phi) \cos(\delta_1 \Phi) + \beta_{i2} \cosh(\gamma_1 \Phi) \sin(\delta_1 \Phi) + \\ + \beta_{i3} \sinh(\gamma_2 \Phi) \cos(\delta_2 \Phi) + \beta_{i4} \cosh(\gamma_2 \Phi) \sin(\delta_2 \Phi) , \tag{2.241}$$

for $i = 0, 2, 4, 6$; and

$$g_i(\Phi) = \beta_{i1}\cosh(\gamma_1\Phi)\cos(\delta_1\Phi) + \beta_{i2}\sinh(\gamma_1\Phi)\sin(\delta_1\Phi) + \\ + \beta_{i3}\cosh(\gamma_2\Phi)\cos(\delta_2\Phi) + \beta_{i4}\sinh(\gamma_2\Phi)\sin(\delta_2\Phi)\,, \tag{2.242}$$

for $i = 1$, 3, 5 and 7.

These relations may be expressed in the vector form for $i = 0$, 2, 4 and 6 as follows

$$f_i(\Phi) = (\alpha_{i1}, \alpha_{i2}, \alpha_{i3}, \alpha_{i4})\{\xi_e\}\,, \tag{2.243}$$

$$g_i(\Phi) = (\alpha_{i1}, \alpha_{i2}, \alpha_{i3}, \alpha_{i4})\{\xi_0\} \tag{2.244}$$

and for $i = 1$, 3, 5 and 7 as

$$f_i(\Phi) = (\alpha_{i1}, \alpha_{i2}, \alpha_{i3}, \alpha_{i4})\{\xi_0\}\,, \tag{2.245}$$

$$g_i(\Phi) = (\alpha_{i1}, \alpha_{i2}, \alpha_{i3}, \alpha_{i4})\{\xi_e\}\,, \tag{2.246}$$

where

$$\{\xi_e\} = \begin{Bmatrix} \cosh(\gamma_1\Phi)\cos(\delta_1\Phi) \\ \sinh(\gamma_1\Phi)\sin(\delta_1\Phi) \\ \cosh(\gamma_2\Phi)\cos(\delta_2\Phi) \\ \sinh(\gamma_2\Phi)\sin(\delta_2\Phi) \end{Bmatrix}, \tag{2.247}$$

and

$$\{\xi_0\} = \begin{Bmatrix} \sinh(\gamma_1\Phi)\cos(\delta_1\Phi) \\ \cosh(\gamma_1\Phi)\sin(\delta_1\Phi) \\ \sinh(\gamma_2\Phi)\cos(\delta_2\Phi) \\ \cosh(\gamma_2\Phi)\sin(\delta_2\Phi) \end{Bmatrix}. \tag{2.248}$$

The couplings of the β-coefficients to their α-equivalents is performed using the modified equation (2.237)

$$\boldsymbol{\beta}_i\mathbf{D}^2 + vq_n^2\mathbf{I} = \boldsymbol{\alpha}_i(\tau_1\mathbf{D} + \tau_2\mathbf{D}^3 + \tau_3\mathbf{D}^5)\,, \tag{2.249}$$

or

$$\boldsymbol{\beta}_i\mathbf{K} = \boldsymbol{\alpha}_i\mathbf{L} \tag{2.250}$$

and

$$\boldsymbol{\beta}_i = \boldsymbol{\alpha}_i\mathbf{L}\mathbf{K}^{-1}\,, \tag{2.251}$$

where $\mathbf{I}$ is the identity matrix, the definitions of $\mathbf{K}$ and $\mathbf{L}$ are clear, and $\mathbf{D}$ is given by

$$\mathbf{D} = \begin{bmatrix} \gamma_1 & -\delta_1 & 0 & 0 \\ \delta_1 & \gamma_1 & 0 & 0 \\ 0 & 0 & \gamma_2 & -\delta_2 \\ 0 & 0 & \delta_2 & \gamma_2 \end{bmatrix}. \tag{2.252}$$

The 16 conditions the even functions must satisfy in order to deal with equations (2.235) are concisely stated in matrix form as

$$\begin{bmatrix} f_0(0) & P^2[f_0(0)] & P^4[f_0(0)] & P[g_0(0)] \\ f_2(0) & P^2[f_2(0)] & P^4[f_2(0)] & P[g_2(0)] \\ f_4(0) & P^2(f_4(0)] & P^4[f_4(0)] & P[g_4(0)] \\ f_6(0) & P^2[f_6(0)] & P^4[f_6(0)] & P[g_6(0)] \end{bmatrix} = \mathbf{I}. \tag{2.253}$$

The 16 conditions for the odd functions are

$$\begin{bmatrix} P[f_1(0)] & P^3[f_1(0)] & g_1(0) & P^2[g_1(0)] \\ P[f_3(0)] & P^3(f_3(0)] & g_3(0) & P^2[g_3(0)] \\ P[f_5(0)] & P^3[f_5(0)] & g_5(0) & P^2(g_5(0)] \\ P[f_7(0)] & P^3[f_7(0)] & g_7(0) & P^2[g_7(0)] \end{bmatrix} = \mathbf{I}. \tag{2.254}$$

Treating of the even and odd functions separately allows the reduction of the size of the above matrix statements. These statements in terms of the $f_i(\Phi)$ and $g_i(\Phi)$ functions may be conveniently rephrased in terms of the α-coefficients. When considering the even functions as examples, $f_0(0)$ means

$$[\alpha_{01}, \alpha_{02}, \alpha_{03}, \alpha_{04}] \begin{bmatrix} 1 \\ 0 \\ 0 \\ 0 \end{bmatrix} = 1, \tag{2.255}$$

$P^2[f_0(0)]$ means

$$\boldsymbol{\alpha}_0 \mathbf{D}^2 \begin{bmatrix} 1 \\ 0 \\ 1 \\ 0 \end{bmatrix} = 0, \tag{2.256}$$

$P^4[f_0(0)]$ means

$$\boldsymbol{\alpha}_0 \mathbf{D} \begin{bmatrix} 1 \\ 0 \\ 1 \\ 0 \end{bmatrix} = 0, \tag{2.257}$$

and, finally, $P[g_0(0)]$ means

$$\boldsymbol{\alpha}_0 \mathbf{D} \begin{bmatrix} 1 \\ 0 \\ 0 \\ 0 \end{bmatrix} = 0. \tag{2.258}$$

Repeating this analysis for the remaining 12 expressions gives

$$\begin{bmatrix} a_{01} & a_{02} & a_{03} & a_{04} \\ a_{21} & a_{22} & a_{23} & a_{24} \\ a_{41} & a_{42} & a_{43} & a_{44} \\ a_{61} & a_{62} & a_{63} & a_{64} \end{bmatrix} \mathbf{Q} = \mathbf{I}, \tag{2.259}$$

where $\mathbf{Q}$ has as its columns

$$\begin{bmatrix} 1 \\ 0 \\ 1 \\ 0 \end{bmatrix}, \quad \mathbf{D}^2 \begin{bmatrix} 1 \\ 0 \\ 1 \\ 0 \end{bmatrix}, \quad \mathbf{D}^4 \begin{bmatrix} 1 \\ 0 \\ 1 \\ 0 \end{bmatrix}, \quad \mathbf{DLK}^{-1} \begin{bmatrix} 1 \\ 0 \\ 1 \\ 0 \end{bmatrix}. \tag{2.260}$$

Hence, for the even functions

$$\mathbf{A} = \mathbf{Q}^{-1} \tag{2.261}$$

and for the odd functions

$$\mathbf{A} = \mathbf{U}^{-1}, \tag{2.262}$$

where $\mathbf{U}$ has as its columns

$$\mathbf{D} \begin{bmatrix} 1 \\ 0 \\ 1 \\ 0 \end{bmatrix}, \quad \mathbf{D}^3 \begin{bmatrix} 1 \\ 0 \\ 1 \\ 0 \end{bmatrix}, \quad \mathbf{LK}^{-1} \begin{bmatrix} 1 \\ 0 \\ 1 \\ 0 \end{bmatrix}, \quad \mathbf{D}^2\mathbf{LK}^{-1} \begin{bmatrix} 1 \\ 0 \\ 1 \\ 0 \end{bmatrix}, \tag{2.263}$$

with $\mathbf{A}$ as the matrix of coefficients.

At this point all the functions $f_i(\Phi)$ and $g_i(\Phi)$ defined in equations (2.232) and (2.236) are determined, when satisfying the conditions (2.235). Hence, the state vector

$$\begin{aligned} \boldsymbol{k}_n = \{&\eta_n(\Phi), \quad P[\eta_n(\Phi)], \quad P^2[\eta_n(\Phi)], \quad P^3[\eta_n(\Phi)], \\ &P^4[\eta_n(\Phi)], \quad \beta_n(\Phi), \quad P[\beta_n(\Phi)], \quad P^2[\beta_n(\Phi)]^{\mathrm{T}}, \end{aligned} \tag{2.264}$$

can be determined for any value of the argument Φ.

The couplings of two neighbouring state vectors at $\Phi = \Theta_j$ is then given by

$$\boldsymbol{k}_j^r = \mathbf{R}_j \boldsymbol{k}_j^l, \tag{2.265}$$

where the elements of $\mathbf{R}_j$ are given by

$$r_{k1} = [a_{i1}, a_{i2}, a_{i3}, a_{i4}]\,\mathbf{D}^{k-1}[\xi(\Theta_j)], \tag{2.266}$$

for $i = l + 1$, $k = 1, 2, 3, 4$ and 5, and

$$r_{kl} = [a_{i1}, a_{i2}, a_{i3}, a_{i4}]\,\mathbf{D}^{k-6}\mathbf{LK}^{-1}[\xi(\Theta_j)], \tag{2.267}$$

for $i = l + 1$, $k = 6, 7$ and 8, and where

$$[\xi(\Theta_j)] = [\xi_e(\Theta_j)], \quad \text{for even values of } l + k, \tag{2.268}$$

$$[\xi(\Theta_j)] = [\xi_0(\Theta_j], \quad \text{for odd values of } l + k. \tag{2.269}$$

In accordance with Figure 47, the studied state vector is then modified according to

$$\boldsymbol{k}_n = \left[\alpha_n, \beta_n, \eta_n, \frac{1}{a}\varphi_n, M_{\Phi,n}, V_{\Phi,n}, N_{\Phi,n}, N_{\Phi y,n}\right]^{\mathrm{T}}. \tag{2.270}$$

The first component of the present mixed displacement and force type state vector can be expressed, referred to equation (2.219), as

$$\alpha_n(\Phi) = \frac{l}{\nu q_n}[P(\beta_n) - (1 + k q_n^4)\,\eta_n + 2kq_n^2 P^2(\eta_n) - kP^4(\eta_n)]. \tag{2.271}$$

The terms β_n and η_n of the state vector were derived earlier. φ_n is the slope of the function η_n. Further terms of the state vector are given by

$$M_{\Phi,n} = \frac{D}{a^2}[P^2(\eta_n) - \nu q_n^2 \eta_n], \tag{2.272}$$

$$V_{\Phi,n} = \frac{D}{a^3}[P^3(\eta_n) - (2 - \nu)\,q_n^2 P(\eta_n)], \tag{2.273}$$

$$N_{\Phi,n} = \frac{D}{a^3}[q_n^4 \Phi_n - 2q_n^2 P^2(\eta_n) + P^4(\eta_n)], \tag{2.274}$$

$$N_{\Phi y,n} = \frac{Eh}{2a(1+\nu)}[P(\alpha_n) + q_n\beta_n] = $$
$$= \frac{Eh}{a(1+\nu)^2}\left[\frac{1}{q_n}P^2(\beta_n) + \nu q_n\beta_n - \frac{1}{q_n}P(\eta_n)\right]. \tag{2.275}$$

The conversion matrix is set up so that

$$\boldsymbol{k}_n = \mathbf{B}_n \bar{\boldsymbol{k}}_n, \tag{2.276}$$

where the subscript n associated with $\mathbf{B}$ indicates that this matrix is to be computed from the physical constants of the n-th shell segment. The details of the matrix $\mathbf{B}_n$ are shown in the terms D-1 and D-2 of the CTM. The $\boldsymbol{k}$ vectors at the two opposite edges of the n-th shell segment are then related by equation

$$\boldsymbol{k}_n^{\mathrm{l}} = \mathbf{B}_n \mathbf{R}_n \mathbf{B}_n^{-1} \boldsymbol{k}_{n-1}^{\mathrm{r}} = \mathbf{BK}_n \boldsymbol{k}_{n-1}^{\mathrm{r}}, \tag{2.277}$$

where $\mathbf{BK}$ is the sought transfer matrix of the studied problem. The corresponding nodal matrix is summed up in the term D-3 of the CTM.

The incorporation of the present matrices into the spatial shell grids is performed in analogy with the aforementioned analysis. The schemes of generalized transfer and nodal matrices together with the corresponding matrices of initial parameters and boundary conditions are constructed similarly as in the terms B-17 and B-18 or B-23 and B-24 of the CTM.

The present transfer matrix of the shell substructure, developed in accordance with the analysis in reference [49], is convenient for effective numerical applications when solving the aforementioned spatial grids with shell members. Similar solution methods may be employed when using another transfer matrices for shell members developed, as for example in reference [63].

2.8. Corrugated Sheet Substructure

Corrugated sheets are typical members of light-weight slender structural configurations. They consist of thin plate and wall sheets or folds for systematic analysis of which the set of modified matrix expressions derived in foregoing sections or in references [49, 63, 103] may be applied.

For simplified checking of results obtained by such refined analyses will be subsequently presented an approximative approach for the solution of corrugated sheet substructures. This approach allows the numerical verification of results obtained by more advanced approximations derived in the foregoing sections.

It can be shown, that almost all corrugated or folded plate substructures may be assumed to act as simple beam for each individual plane member, transversely as well as longitudinally [32, 114]. This approach lends itself

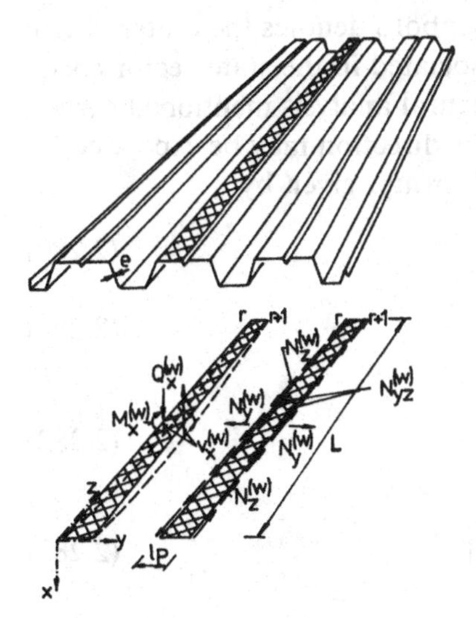

Fig. 48. State components on the studied corrugated sheet substructure.

readily to a systematic arrangement and a continuous control when applying the transfer matrix method.

Figure 48 shows a part of a corrugated sheet simply supported at the ends $z = 0$ and $z = l_K$. The thickness of each plate is assumed to be constant. A thickness variation in the transverse direction may be handled approximately by assuming the thickness to vary in steps. At each change of thickness an imaginary plate edge is inserted. The torsional rigidity of the plate is neglected.

The analysis is carried out in two steps. In the first step each plate is supposed to be supported and clamped, or occasionally simply supported at the corrugated folds. Hence a supporting force and an unbalanced moment arise at each fold. These forces and moments are calculated by the elementary beam statics. In the second step the substructure is loaded by the line loads and moments at the folds, of opposite direction to those calculated in the first step. The complete solution is then obtained by superposition of the two step solutions.

When assuming the normal and tangential influences in the arbitrary node of the corrugated sheet with nomenclature as shown in Figure 48, the corresponding state vector is given by

$$\boldsymbol{k}_k^{(w)}(s,t) = (N_y^{(w)}, N_{yz}^{(w)}, N_z^{(w)}, v_y^{(w)}, v_x^{(w)}, \varphi_x^{(w)}, M_x^{(w)}, Q_x^{(w)}, 1)^{\mathrm{T}}, \tag{2.278}$$

where the tangential group of forces which act upon the plate are the members $N_y^{(w)}$ and $N_{yz}^{(w)}$ with associated displacement $v_y^{(w)}$ and a longitudinal force $N_z^{(w)}$ given by

$$N_z^{(w)}(s,t) = Eh\varepsilon_z^{(w)}, \tag{2.279}$$

with sheet thickness h and strain $\varepsilon_z^{(w)}$. The symbol s denotes the centre-line of the sheet wall. The normal components associated in the state vector correspond with standard components of the flexural analysis mentioned earlier.

The variation of state components in the z-direction may be expressed by the quantities of the tangential and normal group, given by

$$N_y^{(w)}(l_P,t) = N_y^{(w)}(0,t), \tag{2.280}$$

$$N_{yz}^{(w)}(l_P,t) = N_{yz}^{(w)}(0,t)\left(1 - \frac{2z}{L}\right), \tag{2.281}$$

$$N_z^{(w)}(l_P,t) = N_z^{(w)}(0,t)\left(\frac{4z}{L} - \frac{4z^2}{L^2}\right), \tag{2.282}$$

$$v_y^{(w)}(l_P,t) = v_y^{(w)}(0,t)\ \frac{16z}{5L}\left(1 - 2\frac{z^2}{L^2} + \frac{z^3}{L^3}\right). \tag{2.283}$$

It is known from the theory of cylindrical shells that this distribution of forces constitutes a very good approximation to the actual elastic behaviour [33].

Neglecting the deformation of the plate in the y-direction and assuming N_z to vary linearly with y, one obtains from elementary beam statics the following relations for the tangential and normal behaviour

$$N^{(w)}_{y,r+1} = N^{(w)}_{y,r} + \frac{2l_P}{L} N^{(w)}_{yz,r} - \frac{4l_P^2}{L^2} N^{(w)}_{z,r} - \frac{64Ehl_P^3}{5L^4} v^{(w)}_{y,r}, \tag{2.284}$$

$$N^{(w)}_{yz,r+1} = N^{(w)}_{yz,r} - \frac{4l_P}{L} N^{(w)}_{z,r} - \frac{96Ehl_P^2}{5L^3} v^{(w)}_{y,r}, \tag{2.285}$$

$$N^{(w)}_{z,r+1} = N^{(w)}_{z,r} + \frac{48Ehl_P}{5L} v^{(w)}_{y,r}, \tag{2.286}$$

$$v^{(w)}_{y,r+1} = v^{(w)}_{y,r}, \tag{2.287}$$

$$v^{(w)}_{x,r+1} = v^{(w)}_{x,r} + l_P \varphi^{(w)}_{x,r} - \frac{l_P^2}{2EF_{xx}} M^{(w)}_{x,r} - \frac{l_P^3}{6EF_{xx}} Q^{(w)}_{x,r}, \tag{2.288}$$

$$\varphi^{(w)}_{x,r+1} = \varphi^{(w)}_{x,r} - \frac{l_P}{EF_{xx}} M^{(w)}_{x,r} - \frac{l_P^2}{2EF_{xx}} Q^{(w)}_{x,r}, \tag{2.289}$$

$$M^{(w)}_{x,r+1} = M^{(w)}_{x,r} + l_P Q^{(w)}_{x,r}, \tag{2.290}$$

$$Q^{(w)}_{x,r+1} = Q^{(w)}_{x,r}. \tag{2.291}$$

These relations are contained in the transfer matrix of the studied problem, which is summed up in the term E-1 of the CTM when applying the substitutions $E = E_p$ and $F_{xx} = J_{p,x}$. Corresponding nodal matrix is specified in the term E-2 of the CTM.

Hence, it is of particular importance to make the matrix elements as simple as possible. This is accomplished by separating multipliers from stress resultants and displacements. The reduced quantities obtained in this manner are denoted by the sign ˆ.

The most favourable choice of external multipliers for the individual components of the modified state vector were found to be

$$\hat{N}^{(w)}_y = N^{(w)}_y, \tag{2.292}$$

$$\hat{N}^{(w)}_{yz} = \frac{2s_k}{L} N^{(w)}_{yz}, \tag{2.293}$$

$$\hat{N}^{(w)}_z = \frac{4s_k^2}{L^2} N^{(w)}_z, \tag{2.294}$$

$$\hat{v}_y^{(w)} = \frac{192}{5} \frac{E t_k s_k^3}{L^4} v_y^{(w)}, \tag{2.295}$$

$$\hat{v}_x^{(w)} = \frac{192}{5} \frac{E t_k s_k^3}{L^4} v_x^{(w)}, \tag{2.296}$$

$$\hat{\varphi}_x^{(w)} = \frac{192}{5} \frac{E t_k s_k^4}{L^4} \varphi_x^{(w)}, \tag{2.297}$$

$$\hat{M}_x^{(w)} = \frac{1}{s_k} M_x^{(w)}, \tag{2.298}$$

$$\hat{Q}_x^{(w)} = Q_x^{(w)}, \tag{2.299}$$

$$1 = 1, \tag{2.300}$$

where t_k and s_k are the arbitrarily chosen dimensions. If the substructure contains several equal plates, t_k and s_k are most conveniently taken as the thickness and width of one plate. Furthermore, a parameter

$$\alpha = \frac{7.2(2s_k)^6}{t_k^2 L^4}, \tag{2.301}$$

has been introduced.

The transfer matrix for such modified state vectors is given in the term E-3 of the CTM.

When two plates meet at a fold, the total transformation of reduced quantities across the fold is performed by the corresponding transformation matrix summed up in the term E-4 of the CTM.

The incorporation of the derived matrix expressions into the algorithm of the transfer matrix method is performed in accordance with theoretical considerations presented in aforementioned sections of this book.

2.9. Direct Time-Integration Methods

In the foregoing sections the attention was focused on the problems of the implementation of the TMM for the spatial discretization of structural configurations. The present concepts allow the application of mixed schemes of spatial discretization which are suitable for advanced linear and nonlinear analyses of complex structural simulations.

However, the modern problems of mechanics require the simultaneous discretization in time, using analogous mixed time-integration schemes. The principles of direct time-integration schemes, suitable for the present problems, are known [5, 6, 119] and will not be dealt with here. The basic concepts of implementation of direct time-integration methods and especially of the Wilson technique into algorithm of the transfer matrix method were explained in Sections 1.10 and 2.2 of this book.

Subsequently there will be explained the implementation of the step-by-step and Newmark methods of direct time integration into the algorithm of transfer matrix method. In accordance with the principles of substructuring in space and time, these methods may be combined in various substructures of regions as well as in various time intervals of linear and nonlinear analyses. These operations correspond with the implementation of the Wilson technique described in Section 2.2.

a) Step-By-Step Technique

The flow-chart of this concept is given by the following operations:

A. Initial stage of solution.

1. Calculation of displacements $v_0(t)$ in all nodes of spatial simulation.

2. The choice of the time step Δt_0 and calculation of initial velocities and accelerations in all nodes

$$\dot{v}_0(t) = v_0(t)/\Delta t_0, \quad \ddot{v}_0(t) = \dot{v}_0(t)/\Delta t_0. \tag{2.302}$$

3. The choice of the time step Δt and formulation of effective stiffness $\tilde{k}(t)$ as

$$\tilde{k}(t) = k(t) + \frac{6}{\Delta t^2} m + \frac{3}{\Delta t} c(t), \tag{2.303}$$

with mass and damping parameters m and $c(t)$, respectively. Incorporation of these parameters into the algorithm of the TMM is performed in accordance with explanation in Section 2.2

B. For each time-integration step.

4. The calculation of the modified load vector in time $t + \Delta t$

$$\Delta\tilde{p}(t) = \Delta p(t) + m\left(\frac{6}{\Delta t}\dot{v}_0(t) + 3\ddot{v}_0(t)\right) + c(t)\left(3\dot{v}_0(t) + \frac{\Delta t}{2}\ddot{v}_0(t)\right). \tag{2.304}$$

Modified loads are put into the corresponding terms of the loading column of nodal matrices in each point of the applied discrete simulation.

5. Using the algorithm of the transfer matrix method with application of effective stiffness and modified load parameters the increments of deformations $\Delta v(t)$ in all nodes of the simulation are calculated.

6. The calculation of increments of velocities and accelerations in all nodal points

$$\Delta\dot{v}(t) = \frac{3}{\Delta t}\Delta v(t) - 3\dot{v}_0(t) - \frac{\Delta t}{2}\ddot{v}_0(t), \tag{2.305}$$

$$\Delta\ddot{v}(t) = \frac{6}{\Delta t^2}\Delta v(t) - \frac{6}{\Delta t}\dot{v}_0(t) - 3\ddot{v}_0(t). \tag{2.306}$$

7. The calculation of resulting displacements, velocities and accelerations

$$v(t + \Delta t) = v_0(t) + \Delta v(t) = v_1(t), \tag{2.307}$$

$$\dot{v}(t + \Delta t) = \dot{v}_0(t) + \Delta\dot{v}(t) = \dot{v}_1(t), \tag{2.308}$$

$$\ddot{v}(t + \Delta t) = \ddot{v}_0(t) + \Delta\ddot{v}(t) = \ddot{v}_1(t). \tag{2.309}$$

C. Repetition of the calculation for following time steps with initial displacements, velocities and accelerations $v_1(t)$, $\dot{v}_1(t)$ and $\ddot{v}_1(t)$, respectively.

b) Newmark Technique

The Newmark integration method is implemented into the algorithm of the TMM as follows:

A. Initial stage of solution.

1. Calculation of $v_0(t)$, $\dot{v}_0(t)$ and $\ddot{v}_o(t)$ for given loads.

2. The choice of time step Δt, parameters $\delta = 0.5$ and $\alpha = 1/6$. Calculation of integration constants

$$a_0 = \frac{1}{\alpha\Delta t^2}, \qquad a_1 = \frac{\delta}{\alpha\Delta t}, \qquad a_2 = \frac{1}{\alpha\Delta t},$$

$$a_3 = \frac{1}{2\alpha} - 1, \qquad a_4 = \frac{\delta}{\alpha} - 1, \qquad a_5 = \frac{\Delta t}{2}\left(\frac{\delta}{\alpha} - 2\right),$$

$$a_6 = \Delta t(1 - \delta), \qquad a_7 = \delta\Delta t. \tag{2.310}$$

3. The formulation of the effective stiffness

$$\tilde{k}(t) = k(t) + a_0 m + a_1 c(t) \tag{2.311}$$

and incorporation into corresponding transfer or nodal matrices.

B. For each time-integration step.

4. The calculation of the modified load vector in time $t + \Delta t$

$$\Delta\tilde{p}(t) = \Delta p(t) + m(a_0 v_0(t) + a_2\dot{v}_0(t) + a_3\ddot{v}_0(t)) + \\ + c(t)(a_1 v_0(t) + a_4\dot{v}_0(t) + a_5\ddot{v}_0(t)). \tag{2.312}$$

5. The calculation of resulting deformations, accelerations and velocities

$$v(t + \Delta t) = v_0(t) + \Delta v(t), \tag{2.313}$$

$$\dot{v}(t + \Delta t) = \dot{v}_0(t) + a_6\ddot{v}_0(t) + a_7\ddot{v}_0(t + \Delta t). \tag{2.314}$$

$$\ddot{v}(t + \Delta t) = a_0(\Delta v(t)) - a_2\dot{v}_0(t) - a_3\ddot{v}_0(t), \tag{2.315}$$

C. Repetition of calculation for following time steps.

2.10. Problem of Numerical Difficulties

Numerical difficulties may occur when applying the transfer matrix method for analysing some cases of static and dynamic behaviour of complex structural configurations. Numerical instabilities may appear when using the transfer matrices which have hyperbolic functions in diagonal or non-diagonal terms. The cumulation of little differences of great numbers in a calculation cycle of the transfer matrix method may appear further when solving the configurations with very rigid supports or structural couplings. For small time steps the numerical difficulties may be expected when solving the dynamic response of structural configurations by application of direct time-integration methods combined with the transfer matrix method.

There are some ways for eliminating or reducing such phenomena [24, 59, 125]. Effective is the application of suitable multipliers allowing the non-dimensional expression of individual terms of the transfer matrices, as shown in Section 2.8. To a limited extent the double or higher precision of the computer calculation may also be adopted. For the elimination of the numerical instability in nodes of rigid supports or couplings, the elimination of unknown parameters in such points can be applied in a manner similar to Gaussian elimination,First of all the initial parameters having the greatest coefficients in the corresponding equations are eliminated. In the computer program the checking subroutines must be included, allowing the control of initial parameters with maximum coefficients in nodes of rigid supports or couplings. These parameters are in the subsequent calculation substituted by the unknown reaction components in such nodes.

An effective technique for overcoming numerical difficulties is to combine the *Riccati transformation* with the TMM. An important aspect of this combination is that less computer storage is required and improved computational time is achieved [63]. The principles of this concept will be explained below.

The transfer of state variables from station i to the station $i + 1$ across the analysed substructure, for a member with n state variables can be written as

$$\boldsymbol{s}_{i+1} = \mathbf{U}_i\boldsymbol{s}_i + \boldsymbol{F}_i, \tag{2.316}$$

where $\boldsymbol{s}$ is the vector of state variables, $\mathbf{U}$ is an $(n \times n)$ transfer matrix and $\boldsymbol{F}$

is the vector of loading or inducing functions. Let equation (2.316) be modified so that

$$\begin{bmatrix} \boldsymbol{f} \\ \boldsymbol{e} \end{bmatrix}_{i+1} = \begin{bmatrix} \mathbf{U}_{11} & \mathbf{U}_{12} \\ \mathbf{U}_{21} & \mathbf{U}_{22} \end{bmatrix}_i \begin{bmatrix} \boldsymbol{f} \\ \boldsymbol{e} \end{bmatrix}_i + \begin{bmatrix} \boldsymbol{F}_f \\ \boldsymbol{F}_e \end{bmatrix}_i, \tag{2.317}$$

where $\boldsymbol{f}$ contains the $n/2$ state variables which are zero at the left-hand boundary and $\boldsymbol{e}$ contains the remaining $n/2$ complementary state variables. The $\mathbf{U}_{ij}$ are the $(n/2 \times n/2)$ submatrices, $\boldsymbol{F}_f$ is a vector of loading function terms corresponding to the f state variables and $\boldsymbol{F}_e$ contains the forcing terms for the e state variables.

A generalized Riccati transformation at station i can be defined by

$$\boldsymbol{f}_i = \mathbf{S}_i \boldsymbol{e}_i + \boldsymbol{P}_i, \tag{2.318}$$

which relates the f state variables to the e state variables. The $(n/2 \times n/2)$ matrix $\mathbf{S}$ is the Riccati transfer matrix and the vector $\boldsymbol{P}$ contains the corresponding loading function terms. Expansion of equation (2.317) gives

$$\boldsymbol{f}_{i+1} = \mathbf{U}_{11,i} \boldsymbol{f}_i + \mathbf{U}_{12,i} \boldsymbol{e}_i + \boldsymbol{F}_{f,i} \tag{2.319}$$

and

$$\boldsymbol{e}_{i+1} = \mathbf{U}_{21,i} \boldsymbol{f}_i + \mathbf{U}_{22,i} \boldsymbol{e}_i + \boldsymbol{F}_{e,i}. \tag{2.320}$$

Substitution of equation (2.318) into equation (2.320) and solving for $\boldsymbol{e}_i$ yields

$$\boldsymbol{e}_i = (\mathbf{U}_{21}\mathbf{S} + \mathbf{U}_{22})_i^{-1} \boldsymbol{e}_{i+1} - (\mathbf{U}_{21}\mathbf{S} + \mathbf{U}_{22})_i^{-1} (\mathbf{U}_{21}\boldsymbol{P} + \boldsymbol{F}_e)_i. \tag{2.321}$$

Use of equations (2.318) and (2.321) to eliminate $\boldsymbol{f}_i$ and $\boldsymbol{e}_i$ from equation (2.319) gives

$$\begin{aligned} \boldsymbol{f}_{i+1} = {} & (\mathbf{U}_{11}\mathbf{S} + \mathbf{U}_{12})_i (\mathbf{U}_{21}\mathbf{S} + \mathbf{U}_{22})_i^{-1} \boldsymbol{e}_{i+1} + (\mathbf{U}_{11}\boldsymbol{P} + \boldsymbol{F}_f)_i - \\ & - (\mathbf{U}_{11}\mathbf{S} + \mathbf{U}_{12})_i (\mathbf{U}_{21}\mathbf{S} + \mathbf{U}_{22})_i^{-1} (\mathbf{U}_{21}\boldsymbol{P} + \boldsymbol{F}_e)_i. \end{aligned} \tag{2.322}$$

Equation (2.322) may be written as

$$\boldsymbol{f}_{i+1} = \mathbf{S}_{i+1} \boldsymbol{e}_{i+1} + \boldsymbol{P}_{i+1}, \tag{2.323}$$

where

$$\mathbf{S}_{i+1} = (\mathbf{U}_{11}\mathbf{S} + \mathbf{U}_{12})_i (\mathbf{U}_{21}\mathbf{S} + \mathbf{U}_{22})_i^{-1} \tag{2.324}$$

and

$$\boldsymbol{P}_{i+1} = (\mathbf{U}_{11}\boldsymbol{P} + \boldsymbol{F}_f)_i - \mathbf{S}_{i+1} (\mathbf{U}_{21}\boldsymbol{P} + \boldsymbol{F}_e)_i. \tag{2.325}$$

Equations (2.324) and (2.325) are the general recursion relations for $\mathbf{S}$ and $\boldsymbol{P}$. Since the left-hand boundary conditions are homogeneous, the initial conditions are

$$\mathbf{S}_0 = \boldsymbol{P}_0 = 0 . \tag{2.326}$$

For present case it is useful to be able to move in the opposite direction from station $i + 1$ to station i. Then

$$\boldsymbol{e}_i = \boldsymbol{T}_i \boldsymbol{e}_{i+1} + \boldsymbol{Q}_i \tag{2.327}$$

and using equation (2.321) there may be derived the expressions

$$\mathbf{T}_i = (\mathbf{U}_{21}\mathbf{S} + \mathbf{U}_{22})_i^{-1} \tag{2.328}$$

and

$$\boldsymbol{Q}_i = -\mathbf{T}_i(\mathbf{U}_{21}\boldsymbol{P} + \boldsymbol{F}_e)_i . \tag{2.329}$$

The symbol $\mathbf{T}$ denotes the $(n/2 \times n/2)$ matrix that transmits the $\boldsymbol{e}$ state variables and $\boldsymbol{Q}$ is the vector containing loading function terms. Equations (2.328) and (2.329) are the general recursion relations for $\mathbf{T}$ and $\boldsymbol{Q}$.

Equations (2.324), (2.325), (2.328), and (2.329) are needed while moving from left to right through the structure and equations (2.327) and (2.323) are solved while moving from right to left. This forward and backward movement in solving equations is equivalent to the Gauss elimination procedure.

The process starts at the left-hand boundary and proceeds from left to right through the member until the right-hand boundary is reached. All of the intermediate values of $\mathbf{S}$, $\boldsymbol{P}$, $\mathbf{T}$ and $\boldsymbol{Q}$ are retained. At the right-hand boundary the $n/2$ known state variables are applied to equation (2.318), a procedure which determines the remaining $n/2$ state variables at that point. Finally, the state variables at each station are determined by moving from right to left through the substructure or total structure. Successive applications of equation (2.327) give $\boldsymbol{e}$ at any station. Equation (2.318) is now used to calculate $\boldsymbol{f}$ at any station. This completes the solution procedure.

For avoiding the numerical difficulties when applying the direct methods of time integration combined with the transfer matrix method, the known techniques of optimum choice of integration operators as shown in Section 2.2d are to be adopted.

Chapter 3

FETM-Method — Nonlinear Approach

3.1. Introductory Remarks

Further development of the transfer matrix techniques is directed towards the streamline, vectorized algorithms and their utilization in advanced nonlinear problems of mechanics. Vectorized algorithms come distinctly to the fore especially with the development of a new generations of vector computers. One of the algorithms of this type is the FETM-method which is accepted as a problem-oriented combination of finite element and transfer matrix techniques [17, 64, 89, 93].

Theoretical analyses in the foregoing chapter were focused on the linear problems of static and dynamic behaviour of structures when applying the transfer matrix method. This chapter details geometrically and physically nonlinear analyses of structural simulations in limit states of their behaviour. The development of reliable and efficient techniques for the handling of such problems is emphasized. A generalized analysis of motion with implementation of the FETM-method and updated Lagrangian formulation is performed. Utilization of mixed schemes of discretization in space and time is put forth. Special numerical techniques for solving nonlinear equations of motion are presented.

Most of the research work within the field of structural simulations aims at developing more economical and functional structures with an acceptable service ability and safety. Advanced computational models have been developed in order to obtain reliable information concerning the behaviour of structures in extremal limit regimes of exploitation. For the analysis of limit states and safety of slender structures the complete nonlinear spatial behaviour must be taken into account.

The solution of the nonlinear response in the limit states of the structure requires the consideration of a number of physical phenomena occurring on a macroscopic level. As most important should be mentioned:

1. geometric imperfections and second-order geometric effects,
2. elastic-plastic material behaviour,
3. local and total stability effects in critical and postcritical regions,

4. nonlinear damping parameters.

All these effects must be taken into account in a linear and nonlinear general interaction.

The solution of these problems is concerned with the generalized dynamic simulation of complex structures consisting of an assemblage of substructures in space and in time. A general substructure synthesis method is adopted in which the motion of each substructure is represented by a given number of substructure matrix expressions as developed in the previous chapter. The coupling process of substructures requires the satisfaction of geometric compatibility conditions at every point of an internal boundary between two adjacent substructures. An internal boundary is considered to consist of a finite set of points and geometric compatibility conditions must be satisfied at each one of these points.

Mixed multigrid schemes of discretization in space and time consist of a problem-oriented variability of substructure sizes and time steps in various regions of the structure as well as in various time intervals. The mixed Newmark—Wilson technique for direct time integration of nonlinear equations of motion is applied in combination with the FETM-method of mixed space discretization when adopting the substructuring of these problems in space and in time.

The adoption of the developed concepts for solving of elastic-plastic, stability and optimization problems of structural simulations is dealt with in the last two sections of this chapter.

3.2. Pseudo-Force Technique

Consider the motion of a flat substructure in three-dimensional *Euclidean space* as shown in Figure 49. The simultaneous position of the set of material particles is called the *configuration* of the substructure. Alternatively, a configuration may also be regarded as a mapping of a substructure into a region of the Euclidean space. A sequence of such mappings defining the configuration at an arbitrary time t defines the motion of the studied substructure.

The deformation path of structure is a continuous family of configurations H_t which takes each substructure from an *initial position* H_0 at time $t = t_0$ to a *final configuration* H_F at $t = t_F$. The kinematics of deformation is described by two additional configurations, the *current configuration* H_1, corresponding to time t_1 and the *neighbouring configuration* H_2 in time t_2. It is assumed that H_2 is close to H_1, i.e., $t_2 = t_1 + \Delta t$, where Δt is a smal quantity.

The *total Lagrangian formulation* describes the motion of structure relative to the initial configuration H_0. In the *updated Lagrangian formulation* adopted herein, the reference state is taken as the current configuration H_1, which is continuously updated throughout the entire deformation process. A new reference frame is established at each stage along the deformation path. It will be demostrated in the next sections that the updated Lagrangian description of motion can be utilized to develop highly efficient solution procedures for the linear and nonlinear discrete analyses of slender structures. A major advantage of the formulation is its simplicity which provides an easy physical interpretation of the generalized nonlinear behaviour of the analysed structural simulations.

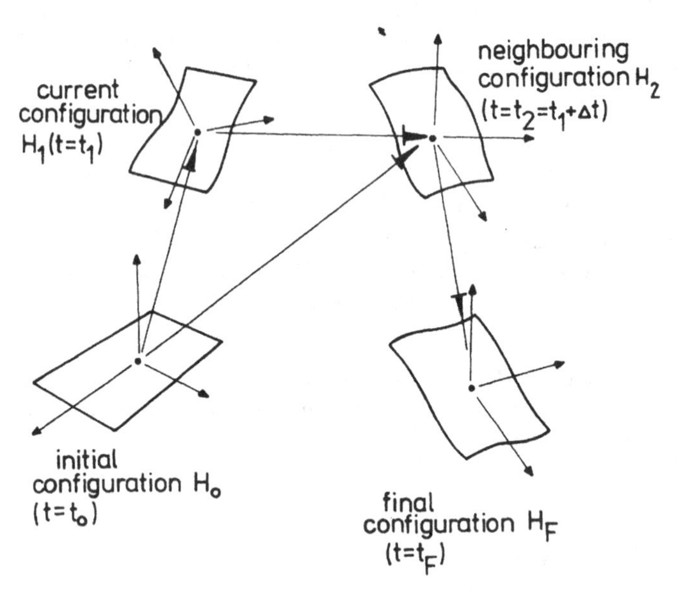

Fig. 49. Description of the deformation path.

In accordance with equations (1.16) and (1.20), the governing incremental equations of motion for a nonlinear structural dynamic system are given in modified form

$$\mathbf{M}_t \Delta \ddot{\boldsymbol{r}}_t + \mathbf{C}_t \Delta \dot{\boldsymbol{r}}_t + \mathbf{P}_t \Delta \boldsymbol{r}_t = \Delta \boldsymbol{R}_t + \Delta \boldsymbol{Q}_t, \tag{3.1}$$

where $\mathbf{P}_t \Delta \boldsymbol{r}_t$ is the vector of internal, deformation-dependent nonlinear forces, $\Delta \boldsymbol{Q}_t$ is the increment of effective plastic loads and $\mathbf{C}_t \Delta \dot{\boldsymbol{r}}_t$ is the vector of deformation and stress-state-dependent nonlinear damping forces (see Section 1.5). All other symbols have been explained earlier.

The pseudo-force technique [16, 41, 66], as applied here, is defined by

$$\mathbf{P}_t \Delta \boldsymbol{r}_t = \mathbf{K}_t \Delta \boldsymbol{r}_t + \mathbf{N}_t \Delta \boldsymbol{r}_t - \Delta \boldsymbol{V}_{t+\Delta t} + \Delta \boldsymbol{Q}_t, \tag{3.2}$$

where $\mathbf{N}_t \Delta \boldsymbol{r}_t$ is the vector of nonlinear terms (pseudo-forces) and $\Delta \boldsymbol{V}_{t+\Delta t}$ is the local approximation error defined in equation (1.20). In the application of the pseudo-force technique, the term $\mathbf{P}_t \Delta \boldsymbol{r}_t$ is placed on the right-hand side of equation (3.1) and the vector of nonlinear terms is treated as a pseudo-force vector. At each time step an estimate of $\mathbf{N}_t \Delta \boldsymbol{r}_t$ is computed and iterations are performed until $\Delta \boldsymbol{V}_{t+\Delta t}$ becomes sufficiently small when compared to a prescribed tolerance norm. As an estimate for $\mathbf{N}_t \Delta \boldsymbol{r}_t$ for the first iteration at time step t an extrapolated value from previous solutions can be used, i.e.,

$$\mathbf{N}_t \Delta \boldsymbol{r}_t = (1 + \alpha) \mathbf{N}_{t-\Delta t} \Delta \boldsymbol{r}_{t-\Delta t} - \alpha \mathbf{N}_{t-2\Delta t} \Delta \boldsymbol{r}_{t-2\Delta t}, \tag{3.3}$$

where α is an extrapolation parameter ranging from 0 to 1. The major advantage of this technique is the fact, that the effective stiffness contained in the flow-chart of direct time-integration schemes, as explained in Sections 1.2 and 2.10, need to be decomposed once, and only the right-hand sizes are modified to account for nonlinear terms.

Because of the large computational effort required for the nonlinear analysis, it is desirable to seek a strategy of optimal numerical calculations in which the number of calculation operations is minimized. A general solution strategy for nonlinear problems may be defined in terms of a number of control parameters which specify the linearization techniques, the frequency with which the effective stiffness matrix is reformulated, convergence tolerances and limits on the maximum number of iterations and adaptively change the time step size in order to minimize the computational effort. There are three typical macro-operations of nonlinear structural analyses, defined by

— solution procedures for nonlinear systems of equations,
— discretization procedures and the element technology,
— adaptive programming procedures, automatic choice of methods and of time and load steps.

Each of these topics can be thought of as consisting of a sequence of micro-operations on matrices and vectors. Some of the micro-operations are performed once, others have to be performed repeatedly at each time step. The operations of the first of the mentioned items were discussed in Section 1.7. The operations of both remaining items are examined in detail in following sections.

3.3. Mixed Discretization in Space

The actual geometry of the structural configuration is simulated over a multigrid space mesh of elastic-plastic microelements and macroelements as shown in Figure 50. In accordance with the analyses in Chapter 2, other types

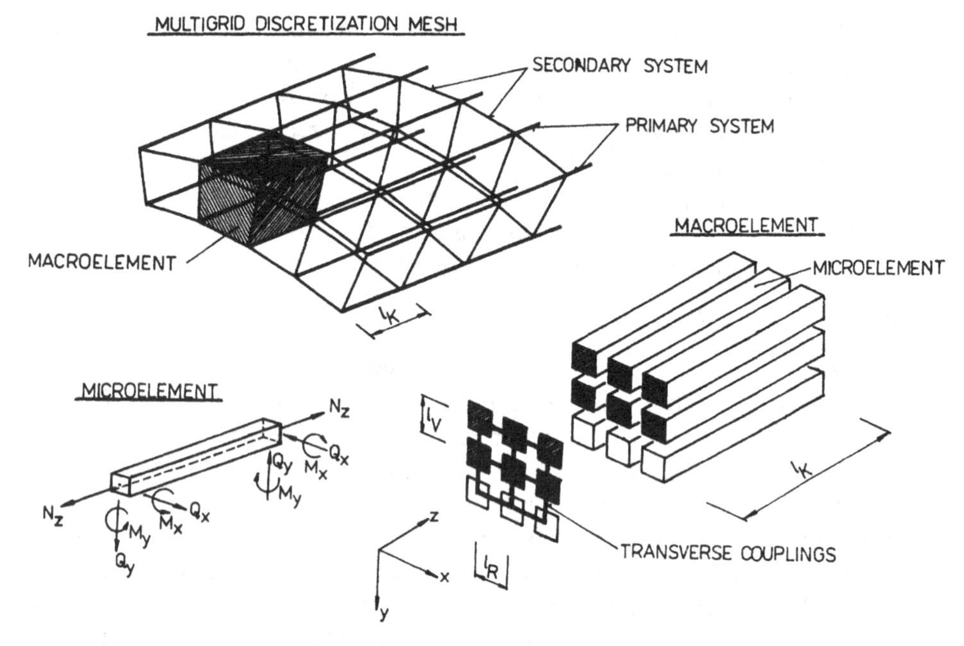

Fig. 50. The assumed multigrid space mesh.

of substructures may be incorporated into such a grillage mesh in order to simulate the actual updated geometry of all particles of the nonlinear structure. The assumed discrete model allows the simulation of the general anisotropy of structural and material parameters which are present in large deformation and elastic-plastic regions of the nonlinear analysis. Different types of nonlinearities are dealt with using various systems of the applied discrete simulation. In the updated Lagrangian formulation of motion, accomplished using local element coordinate systems that translate and rotate with the large deformations of structure, the major rigid-body geometric nonlinearities are embodied in the coordinate transformations of the microelement mesh. The only nonlinear effects that occur on the microelement level are due to its geometric stiffness. Otherwise, the effects of physical nonlinearities, e.g., the nonlinear damping and elastic-plastic influences, are analysed on the level of the macroelement simulation mesh. Both systems are interactively coupled and may be arbitrarily varied and combined. The lumped mass representation of inertial parameters is applied in the present model.

The proposed multigrid simulation uses the dynamically variable size of micro- and macroelements in various regions of the structure and in various time and load step levels of the nonlinear analysis.

3.4. Algorithm of FETM-Method

The generalized transfer hypermatrices of the FETM-method when applied for the analysis of the proposed multigrid simulation are constructed over the diagonal set-up of transfer matrices of macroelements in accordance with Figure 50. The transfer matrices of macroelements as typical substructures consist of the diagonal assembly of transfer submatrices of individual microelements. Transfer submatrices are derived over the inverse transformations of linear and nonlinear stiffness matrices of the assumed microelements.

Consider the microelement shown in Figure 50 with force—deformation relations defined in the known stiffness matrices $\mathbf{K}_L$ and $\mathbf{K}_G$ (see equation 1.41) of its linear and nonlinear behaviour. The corresponding subvectors of force $\boldsymbol{p}$ and displacement $\boldsymbol{d}$ components are expressed as

$$\boldsymbol{p}_{i-1} = (S_1, S_2, S_3, S_4, S_5, S_6)^T, \tag{3.4}$$

$$\boldsymbol{p}_i = (S_7, S_8, S_9, S_{10}, S_{11}, S_{12})^T, \tag{3.5}$$

$$\boldsymbol{d}_{i-1} = (u_1, u_2, u_3, u_4, u_5, u_6)^T, \tag{3.6}$$

$$\boldsymbol{d}_i = (u_7, u_8, u_9, u_{10}, u_{11}, u_{12})^T. \tag{3.7}$$

Taking account of these substitutions, the force-deformation couplings are given by

$$\begin{bmatrix} \boldsymbol{p}_{i-1} \\ \boldsymbol{p}_i \end{bmatrix} = \begin{bmatrix} \mathbf{A} & \mathbf{B} \\ \mathbf{C} & \mathbf{D} \end{bmatrix} \begin{bmatrix} \boldsymbol{d}_{i-1} \\ \boldsymbol{d}_i \end{bmatrix}, \tag{3.8}$$

with adopted stiffness matrix $\mathbf{K}_I$

$$\mathbf{K}_I = \mathbf{K}_L + \mathbf{K}_G = \begin{bmatrix} \mathbf{A} & \mathbf{B} \\ \mathbf{C} & \mathbf{D} \end{bmatrix}. \tag{3.9}$$

The solution and modifications of the system (3.8) yield the following hybrid couplings of state components in initial and end points of the analysed microelement

$$\begin{bmatrix} \boldsymbol{d}_i \\ \boldsymbol{p}_i \end{bmatrix} = \begin{bmatrix} -\mathbf{B}^{-1}\mathbf{A} & \mathbf{B}^{-1} \\ \mathbf{C} - \mathbf{D}\mathbf{B}^{-1}\mathbf{A} & \mathbf{D}\mathbf{B}^{-1} \end{bmatrix} \begin{bmatrix} \boldsymbol{d}_{i-1} \\ \boldsymbol{p}_{i-1} \end{bmatrix} = \mathbf{BKS} \begin{bmatrix} \boldsymbol{d}_{i-1} \\ \boldsymbol{p}_{i-1} \end{bmatrix}, \tag{3.10}$$

where **BKS** is the sought transfer submatrix of the microelement. The submatrix **BKS** transfers the state vectors from the initial into the end points of the longitudinal discretization of the analysed microelement simulation.

The nodal matrices perform the nodal spatial couplings for individual microelements of the studied macroelement simulation as shown in Figure 50. The nodal matrices are derived over the direct transformations of linear and nonlinear stiffness matrices of microelement transverse couplings (the members of secondary system as shown in Figure 50). The nodal matrices also include into the calculation cycle of the FETM-method, the variable coordinate transformations of the updated Lagrangian formulation of motion. The inertial parameters and external supports or couplings may also be incorporated into the corresponding nodal matrices.

The spatial macroelement constructed by such matrix operations represents a typical substructure of the proposed multigrid simulation. The coupling of individual macroelements is performed over transfer and nodal hypermatrices defined by the characteristic set-up of their transfer and nodal matrix equivalents.

The matrix and hypermatrix expressions for initial parameters and boundary conditions as well as the governing equations of the FETM-method are defined in accordance with principles of the transfer matrix method as explained in Chapter 2.

These concepts may also be adopted for the derivation of other types of macroelement substructures (flat members, shells, three-dimensional solids, etc.) into proposed multigrid space simulations of nonlinear problems. For the derivation of such substructures the known stiffness matrices of the finite element method may be applied. For example, in nonlinear analyses the nodal point degrees of freedom for a flat substructure are separated into two groups, namely those associated with inplane displacements and those associated with bending displacements. Details of derivation of such plate bending element can be found for example in reference [12]. A general quadrilateral membrane element is needed to match such a plate bending element. The use of refined elements utilizing polynomials of higher order will be avoided because stiffness evaluations of such elements will be extremely time-consuming when including the nonlinear geometric effects. Therefore an element with only two inplane degrees of freedom at each corner node should be used for the incorporation into the algorithm of the FETM-method. Such an element is, for example, *the Zienkiewicz—Irons quadrilateral* described in reference [124]. The incremental form of stiffness matrices of these elements corresponding to the theory of large deformations, which accounts for the geometrically nonlinear effects at the substructure level, will be used for deriving the corresponding nonlinear macroelement substructure performed in the Section 3.6. The used elements have six degrees of freedom per nodal point; three translations and three rotations. Finite rotations in space are not vectorially additive quantities; however this problem is tackled by distinguishing between deformational and rigid body rotations and accumulating the former incrementally along the deformation path.

3.5. Vectorization Technique

The process of organizing the calculation within a numerical flow-chart of the FETM-method is further referred to as a vectorization of the algorithm. This is used as the concept of transformation of hypermatrices into corresponding *Jordan permutative equivalents* [20, 109], which allows the total vectorization of the programmatory code.

Let us introduce the phenomena of *Jordan block* and of the *matrix in Jordan form*. If Q is the complex number and m is the natural number, the Jordan block $\mathbf{J}_m(Q)$ is expressed in matrix form as

$$\mathbf{J}_m(Q) = \begin{bmatrix} Q & 1 & 0 & 0 & \cdots \\ 0 & Q & 1 & 0 & \\ 0 & 0 & Q & 1 & \\ \vdots & & & & \end{bmatrix}. \tag{3.11}$$

The matrix in Jordan form when it has a diagonal block structure with partial Jordan blocks is given by

$$\mathbf{J} = \begin{bmatrix} \mathbf{J}_{M_1}(Q) & & & \\ & \mathbf{J}_{M_2}(Q) & & \\ & & \mathbf{J}_{M_3}(Q) & \\ & & & \ddots \end{bmatrix}. \tag{3.12}$$

The validity of the *Jordan lemma*, stating that each complex matrix in a complex zone is similar to the corresponding matrix in Jordan form, is assumed. If the diagonal block matrix is denoted by $\mathbf{A} = \mathrm{diag}(\mathbf{A}_1, \mathbf{A}_2, \mathbf{A}_3, \cdots, \mathbf{A}_m)$ and $(k_1, k_2, k_3, \cdots, k_m)$ is the permutation of indexes $(1, 2, 3, \ldots, m)$, then for the matrix $\mathbf{A} = \mathrm{diag}(\mathbf{A}_{k_1}, \mathbf{A}_{k_2}, \mathbf{A}_{k_3}, \cdots, \mathbf{A}_{k_m})$ the following is valid

$$\mathbf{A} = \mathbf{P}^{-1}\mathbf{A}\mathbf{P}, \tag{3.13}$$

with permutative matrix $\mathbf{P}$, which corresponds to the permutation $(k_1, k_2, k_3, \cdots, k_m)$. The matrix $\mathbf{P}$ has m-block rows and m-block columns. The block have dimensions from A_1 until A_m. In the first column and k-th row of the matrix $\mathbf{P}$ the unit matrix is positioned and the remaining elements are zero. In the second column and k_2-th row of the matrix $\mathbf{P}$ the unit matrix is located and the remaining members are again zero. These operations are continued until the full matrix $\mathbf{P}$ is constructed. Such operations are denoted as simultaneous permutations of rows and columns of the matrix $\mathbf{A}$. The scheme of the obtained modified diagonal form of transfer and nodal hypermatrices of the FETM-method is as shown in Figure 51. Long diagonal vectors are used in the flow-chart of the FETM-method in order to avoid the storage of zero values and computation with them. For such diagonal vectorization only the nonzero elements are stored, along with a bit vector which identifies their location. Diagonals with all zero elements are neither stored nor operated upon.

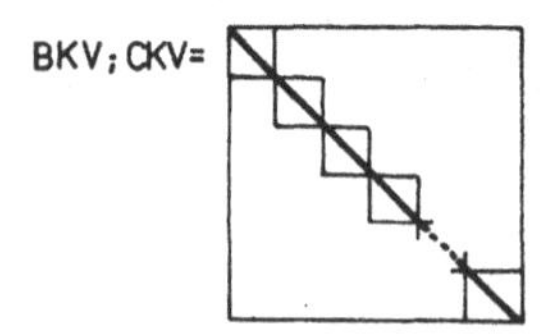

Fig. 51. Vectorized hypermatrices of the FETM-method.

3.6. Nonlinear Substructures

First of all, the attention is concerned with the derivation of matrix expressions of the FETM-method for a nonlinear space grillage substructure as shown in Figure 52. As mentioned earlier, such a substructure is a principal part of the spatial multigrid simulation mesh, adopted for solving nonlinear problems when considering the general variability of physical parameters and interactive couplings of individual nonlinear effects.

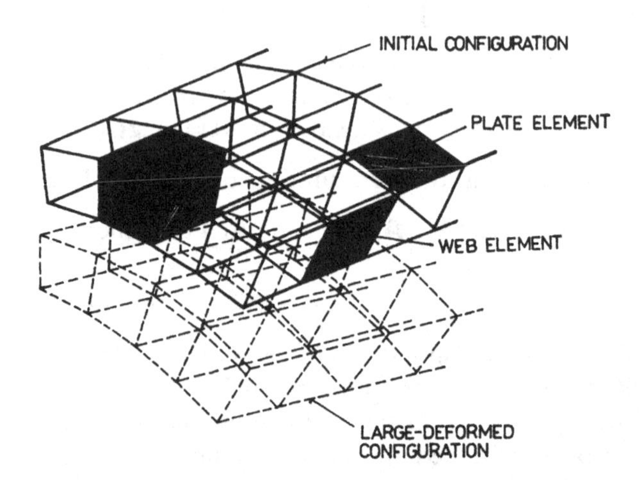

Fig. 52. The large deformed nonlinear space mesh configuration.

The one-dimensional microelement is studied with state parameters and pseudo-forces $\mathbf{N}_i$ resulting from the large deformation analysis as depicted in Figure 53. The incremental stiffness matrix $\mathbf{K}_I$ for coupling of force $\boldsymbol{p}_i$ and deformation $\boldsymbol{d}_i$ state subvectors for such microelement is given by

$$\boldsymbol{p}_i = (\mathbf{K}_\mathrm{L} + \mathbf{K}_\mathrm{G})\boldsymbol{d}_i = \mathbf{K}_\mathrm{I}\boldsymbol{d}_i\,, \tag{3.14}$$

$$\boldsymbol{p}_i = (S_1, S_2, S_3, S_4, S_5, S_6, \;\vdots\; S_7, S_8, S_9, S_{10}, S_{11}, S_{12})^\mathrm{T}\,, \tag{3.15}$$

$$\boldsymbol{d}_i = (u_1, u_2, u_3, u_4, u_5, u_6, \;\vdots\; u_7, u_8, u_9, u_{10}, u_{11}, u_{12})^\mathrm{T}\,, \tag{3.16}$$

$$\mathbf{K}_\mathrm{L} = \begin{bmatrix}
AB &&&&&&&&&&& \\
0 & AC &&&&&&&&&& \\
0 & 0 & AD &&&&&&&&& \\
0 & 0 & 0 & AE && \text{symmetry} &&&&&& \\
0 & 0 & -AH & 0 & AF &&&&&&& \\
0 & AI & 0 & 0 & 0 & AG &&&&&& \\
-AB & 0 & 0 & 0 & 0 & 0 & AB &&&&& \\
0 & -AC & 0 & 0 & 0 & -AI & 0 & AC &&&& \\
0 & 0 & -AD & 0 & AH & 0 & 0 & 0 & AD &&& \\
0 & 0 & 0 & -AE & 0 & 0 & 0 & 0 & 0 & AE && \\
0 & 0 & -AH & 0 & AK & 0 & 0 & 0 & AH & 0 & AF & \\
0 & AI & 0 & 0 & 0 & AL & 0 & -AI & 0 & 0 & 0 & AG
\end{bmatrix}, \tag{3.17}$$

$$\mathbf{K}_\mathrm{G} = \begin{bmatrix}
0 &&&&&&&&&&& \\
0 & BB &&&&&&&&&& \\
0 & 0 & 0 &&&&&&&&& \\
0 & 0 & 0 & 0 &&& \text{symmetry} &&&&& \\
0 & 0 & 0 & 0 & 0 &&&&&&& \\
0 & BE & 0 & 0 & 0 & BC &&&&&& \\
0 & 0 & 0 & 0 & 0 & 0 & 0 &&&&& \\
0 & -BD & 0 & 0 & 0 & -BE & 0 & BD &&&& \\
0 & 0 & 0 & 0 & 0 & 0 & 0 & 0 & 0 &&& \\
0 & 0 & 0 & 0 & 0 & 0 & 0 & 0 & 0 & 0 && \\
0 & 0 & 0 & 0 & 0 & 0 & 0 & 0 & 0 & 0 & 0 & \\
0 & BE & 0 & 0 & 0 & -BF & 0 & -BE & 0 & 0 & 0 & BC
\end{bmatrix}, \tag{3.18}$$

with substitutions

$$\begin{aligned}
AB &= EF_c/l_\mathrm{K}\,, & BB &= 6N_i/(5l_\mathrm{K})\,, \\
AC &= 12EF_{xx}/(l_\mathrm{K}^3(1+\Phi_y))\,, & BC &= 2N_il_\mathrm{K}/15\,, \\
AD &= 12EF_{yy}/(l_\mathrm{K}^3(1+\Phi_x))\,, & BD &= 6N_i/(5l_\mathrm{K})\,, \\
AE &= GJ_\mathrm{T}/l_\mathrm{K}\,, & BE &= N_i/10\,, \\
AF &= (4+\Phi_x)EF_{yy}/(l_\mathrm{K}(1+\Phi_x))\,, & BF &= N_il_\mathrm{K}^2/30\,, \\
AG &= (4+\Phi_y)EF_{xx}/(l_\mathrm{K}(1+\Phi_y))\,, && \\
AH &= 6EF_{yy}/(l_\mathrm{K}^2(1+\Phi_x))\,, &&
\end{aligned}$$

$$AI = 6EF_{xx}/(l_K^2(1 + \Phi_y)),$$
$$AK = (2 - \Phi_y)EF_{xx}/(l_K(1 + \Phi_y)),$$
$$AL = (2 + \Phi_x)EF_{yy}/(l_K(1 + \Phi_x)), \quad (3.19)$$

where Φ_i represent the shear deformation parameters.

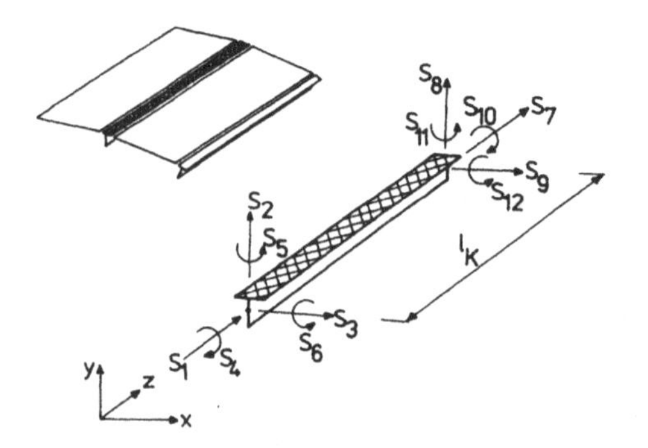

Fig. 53. One-dimensional element.

Similarly, the stiffness matrices for the studied nonlinear microelement may be found when it is subjected to the action of tensile or compressive axial forces. The scheme of the transfer matrix for nonlinear spatial grillage substructure shown in Figure 52 is illustrated in the term F-1 of the CTM. This transfer matrix is constructed over the diagonal set-up of transfer submatrices for the studied one-dimensional microelement derived from the present linear and geometrically nonlinear stiffness matrices in accordance with the theoretical analysis in the Section 3.4. Typical submatrices of such type are summed up in the terms F-2, F-3, F-4, F-5 and F-6 of the CTM. The corresponding nodal matrix and submatrix equivalents for studied grillage substructure are derived similarly as in foregoing sections of this book.

As the next typical substructure, the planar rectangular element as shown in Figure 54 is dealt with. The stiffness matrix expressions for such element, derived in accordance with the conclusions in Section 3.4, are given by

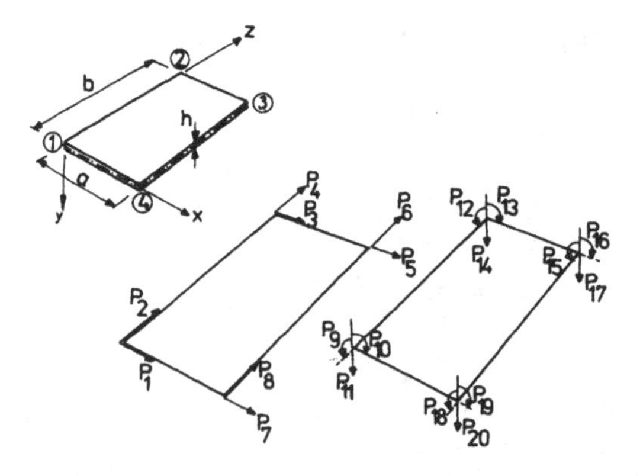

Fig. 54. Planar rectangular element.

$$\boldsymbol{p}_i = (\mathbf{K}_L + \mathbf{K}_G)\,\boldsymbol{d}_i\,, \tag{3.20}$$

$$\boldsymbol{p}_i = (P_1, P_2, \vdots\, P_3, P_4, \vdots\, P_5, P_6, \vdots\, P_7, P_8, \vdots\, P_9, P_{10}, P_{11}, \vdots\, P_{12}, P_{13}, P_{14}, \vdots\, P_{15}, P_{16}, P_{17}, \vdots\, P_{18}, P_{19}, P_{20})^T\,, \tag{3.21}$$

$$\boldsymbol{d}_i = (u_1, u_2, \vdots\, u_3, u_4, \vdots\, u_5, u_6, \vdots\, u_7, u_8, \vdots\, u_9, u_{10}, u_{11}, \vdots\, u_{12}, u_{13}, u_{14}, \vdots\, u_{15}, u_{16}, u_{17}, \vdots\, u_{18}, u_{19}, u_{20})^T\,, \tag{3.22}$$

$$\mathbf{K}_L = \begin{bmatrix} \mathbf{K}_{L,1} & \\ & \mathbf{K}_{L,2} \end{bmatrix}, \quad \mathbf{K}_G = \begin{bmatrix} \mathbf{K}_{G,1} & \\ & \mathbf{K}_{G,2} \end{bmatrix}, \tag{3.23}$$

$$\mathbf{K}_{L,1} = \frac{h}{12} \begin{bmatrix} RA & & & & & & & \\ RC & RB & & & & \text{symmetry} & & \\ RE & RD & RA & & & & & \\ RI & RF & -RC & RB & & & & \\ RK & RI & RG & RC & RA & & & \\ RM & RL & RJ & RH & -RC & RB & & \\ RR & RN & RK & -RI & RE & -RD & RA & \\ RT & RS & -RM & RL & -RI & RF & RC & RB \end{bmatrix}, \tag{3.24}$$

where

$$\begin{aligned}
RA &= 4d_{11}p^{-1} + 4d_{33}p\,, & RJ &= 3d_{21} + 3d_{33}\,,\\
RB &= 4d_{22}p + 4d_{33}p^{-1}\,, & RK &= -4d_{11}p^{-1} + 2d_{33}p\,,\\
RC &= 3d_{21} + 3d_{33}\,, & RL &= 2d_{22}p - 4d_{33}p^{-1}\,,\\
RD &= 3d_{21} - 3d_{33}\,, & RM &= 3d_{21} - 3d_{33}\,,\\
RE &= 2d_{11}p^{-1} - 4d_{33}p\,, & RN &= -3d_{21} - 3d_{33}\,,\\
RF &= -4d_{22}p + 2d_{33}p^{-1}\,, & RR &= -2d_{11}p^{-1} - 2d_{33}p\,,\\
RG &= -2d_{11}p^{-1} - 2d_{33}p\,, & RS &= -2d_{22}p - 2d_{33}p^{-1}\,,\\
RH &= -2d_{22}p - 2d_{33}p^{-1}\,, & RT &= 3d_{33}\,,\\
RI &= -3d_{21} + 3d_{33}\,, & &
\end{aligned} \tag{3.25}$$

in which

$$d_{11} = d_{22} = \frac{(1-\nu)E}{(1+\nu)(1-2\nu)}, \quad d_{21} = d_{12} = \frac{\nu E}{(1+\nu)(1-2\nu)},$$

$$d_{33} = \frac{E}{2(1+\nu)}, \quad p = a/b\,, \tag{3.26}$$

$$
\mathbf{K}_{L,2} = \frac{1}{15ab}\begin{bmatrix}
SA & & & & & & & & & & & \\
-SB & SC & & & & & & & & & & \\
-SD & SE & SF & & & & & & & & & \\
SG & 0 & SH & SA & & & & & & & & \\
0 & SI & SJ & SB & SC & \text{symmetry} & & & & & & \\
-SH & SJ & SM & SD & SE & SF & & & & & & \\
SN & 0 & SO & SP & 0 & SQ & SA & & & & & \\
0 & SR & SS & 0 & ST & SU & SB & SC & & & & \\
SO & -SS & SX & -SQ & -SU & SY & -SD & -SE & SF & & & \\
SP & 0 & -SQ & SN & 0 & -SO & SG & 0 & SH & SA & & \\
0 & ST & SU & 0 & SR & SS & 0 & SI & -SJ & -SB & SC & \\
SQ & -SU & SY & -SO & -SS & SX & -SH & -SJ & SM & SD & -SE & SF
\end{bmatrix}, \tag{3.27}
$$

The following substitutions have been made:

$$
\begin{aligned}
SA &= 20a^2D_y + 8b^2D_{xy}, \\
SB &= 15abD_1, \\
SC &= 20b^2D_x + 8a^2D_{xy}, \\
SD &= 30apD_y + 15bD_1 + 6bD_{xy}, \\
SE &= 30bp^{-1}D_x + 15bD_1 + \\
&\quad + 6aD_{xy}, \\
SF &= 60p^{-2}D_x + 60p^2D_y + \\
&\quad + 30D_1 + 84D_{xy}, \\
SG &= 10a^2D_y - 2b^2D_{xy}, \\
SH &= -30apD_y - 6bD_{xy}, \\
SI &= 10b^2D_x - 8a^2D_{xy}, \\
SJ &= 15bp^{-1}D_x - 15aD_1 - \\
&\quad - 6aD_{xy}, \\
SM &= 30p^{-2}D_x - 60p^2D_y - 30D_1 - \\
&\quad - 84D_{xy},
\end{aligned}
\qquad
\begin{aligned}
SN &= 10a^2D_y - 8b^2D_{xy}, \\
SO &= -15paD_y + 15bD_1 + \\
&\quad + 60bD_{xy}, \\
SP &= 5a^2D_y + 2b^2D_{xy}, \\
SQ &= 15apD_y - 6bD_{xy}, \\
SR &= 10b^2D_x - 2a^2D_{xy}, \\
SS &= 30bp^{-1}D_x + 6aD_{xy}, \\
ST &= 5b^2D_x + 2a^2D_{xy}, \\
SU &= 15bp^{-1}D_x - 6aD_{xy}, \\
SX &= 60p^{-2}D_x + 30p^2D_y - \\
&\quad - 30D_1 - 84D_{xy}, \\
SY &= -30p^{-2}D_x - 30p^2D_y + \\
&\quad + 30D_1 + 84D_{xy},
\end{aligned} \tag{3.28}
$$

with terms

$$
p = a/b, \quad D_x = D_y = D = Eh^3/(12(1 - \nu^2)), \quad D_{xy} = 0{,}5(1 - \nu)D, \quad D_1 = \nu D. \tag{3.29}
$$

$$
\mathbf{K}_{G,1} = \begin{bmatrix}
TA & & & & & & & \\
0 & TB & & & & & & \\
0 & 0 & TC & & & & & \\
0 & 0 & 0 & TD & & \text{symmetry} & & \\
-TE & 0 & 0 & 0 & TE & & & \\
0 & -TF & 0 & 0 & 0 & TF & & \\
0 & 0 & -TG & 0 & 0 & 0 & TG & \\
0 & 0 & 0 & -TH & 0 & 0 & 0 & TH
\end{bmatrix}, \tag{3.30}
$$

where

$$
\begin{aligned}
TA &= N_{x,1}/(2a), & TE &= N_{x,3}/(2a),\\
TB &= N_{y,1}/(2b), & TF &= N_{y,3}/(2b),\\
TC &= -N_{x,2}/(2a), & TG &= -N_{x,4}/(2a),\\
TD &= -N_{y,2}/(2b), & TH &= -N_{y,4}/(2b),
\end{aligned}
\tag{3.31}
$$

with $N_{x,i}$ and $N_{y,i}$ as the membrane pseudo-forces per unit width at the element centre.

$$
\mathbf{K}_{G,2} = \begin{bmatrix}
VA \\
0 & VB \\
0 & 0 & VC \\
-VD & 0 & 0 & VD \\
0 & VM & 0 & 0 & VE & & & & \text{symmetry} \\
0 & 0 & 0 & 0 & 0 & VF \\
-VN & 0 & 0 & -VN & 0 & 0 & VG \\
0 & VP & 0 & 0 & -VH & 0 & 0 & VH \\
0 & 0 & -VI & 0 & 0 & 0 & 0 & 0 & VI \\
VR & 0 & 0 & -VR & 0 & 0 & -VJ & 0 & 0 & VJ \\
0 & -VK & 0 & 0 & -VS & 0 & 0 & VS & 0 & 0 & VK \\
0 & 0 & 0 & 0 & 0 & -VL & 0 & 0 & 0 & 0 & 0 & VL
\end{bmatrix},
\tag{3.32}
$$

where

$$
\begin{aligned}
VA &= N_{xx,1}b/(3a), & VJ &= N_{xx,4}b/(3a),\\
VB &= N_{yy,1}a/(3b), & VK &= N_{yy,4}a/(3b),\\
VC &= N_{xy,1}/2, & VL &= N_{xy,4}/2,\\
VD &= N_{xx,2}b/(3a), & VM &= N_{yy,2}a/(6b),\\
VE &= N_{yy,2}a/(3b), & VN &= N_{xx,3}b/(6a),\\
VF &= N_{xy,2}/2, & VP &= N_{yy,3}a/(6b),\\
VG &= N_{xx,3}b/(3a), & VR &= N_{xx,4}b/(6a),\\
VH &= N_{yy,3}a/(3b), & VS &= N_{yy,4}a/(6b),\\
VI &= N_{xy,3}/2,
\end{aligned}
\tag{3.33}
$$

where $N_{xx,i}$, $N_{yy,i}$ and $N_{xy,i}$ are the membrane pseudo-forces per unit width at the element centre.

Once the vector of nodal displacements $\boldsymbol{d}_i$ is known, the product $\mathbf{K}_L\ \boldsymbol{d}_i$ will give the corresponding stress resultants due to these deformations, which yield the membrane forces acting at the element centre. When the vector of nodal displacements $\boldsymbol{d}_i$ is known in rectangular Cartesian coordinates for the element, the matrix product $\mathbf{K}_I\ \boldsymbol{d}_i$ will readily give the nodal stress resultants referred to in this coordinate system. If, however, current axial and shear forces referred to element axes are wanted, a transformation according the current state of deformation will be necessary.

Present planar quadrilateral elements may also be used for the analysis of shells, for which groups of four surface points lying in one and the same plane can be conveniently found. This is only possible if the corner nodes of a quadrilateral are located on the two lines of principal curvatures of the shell surface. The use of quadrilateral elements is therefore restricted to shells for which it is possible to determine a coordinate system coinciding with the network of the lines of principal curvatures. Such types of substructures are the shells with cylindrical or quasi-cylindrical shapes dealt with in this book. In contrast to shell elements, the flat elements produce no interior couplings between stretching and flexural behaviour, this interaction being established at the nodal points only via coordinate transformations. Flat elements present no problems with regard to the correct representation of rigid-body motions, since these modes are exactly described by polynomial functions.

The transfer submatrices and matrices for the present planar elements are constructed in accordance with the operations defined by equation (3.10) and theoretical conclusions specified in Section 3.4.

When studying the substructures consisting of an assembly of the aforementioned one-dimensional and planar elements, the appertaining nodal matrices include into the calculation cycle:

1. parameters of interaction of the members of the primary and secondary system of the analysed substructure,
2. couplings of one-dimensional elements with planar members,
3. coordinate transformations taking account of the updated Lagrangian formulation,
4. nodal inertial parameters of the structure,
5. dead weight, loading and forcing parameters concentrated in nodes of the applied simulation.

These influences wil be systematically dealt with in the following subsections 1—5.

1. First of all the attention is focused on the analysis of interaction of members of primary and secondary system of substructure. State parameters of the secondary system are denoted by the superscript (s). Force state components of the members of the secondary system are expressed as functions of the corresponding nodal displacements regarding the stiffness matrices of stiffener and flat elements which are listed in this section. The coupling of the nodal displacement and force state components in nodal points of longitudinal and transverse one-dimensional members of the studied substructure (Figure 55) can be written as

$$\begin{aligned}
u_1 &= -u_3^{(s)}, & S_1 &= -S_3^{(s),L} - S_3^{(s),R} + S_{1,0},\\
u_2 &= u_2^{(s)}, & S_2 &= S_2^{(s),L} + S_2^{(s),R} + S_{2,0},\\
u_3 &= u_1^{(s)}, & S_3 &= -S_1^{(s),L} + S_1^{(s),R} + S_{3,0},
\end{aligned}$$

$$\begin{aligned} u_4 &= -u_6^{(s)}, & S_4 &= -S_6^{(s),L} - S_6^{(s),R} + S_{4,0},\\ u_5 &= u_5^{(s)}, & S_5 &= S_5^{(s),L} + S_5^{(s),R} + S_{5,0},\\ u_6 &= u_4^{(s)}, & S_6 &= S_4^{(s),L} + S_4^{(s),R} + S_{6,0}, \end{aligned} \tag{3.34}$$

where superscripts L and R denote the state components from the left and right hand directions of analysed node and $S_{i,0}$ are external nodal forces. Considering these couplings, the redistribution of stiffness matrices (equations (3.17) and (3.18)) for transverse one-dimensional members, determines the nodal submatrix CKS in the term F-7 of the **CTM**.

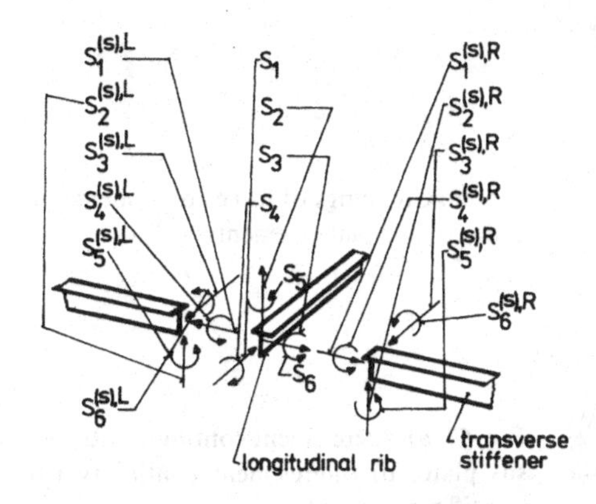

Fig. 55. Nodal couplings of force components of the longitudinal ribs and transverse stiffeners.

The coupling of the nodal displacements and force state components in nodal points of the plate elements (Figure 56) are given by

$$\begin{array}{llll} u_1 = u_7^{(s)}, & P_1 = P_7^{(s)} + P_{1,0}, & u_5 = u_3^{(s)}, & P_5 = P_3^{(s)} + P_{5,0},\\ u_2 = -u_8^{(s)}, & P_2 = -P_8^{(s)} + P_{2,0}, & u_6 = -u_4^{(s)}, & P_6 = -P_4^{(s)} + P_{6,0},\\ u_{11} = u_{20}^{(s)}, & P_{11} = -P_{20}^{(s)} + P_{11,0}, & u_{17} = u_{14}^{(s)}, & P_{17} = P_{14}^{(s)} + P_{17,0},\\ u_{10} = -u_{19}^{(s)}, & P_{10} = -P_{19}^{(s)} + P_{10,0}, & u_{16} = -u_{13}^{(s)}, & P_{16} = -P_{13}^{(s)} + P_{16,0},\\ u_9 = u_{18}^{(s)}, & P_9 = P_{18}^{(s)} + P_{18,0}, & u_{15} = u_{12}^{(s)}, & P_1 = P_{12}^{(s)} + P_{15,0},\\ u_3 = u_5^{(s)}, & P_3 = P_5^{(s)} + P_{3,0}, & u_7 = u_1^{(s)}, & P_7 = P_1^{(s)} + P_{7,0},\\ u_4 = -u_6^{(s)}, & P_4 = -P_6^{(s)} + P_{4,0}, & u_8 = -u_2^{(s)}, & P_8 = -P_2^{(s)} + P_{8,0},\\ u_{14} = u_{17}^{(s)}, & P_{14} = P_{17}^{(s)} + P_{14,0}, & u_{20} = u_{11}^{(s)}, & P_{20} = P_{11}^{(s)} + P_{20,0},\\ u_{13} = -u_{16}^{(s)}, & P_{13} = -P_{16}^{(s)} + P_{13,0}, & u_{19} = -u_{10}^{(s)}, & P_{19} = -P_{10}^{(s)} + P_{19,0},\\ u_{12} = u_{15}^{(s)}, & P_{12} = P_{15}^{(s)} + P_{12,0}, & u_{18} = u_9^{(s)}, & P_{18} = P_9^{(s)} + P_{18,0}, \end{array} \tag{3.35}$$

Taking account of these expressions and using the redistribution of the corresponding stiffness matrices (equations 3.20 until 3.33) for flat members, determine the sught nodal submatrix **CKS** in the term F-7 of the CTM.

2. When coupling the stiffener (one dimensional) element to planar elements, the continuity of the in-plane displacements along the intersection is

to be preserved. However, the torsional rotation of the stiffener is not compatible with the normal rotation of the plate along the intersection. This problem concerns the coupling of rotation parameters at nodes between stiffeners and planar elements. At node i (Figure 57) the torsional rotation of

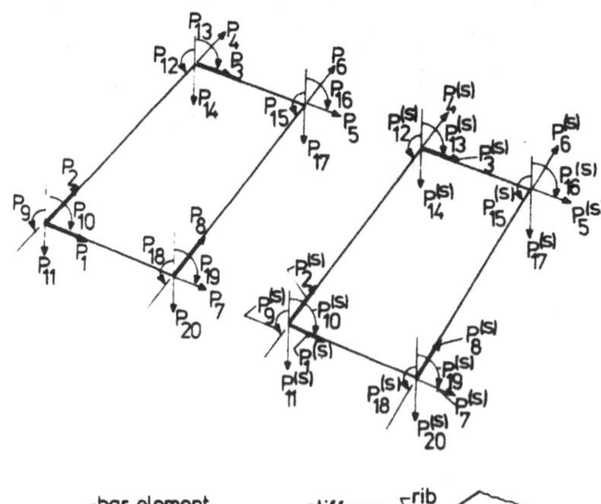

Fig. 56. Couplings of force components in nodes of the plate elements.

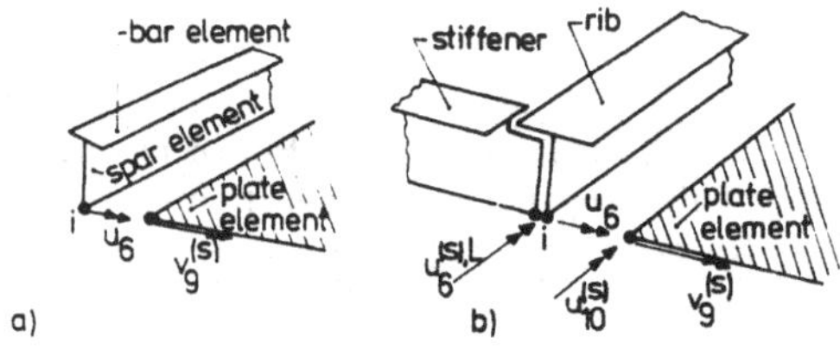

Fig. 57. a) Interelement continuity rib versus plate; b) Interelement continuity rib — stiffener — plate.

the rib and plate rotation about the same axis coincide. However, the end rotation u_6 of the rib cannot be coupled directly to the rotation $u_9^{(s)}$ of the plate due to the shear deformation in the rib. In an effort to include the shear deformation in the rib idealization, a stiffness matrix of a spar element [9, 118] may be used to represent the web Figure 58. The basic element is illustrated in Figure 59. The relationship between Cartesian coordinates x, z and normalized coordinates ξ, η is given by

$$x = \frac{1}{2}\sum_{i=1}^{2}(1 + \xi_i\xi)x_i, \quad z = \frac{1}{4}\sum_{i=1}^{4}(1 + \xi_i\xi)(1 + \eta_i\eta)z_i. \tag{3.36}$$

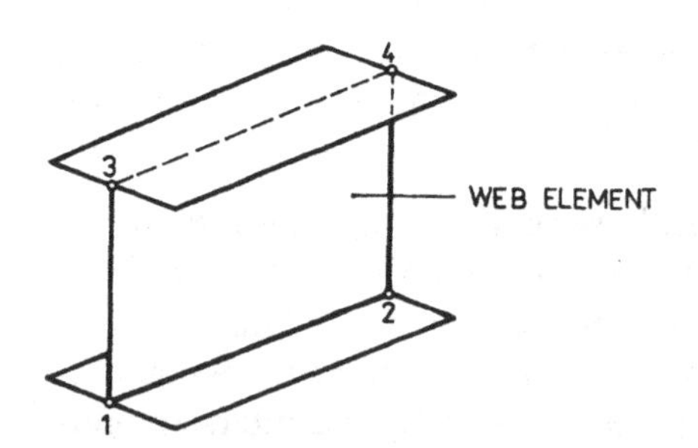

Fig. 58. Web element.

The x-axis in Figure 59 lies in the middle plane of the plate to which the stiffener is attached and does not always coincide with one of the nodal lines 1—2 or 3—4. However, the displacement parameters to be coupled to the plate degrees of freedom should be referred to the line $z = 0$. Therefore, a

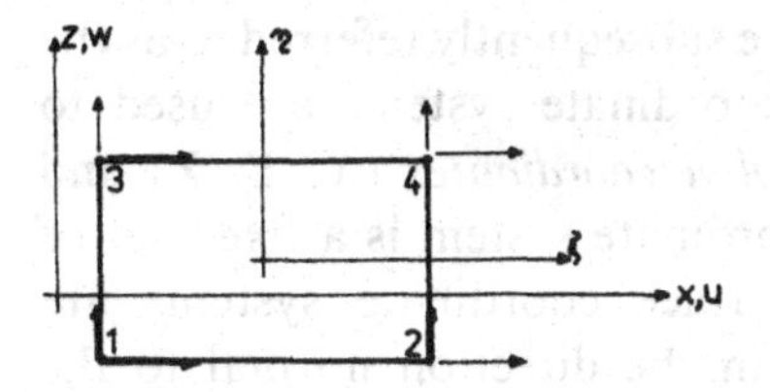

Fig. 59. Spar element.

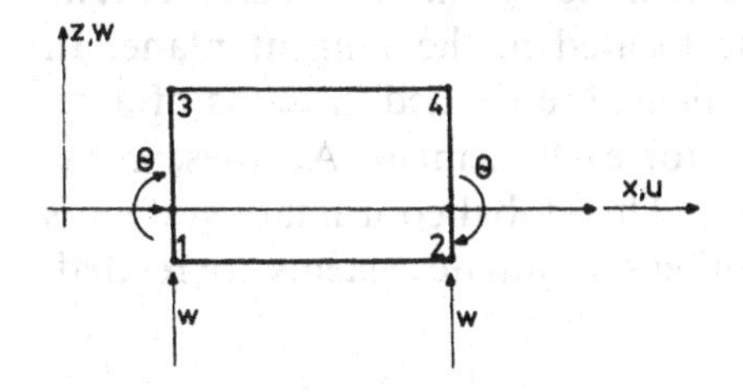

Fig. 60. New parameters for the rib element.

change of parameters is performed from the basic set in Figure 59 to the new set in Figure 60, where the end rotation Θ has been introduced. The transformations between the two sets of variables read

$$\begin{aligned} u_1 &= u_A + z_1 \Theta_A , \\ u_2 &= u_B + z_2 \Theta_B , \\ u_3 &= u_B + z_3 \Theta_B , \\ u_4 &= u_A + z_4 \Theta_A . \end{aligned} \tag{3.37}$$

An attempt is also made to include the effect of torsion in the present joint of rib and plate members. Due to the couplings between these members the torsional problem is very complicated and no scheme for formulating a simple stiffness matrix for this effect is available. The present approximation is based on the theory of pure torsion. The principal effects of Saint-Venant torsional rigidity is taken care of. This description of torsion is also very simple to include into stiffener and plate member couplings since it implies no interactions of axial and transverse deformations. Taking account of these considerations there can be derived individual coupling terms for corresponding fields of nodal submatrices, which are schematically depicted in the term F-7 of the CTM.

3. The assembly of the element transfer submatrices to form the transfer matrix of the complete substructure may be conveniently accomplished by the coordinate transformations and the subsequent superposition of individual transfer submatrices.

Before adding the element matrices into a total matrix relation, it is necessary to transform the element stiffness coefficients so that the translational and rotational degrees of freedom of all elements sharing a common nodal point are expressed in the same coordinate system. These nodal point coordinate systems, which are then used to relate nodal forces and displacements of the complete element assemblage, are subsequently referred to as the base coordinates. Two different types of coordinate systems are used to establish the base coordinates: a set of *global coordinates* (X, Y, Z) and *surface coordinates* (ξ, η, ζ). The global coordinate system is a fixed set of rectangular Cartesian coordinates. The surface coordinate systems are moving sets located such that ζ is always in the direction normal to the current surface of deformed layer at each node of the substructure. The coordinates η and ξ are assumed to be located in the tangent plane. In addition, the properties of individual elements are defined in terms of a set of element coordinates (x, y, z) specified for each element. All these coordinate systems are illustrated in Figure 61. The global coordinate system is taken as the reference frame to which all other coordinate systems are related.

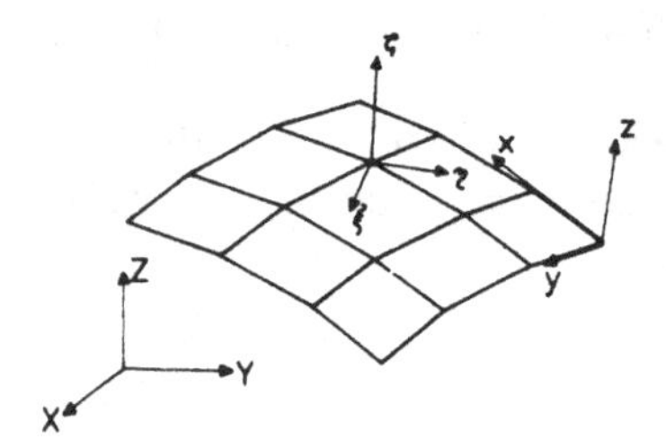

Fig. 61. The applied coordinate systems.

The transformation from one system into another is a purely geometric operation.

The direction cosines of the element coordinate axes relative to the global coordinate system are obtained by decomposing the x, y and z-axes into global directions X, Y and Z, via

$$\begin{bmatrix} X \\ Y \\ Z \end{bmatrix} = \begin{bmatrix} \cos(X,x) & \cos(X,x) & \cos(X,z) \\ \cos(Y,x) & \cos(Y,y) & \cos(Y,z) \\ \cos(Z,x) & \cos(Z,y) & \cos(Z,z) \end{bmatrix} \begin{bmatrix} x \\ y \\ z \end{bmatrix}, \tag{3.38}$$

or

$$\boldsymbol{X} = \mathbf{R}\boldsymbol{x}, \tag{3.39}$$

where (X, x) denotes the angle between positive X-axis and positive x-axis, etc.

A similar relation may be established between global and surface directions, viz.

$$\begin{bmatrix} X \\ Y \\ Z \end{bmatrix} = \begin{bmatrix} \cos(X,\xi) & \cos(X,\eta) & \cos(X,\zeta) \\ \cos(Y,\xi) & \cos(Y,\eta) & \cos(Y,\zeta) \\ \cos(Z,\xi) & \cos(Z,\eta) & \cos(Z,\zeta) \end{bmatrix} \begin{bmatrix} \xi \\ \eta \\ \zeta \end{bmatrix}, \tag{3.40}$$

or in matrix notation

$$\boldsymbol{X} = \mathbf{V}\xi. \tag{3.41}$$

The transformations defined by equations (3.39) and (3.41) are orthogonal, i.e. $\mathbf{R}^{-1} = \mathbf{R}^{\mathrm{T}}$ and $\mathbf{V}^{-1} = \mathbf{V}^{\mathrm{T}}$. From the above definitions it is possible to arrive at a relation between surface and element coordinates. Inverting equation (3.41) and substituting equation (3.39) yields

$$\xi = \mathbf{V}^{\mathrm{T}}\mathbf{X} = \mathbf{V}^{\mathrm{T}}\mathbf{R}\boldsymbol{x} = \mathbf{W}\boldsymbol{x}, \tag{3.42}$$

where

$$\mathbf{W} = \mathbf{V}^{\mathrm{T}}\mathbf{R}. \tag{3.43}$$

The local and global systems are related by equation

$$\boldsymbol{x} = \mathbf{R}^{\mathrm{T}}(\boldsymbol{X} - \boldsymbol{X}_i), \tag{3.44}$$

where $\boldsymbol{X}_i$ contains the global coordinates of local node of reference number i. From this equation the element coordinates of an arbitrary point within the substructure may be obtained once the global coordinates $\boldsymbol{X}$ and the direction cosines $\boldsymbol{R}$ have been specified. Similarly as aforementioned influences also present coordinate transformations will be incorporated into nodal submatrix schematically shown in F-7 of the CTM.

4. The inertia forces are proportional to the mass and acceleration of the elements and directed oppositely to the acceleration. Inertia influences are lumped in nodal points of the applied discrete simulation. Inertia forces appear in the corresponding terms of the nodal matrix in F-7 of the CTM, as the product of mass parameters and second time derivative of the vector of displacement coefficients or nodal deformations.

5. Dead weight and external loadings are treated by replacing the distributed volume and surface loads by equivalent nodal forces obtained from tributary area considerations. For the vector of external nodal loads is preserved the last (loading) column of the generalized nodal hypermatrix as shown in F-8 of the CTM.

Combining all of present influences determines the generalized nodal hypermatrix **CK** schematically depicted in F-8 of the CTM. The matrix **CK** has dimension $(12n + 20m + 1, 12n + 20m + 1)$, with n being the number of longitudinal ribs and m the number of applied plate elements. The vector of external forcing loads ***PK*** is positioned in the last (unit) column of the matrix **CK**. The vector ***PK*** may contain the parameters of initial conditions (displacements or imperfections), too. The scheme of the nodal submatrices **C1**, **C2**, **C3** and **C4** with dimensions (32, 32) (maxielement = one-dimensional rib + plate member) is illustrated in F-7 of the CTM. Taking account of position symbols shown in Figure 62, the typical terms of submatrices **C1**,

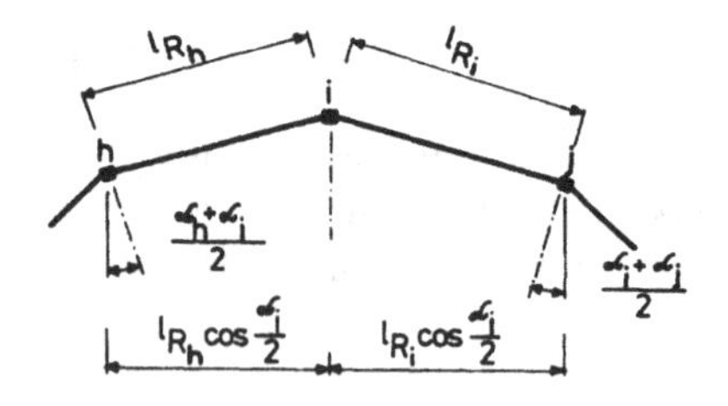

Fig. 62. Position symbols for nodal submatrices.

C2, **C3** and **C4**, for example for one-dimensional members, may be expressed as

$$C1(2,1) = \frac{E^{(s)}F_c^{(s)}}{l_{R,hi}} + PP1 - m_1\omega^2,$$

$$C2(2,1) = \frac{E^{(s)}F_c^{(s)}}{l_{R,ih}} + PP2,$$

$$C3(2,1) = \frac{E^{(s)}F_c^{(s)}}{l_{R,ij}} + PP3,$$

$$C4(2,1) = \frac{E^{(s)}F_c^{(s)}}{l_{R,ij}} - PP4,$$

$$C1(5,4) = -\frac{1}{\cos(\alpha_i/2)}\left[G^{(s)}\left(\frac{J^{(s)}_{T,hi}}{l_{R,hi}} + \frac{J^{(s)}_{T,ij}}{l_{R,ij}}\right) + G^{(p)}\left(\frac{J^{(p)}_{T,hi}}{l_{R,hi}} + \frac{J^{(p)}_{T,ij}}{l_{R,ij}}\right)\right] - m_2\omega^2,$$

$$C2(5,4) = \frac{\cos((\alpha_h + \alpha_i)/2)}{l_{R,ij}\cos(\alpha_i/2)}[G^{(s)}J^{(s)}_{T,hi} + G^{(p)}J^{(p)}_{T,hi}],$$

$$C3(5,4) = \frac{\cos((\alpha_h + \alpha_i)/2)}{l_{R,ij}\cos(\alpha_i/2)}[G^{(s)}J^{(s)}_{T,ij} + G^{(p)}J^{(p)}_{T,ij}],$$

$$C4(5,4) = 2C1(5,4),$$

$$C1(6,3) = \cos^2(\alpha_i/2)[12E^{(s)}(J^{(s)}_{1,hi}/l^3_{R,hi} + J^{(s)}_{1,ij}) + PP5) - m_3\omega^2,$$

$$C2(6,3) = -\cos(\alpha_h/2)\cos(\alpha_i/2)\frac{1}{l^3_{R,hi}}(12E^{(s)}J^{(s)}_{1,hi} + PP6),$$

$$C3(6,3) = -\cos(\alpha_i/2)\cos(\alpha_j/2)\frac{1}{l^3_{R,ij}}(12E^{(s)}J^{(s)}_{1,ij} + PP7),$$

$$C4(6,3) = -2C1(6,3),$$

$$C1(6,11) = -\cos(\alpha_i/2)\frac{1}{l_{R,ij}}(6E^{(s)}J^{(s)}_{1,ij} + PP8),$$

$$C2(6,11) = \cos(\alpha_i/2)\frac{1}{l^2_{R,hi}}(6E^{(s)}J^{(s)}_{1,ij} + PP9),$$

$$C3(6,11) = -\cos(\alpha_i/2)\frac{1}{l^2_{R,ij}}(6E^{(s)}J^{(s)}_{1,ij} + PP10),$$

$$C4(6,11) = \cos(\alpha_i/2)\frac{1}{l^2_{R,ij}}PP11,$$

$$Ci(9,8) = Ci(5,4),$$

$$C1(10,7) = \cos^2(\alpha_i/2)\left[12E^{(s)}\left(\frac{J^{(s)}_{2,hi}}{l^3_{R,hi}} + \frac{J^{(s)}_{2,ij}}{l^3_{R,ij}}\right) + PP12\right] - m_4\omega^2,$$

$$C2(10,7) = -\cos(\alpha_h/2)\cos(\alpha_i/2)\frac{1}{l^3_{R,hi}}(12E^{(s)}J^{(s)}_{2,ij} + PP13),$$

$$C3(10,7) = -\cos(\alpha_i/2)\cos(\alpha_j/2)\frac{1}{l^3_{R,ij}}(12E^{(s)}J^{(s)}_{2,ij} + PP14),$$

$$C4(10,7) = -2C1(10,7),$$

$$C1(10,11) = -\cos(\alpha_i/2)\frac{1}{l_{R,ij}}(6E^{(s)}J^{(s)}_{2,ij} + PP15),$$

$$C2(10,11) = \cos(\alpha_i/2)\frac{1}{l^2_{R,hi}}(6E^{(s)}J^{(s)}_{2,hi} + PP16),$$

$$C3(10,11) = -\cos(\alpha_i/2)\frac{1}{l^2_{R,ij}}(6E^{(s)}J^{(s)}_{2,ij} + PP17),$$

$$C4(10,11) = \cos(\alpha_i/2)\frac{1}{l^2_{R,ij}}PP18,$$

$$C1(12,3) = \cos(\alpha_i/2)[6E^{(s)}(J^{(s)}_{1,hi}/l^2_{R,hi} - J^{(s)}_{1,ij}/l^2_{R,ij}) + PP19],$$

$$C2(12,3) = -\cos(\alpha_h/2)(6E^{(s)}J^{(s)}_{1,hi}/l^2_{R,hi} + PP20),$$

$$C3(12,3) = \cos(\alpha_i/2)(6E^{(s)}J^{(s)}_{1,ij}/l^2_{R,ij} + PP21),$$

$$C4(12,3) = \cos(\alpha_i/2)\frac{1}{l^2_{R,ij}}PP22,$$

$$
\begin{aligned}
C1(12,7) &= \cos(\alpha_i/2)[6E^{(s)}(J^{(s)}_{2,hi}/l^2_{R,hi} - J^{(s)}_{2,ij}/l^2_{R,ij}) + PP23]\,,\\
C2(12,7) &= -\cos(\alpha_h/2)(6E^{(s)}J^{(s)}_{2,hi}/l^2_{R,hi} + PP24)\,,\\
C3(12,7) &= \cos(\alpha_i/2)(6E^{(s)}J^{(s)}_{2,ij}/l^2_{R,ij} + PP25)\,,\\
C4(12,7) &= \cos(\alpha_i/2)\frac{1}{l^2_{R,ij}}PP26\,,\\
C1\,(12,11) &= 4E^{(s)}(J^{(s)}_{1,hi}/l_{R,hi} + J^{(s)}_{1,ij}/l_{R,ij}) + PP27\,,\\
&\quad + 4E^{(s)}(J^{(s)}_{2,hi}/l_{R,hi} + J^{(s)}_{2,ij}/l_{R,ij})PP28 - m_5\omega^2\,,\\
C2(12,11) &= -2E^{(s)}J^{(s)}_{1,hi}/l_{R,hi} + PP29 - 2E^{(s)}J^{(s)}_{2,hi}/l_{R,hi} + PP30\,,\\
C3(12,11) &= -2E^{(s)}J^{(s)}_{1,ij}/l_{R,ij} + PP31 - 2E^{(s)}J^{(s)}_{2,ij}/l_{R,ij} + PP32\,,\\
C4(12,11) &= -8E^{(s)}J^{(s)}_{1,hi}/l_{R,hi} + PP33 - 8E^{(s)}J^{(s)}_{2,hi}/l_{R,hi} + PP34 - m_6\omega^2\,,\\
&\vdots\\
&\text{etc.} \qquad\qquad (3.45)
\end{aligned}
$$

In these expressions the parameters $G^{(s)}, E^{(s)} F^{(s)}_c$, $J^{(s)}_1 = F^{(s)}_{xx,1}$, $J^{(s)}_2 = F^{(s)}_{xx,2}$, $J^{(s)}_T$ denote elasticity moduli, area, moments of inertia and torsional constants of the rib member, respectively. $G^{(p)}$, $J^{(p)}_T$ and PPi are corresponding shear elasticity modulus, torsional rigidity and plate constants of the flat member, m_i are the parameters of mass terms in nodal points and ω are the corresponding frequency parameters.

In the algorithm of the FETM-method the boundary conditions are specified by two types of matrix formulations. The first one, the *hypermatrix of initial parameters* **HL**, formulates the initial parameters on basis of the boundary conditions of individual micro and macroelements in the initial section of the analysed structural simulation. Preliminary unknown initial parameters, after repeated multiplication with corresponding transfer and nodal hypermatrices, are numerically determined by boundary conditions in the end section of the analysed simulation. These conditions are specified by *hypermatrix of boundary conditions* **RP**. Both defined types of hypermatrices, together with the corresponding matrices for alternatively clamped boundary conditions in initial and end sections of the shell structure constructed of orthotropic ribs and planar members are summed up in the terms F-9, F-10, F-11 and F-12 of the CTM. The applied boundary conditions simulate the influence of rigid end transverse stiffeners. However, the present matrices may be modified for the simulation of arbitrarily variable boundary conditions.

The governing equation of the algorithm of the FETM-method for the determination of the complex matrix of initial parameters **UV** is then given by

$$\mathbf{UV} = \mathbf{RP}\left(\prod_{1}^{s}(\mathbf{CK}\,\mathbf{BK})\,\mathbf{HL}\right) \Rightarrow \boldsymbol{k}_0, \tag{3.46}$$

where $\boldsymbol{k}_0$ is the vector of complex boundary conditions in initial section of applied discrete simulation and s is the number of sections in longitudinal direction. The state vectors in the nodes of the simulation are given by

$$\boldsymbol{k}_j = \prod_{1}^{j}(\mathbf{CK}\,\mathbf{BK})\,\boldsymbol{k}_0. \tag{3.47}$$

The computation of stress resultants and stress couples within each discrete element is accomplished in a straightforward manner when the total nodal displacements associated with membrane and bending action of stiffeners and plate members have been computed. The stresses are referred to a set of moving coordinate axes which vary from one element to another. Hence, the actual orientation of the element coordinate systems must be known for an accurate interpretation of the results. For design purposes it may be convenient to compute the principal stresses and corresponding directions within each element. The fundamental step in the updated Lagrangian description of motion is the updating of nodal coordinates for each new configuration. Thus, once the incremental equilibrium equations have been solved, the new nodal coordinates $\mathbf{X}^{(2)}$ in the new configuration are given by

$$\mathbf{X}^{(2)} = \mathbf{X}^{(1)} + \Delta\mathbf{U}\,, \tag{3.48}$$

where $\mathbf{X}^{(1)}$ are the coordinates in the current configuration and $\Delta\mathbf{U}$ are the incremental nodal displacements for the particular nodal point under consideration. The updated nodal coordinates (3.48) provide a unique definition of the deformed facetted shells or three-dimensional grillages. From these updated coordinates the new element direction cosines (3.39) and surface direction cosines (3.41) are established. In this way the corresponding nodal matrices and hypermatrices are continuously updated for each new configuration. As a result, the major geometric nonlinearities are embodied in the congruence transformations when forming the element assemblage. The effect of large rotations on the stiffness of complete structure is similarly introduced via the coordinate transformations when forming the element assemblage.

3.7. Accuracy and Limitations

The discretization error introduced when representing shell structure by flat elements of equal thickness, is a function of local slopes within an element and may be reduced to an arbitrarily small quantity by limiting the element sizes [13, 44, 54]. The geometrically nonlinear effects are embodied in the transformations to the base coordinates. It should be kept in mind that the linear relationship imposes certain limitations on the appropriate element size. The size of individual elements should be so selected that the current deformation can always be regarded as small with respect to local element axes.

The large rotations that frequently occur in thin shell structures present an additional difficulty. This stems from the fact that finite rotations are not true vector quantities; that is they do not comply with the rules of vector transformation. An example of this nonvectorial property of finite rotations can be found in references [50] and [117]. In the concept adopted in this book the rotational degrees of freedom are referred to axes lying in the tangent plane of each of the nodal points. This is a moving reference frame that translates and rotates with the structure during deformation, thereby assuring that rotations are always small with respect to current coordinate axes. An alternative scheme was presented in paper [2], where the nodal rotations are referred to the global coordinate system.

The reformulation of the transfer and nodal matrices and repeated solutions of equilibrium equations are of major concern when the efficiency of present nonlinear formulations is to be considered. The incremental transfer submatrix referred to the microelement coordinates, is obtained as the sum of the linear transfer submatrix and the initial stress transfer submatrix corresponding to large diplacement analysis. The linear element transfer submatrix depends solely on the geometric dimensions of the element. Since the in-plane strains are considered to be small, the linear transfer submatrix will remain approximately constant with respect to local element axes. Hence, the linear element transfer submatrices have to be formed only once (based on the initial geometry) and may be stored on peripheral storage (disc, drum, tape, etc.) for subsequent use. This is particularly convenient for the present formulation where the computation of transfer submatrices requires matrix inversion (see equation 3.10). Finally, it should be emphasized that a considerable saving in computation time can be achieved by taking into account the vectorized structure of corresponding transfer and nodal matrix expressions. All these factors considered, the repeated computation and assembly of new incremental transfer and nodal matrices and internal element forces can be accomplished with relatively moderate computational effort.

3.8. Mixed Discretization in Time

For direct time integration of the present multigrid schemes of the spatial discretization of nonlinear structural simulations, the combination of Newmark and Wilson techniques for mixed discretization in time is adopted. Such *mixed schemes of discretization in space and time* allow an effective simulation of the general variability of physical parameters in various substructuring regions as well as in variable time intervals of nonlinear analyses.

In combined Newmark—Wilson time-integration technique [95], developed specially for the nonlinear response analysis of multigrid spatial simulations as described above, the nodal deformations and velocities in time $t + \Delta\tau$ are taken to be given as

$$\boldsymbol{r}_{t+\Delta\tau} = \boldsymbol{r}_t + \dot{\boldsymbol{r}}_t\Delta_t + (1/2 - \alpha)\ddot{\boldsymbol{r}}_t(\Delta\tau)^2 + \alpha\ddot{\boldsymbol{r}}_{t+\Delta\tau}(\Delta\tau)^2\,, \tag{3.49}$$

$$\dot{\boldsymbol{r}}_{t+\Delta\tau} = \dot{\boldsymbol{r}}_t + (1 - \delta)\ddot{\boldsymbol{r}}_t\Delta\tau + \delta\ddot{\boldsymbol{r}}_{t+\Delta\tau}\Delta\tau\,, \tag{3.50}$$

with variable nonlinear integration operators α and δ and where

$$\Delta\tau = \Theta\Delta t\,, \quad \Theta \geqslant 1.0\,. \tag{3.51}$$

The increments of deformations and velocities in the time interval $\Delta\tau$ are derived from equations (3.49) and (3.50) giving

$$\Delta\dot{\boldsymbol{r}}_t = \dot{\boldsymbol{r}}_{t+\Delta\tau} - \dot{\boldsymbol{r}}_t = \frac{\delta}{\alpha\Delta\tau}\Delta\boldsymbol{r}_t - \frac{\delta}{\alpha}\dot{\boldsymbol{r}}_t - \left(\frac{\delta}{2\alpha} - 1\right)\Delta\tau\ddot{\boldsymbol{r}}_t\,, \tag{3.52}$$

$$\Delta\ddot{\boldsymbol{r}}_t = \ddot{\boldsymbol{r}}_{t+\Delta\tau} - \ddot{\boldsymbol{r}}_t = \frac{1}{\alpha(\Delta\tau)^2}\Delta\boldsymbol{r}_t - \frac{1}{\alpha\Delta\tau}\dot{\boldsymbol{r}}_t - \frac{1}{2\alpha}\ddot{\boldsymbol{r}}_t\,. \tag{3.53}$$

The vectors of effective rigidities and modified loads are specified from the equation of motion (3.1) as

$$\mathbf{K}_t^{\text{eff}} = \frac{1}{\alpha(\Delta\tau)^2}\mathbf{M}_t + \frac{\delta}{\alpha\Delta\tau}\mathbf{C}_t + \mathbf{K}_t\,, \tag{3.54}$$

$$\Delta\boldsymbol{R}_t^{\text{eff}} = \Delta\boldsymbol{R}_t + \mathbf{M}_t\left(\frac{1}{\alpha\Delta\tau}\dot{\boldsymbol{r}}_t + \frac{1}{2\alpha}\ddot{\boldsymbol{r}}_t\right) + \mathbf{C}_t\left[\frac{\delta}{\alpha}\dot{\boldsymbol{r}}_t + \left(\frac{\delta}{2\alpha} - 1\right)\Delta\tau\ddot{\boldsymbol{r}}_t\right] + \Delta\boldsymbol{Q}_t\,. \tag{3.55}$$

The incremental equation of motion in time interval $\Theta\Delta t$ is then modified into the form

$$\mathbf{K}_t^{\text{eff}}\Delta\boldsymbol{r}_t = \Delta\boldsymbol{R}_t^{\text{eff}}\,. \tag{3.56}$$

For solving this equation all numerical concepts specified in this chapter are applied.

The implementation of the present direct time-integration scheme into the algorithm of the FETM-method is given by the following set of operations:

A. Initial stage of solution.

1. Applying the FETM-method, the calculation of deformations $v_0(t)$, velocities $\dot{v}_0(t)$ and accelerations $\ddot{v}_0(t)$ in all nodes for initial configuration of studied structural simulation.

2. The choice of the time step Δt, the initial specification of operators $\delta \gtreqless 1/2$; $\alpha = 1/6$ and $\Delta\tau = \Theta\Delta t$ (the nonlinear equivalents of operators β, γ and $\Delta\tau$ in Section 1.10). The calculation of integration parameters

$$a_0 = \frac{1}{\alpha\Delta\tau^2}, \quad a_1 = \frac{\delta}{\alpha\Delta\tau}, \quad a_2 = \frac{1}{\alpha\Delta\tau},$$
$$a_3 = \frac{1}{2\alpha}, \quad a_4 = \frac{\delta}{\alpha}, \quad a_5 = \left(\frac{\delta}{2\alpha} - 1\right)\Delta\tau. \tag{3.57}$$

3. The formulation of the effective rigidities

$$\tilde{\mathbf{K}}(t) = \mathbf{K}(t) + a_0\mathbf{M}(t) + a_1\mathbf{C}(t), \tag{3.58}$$

with nonlinear mass and damping matrices $\mathbf{M}(t)$ and $\mathbf{C}(t)$. Incorporation into algorithm of the FETM-method is performed in accordance with explanation in Section 2.2.

B. For each time-integration step.

4. The calculation of the modified load vector

$$\tilde{\mathbf{R}}(t) = \mathbf{R}(t) + \mathbf{M}(t)[a_2\dot{\boldsymbol{r}}(t) + a_3\ddot{\boldsymbol{r}}(t)] + \mathbf{C}(t)[a_4\dot{\boldsymbol{r}}(t) + a_5\ddot{\boldsymbol{r}}(t)]. \tag{3.59}$$

3. Using the algorithm of the FETM-method with application of effective stiffness and modified load parameters, the increments of deformations $\Delta\boldsymbol{r}(t)$ in all nodes of simulation are calculated.

6. The calculation of accelerations $\ddot{\boldsymbol{r}}(t + \Delta\tau)$

$$\ddot{\boldsymbol{r}}(t + \Delta\tau) = a_0\Delta\boldsymbol{r}(t) + a_2\dot{\boldsymbol{r}}(t) - a_3\ddot{\boldsymbol{r}}(t) + \ddot{\boldsymbol{r}}(t). \tag{3.60}$$

7. The calculation of acceleration $\ddot{\boldsymbol{r}}(t + \Delta t)$

$$\ddot{\boldsymbol{r}}(t + \Delta t) = \ddot{\boldsymbol{r}}(t) + \frac{1}{\Theta}[\ddot{\boldsymbol{r}}(t + \Delta\tau)] = \ddot{\boldsymbol{r}}_1(t). \tag{3.61}$$

8. The calculation of velocities $\dot{\boldsymbol{r}}(t + \Delta t)$

$$\dot{\boldsymbol{r}}(t + \Delta t) = \dot{\boldsymbol{r}}(t) + (1 - \delta)\ddot{\boldsymbol{r}}(t)\Delta t + \delta\ddot{\boldsymbol{r}}(t + \Delta t)\Delta t = \dot{\boldsymbol{r}}_1(t). \tag{3.62}$$

9. The calculation of dynamic deformations $\boldsymbol{r}(t + \Delta t)$

$$\boldsymbol{r}(t + \Delta t) = \boldsymbol{r}(t) + \dot{\boldsymbol{r}}(t)\Delta t + \left(\frac{1}{2} - \alpha\right)\ddot{\boldsymbol{r}}(t)\Delta t^2 +$$

$$+ \alpha \ddot{r}(t + \Delta t)\Delta t^2 = r_1(t). \qquad (3.63)$$

C. Repetition of calculation for following time steps with initial deformations, velocities and accelerations $r_1(t)$, $\dot{r}_1(t)$ and $\ddot{r}_1(t)$, respectively.

In accordance with the conclusions in Section 1.10, the variability of individual time-integration operators is as follows:

$\delta \gtreqless 1/2$,
$\alpha \gtreqless 0$ — method of central differences,
$\alpha \gtreqless 1/12$ — Fox—Goodwin method,
$\alpha \gtreqless 1/6$ — method of linear acceleration,
$\alpha \gtreqless 1/4$ — method of constant average acceleration,
$\alpha \gtreqless 1/6$ — Wilson Θ-method (for $\Theta \gtreqless 1.5$).

When combining the present parameters there may be obtained various problem-oriented modifications of present mixed concept.

For calculation it is important to consider the accuracy, stability and convergence of present mixed schemes for direct time integration. The computational errors may again be expressed in the terms of an artificial change of the period T and of the reduction of amplitude of vibration. To control the stability, accuracy and convergence of nonlinear numerical analyses, there may be used the length of the time step Δt, the value of the ratio $\Delta t/T$, the period elongation *PE*, amplitude decay *AD* and finally the checking of numerical damping. Present control items have to be performed in accordance with the nonlinear variability of multigrid discretization schemes in space, as well as with the variability of present integration operators for mixed schemes of discretization in time.

The variability of the ratio of the calculated and actual period of vibration for analysed simulation, as the function of the ratio $\Delta t/T$ and of the parameter Θ, is illustrated in Figure 63. The increasing values of the ratio $\Delta t/T$ and of the parameter Θ specify the growing value of the ratio of calculated and actual periods. The values of this ratio for $\Theta = 1.4$ and $\Delta t/T = 0.2$ are distinctly increased, the period is elongated and the accuracy of nonlinear numerical solution fails.

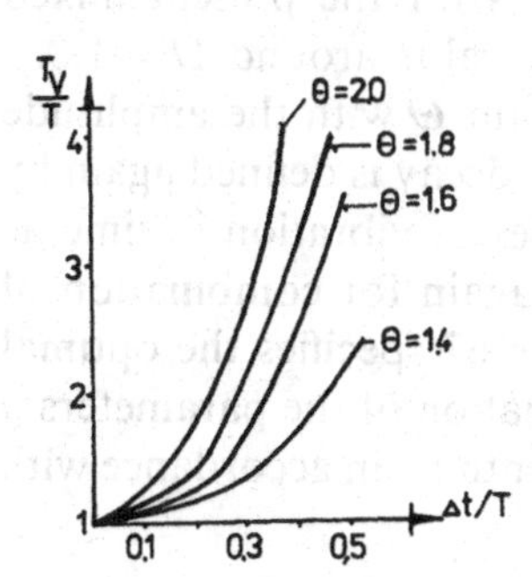

Fig. 63. Period elongation as a function of time step and period for various Θ.

The dependence of the time step, period and amplitude decay (in %) for various parameters Θ, when assuming the zero values of internal and external damping (e.g. studied is the numerical damping), is plotted in Figure 64. In analogy with the aforementioned, these diagrams again confirm the distinct increase of the amplitude decay for $\Theta > 1.4$ and thereby the decrease in accuracy. Similar conclusions can be drawn from Figure 65, illustrating the period elongation related to the parameters $\Delta t/T$ and Θ.

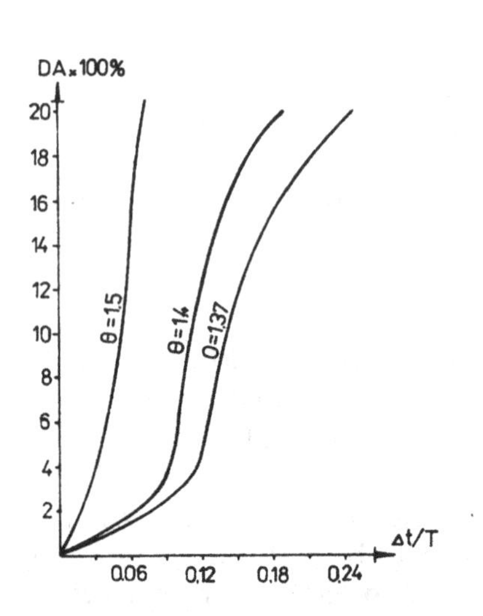

Fig. 64. Amplitude decay as a function of time step and period for various Θ.

Fig. 65. Period elongation as a function of time step and period for various Θ.

The relation for the ratio of calculated (T_V) and actual (T) periods and of the parameter Θ, for values $\alpha = 1/6$ and $\delta = 0.1$ (Newmark—Wilson combination) is shown in Figure 66. For the assumed combination of integration parameters $\alpha = 1/6$ and $\delta = 0.1$ the optimal value of $\Theta = 1.8$ is specified for which the calculated and actual periods of vibration coincides. The value $\Theta = 1.37$, valid for the classic Wilson approach, is for the present mixed Newmark—Wilson approach shifted into a higher value around $\Theta = 1.8$.

The analogous relation of the integration constant Θ with the amplitude decay AD is illustrated in Figure 67. The amplitude decay is defined again by the ratio of calculated A_V and actual A_E amplitudes of vibration in time as shown in Figure 67. The parameter Θ is studied again for combination of values $\alpha = 1/6$ and $\delta = 0.1$. The diagram in Figure 67 specifies the optimal amplitude decay for $\Theta = 1.8$. For assumed combination of the parameters α and δ the accuracy of the solution decreases exponentially in accordance with decreasing values of α and δ.

In the next step the attention will be directed towards finding of an optimal value of the parameter δ when assuming the specified optimum $\Theta = 1.8$ as well as the value $\alpha = 1/6$ which is typical of studied mixed Newmark—Wilson integration scheme. The relation between the integration constant δ and the numerical damping for the freely vibrating and physically nondamped dynamic simulation, for the above combination of parameters $\alpha = 1/6$ and $\Theta = 1.8$ is depicted in Figure 68. The numerical damping specified by amplitude decay AD is zero for $\delta = 0.1$ when assuming the above values of the other two integration parameters. In analogy with the previous results the other values of δ, for this combination of integration parameters, initiate the explosive increase of numerical damping and the loss of numerical accuracy.

The optimization of the integration operators α, δ and Θ is to be performed for each parametric combination of the present mixed schemes of direct time integration. The dependence of the parameter Θ on the period elonga-

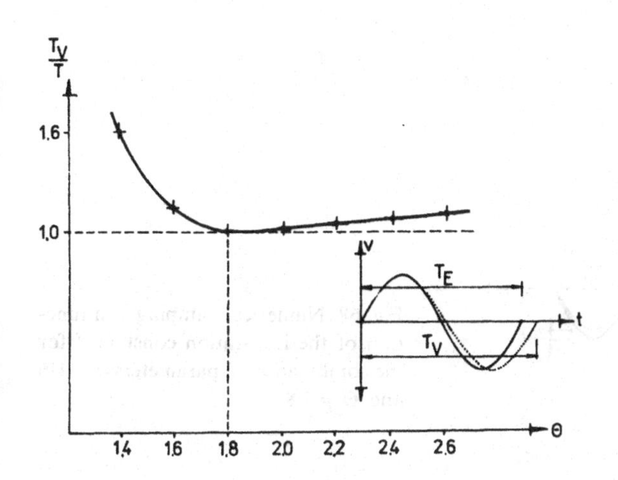

Fig. 66. Ratio of periods T_V/T as a function of Θ for the combination parameters $\alpha = 1/6$ and $\delta = 0.1$.

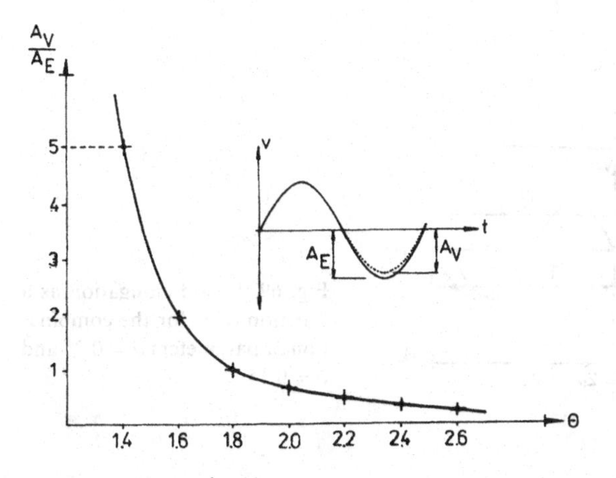

Fig. 67. Amplitude decay as a function of Θ for the combination of parameters $\alpha = 1/6$ and $\delta = 0.1$.

tion for a free nondamped vibration, when assuming the combination $\alpha = 0.25$ and $\delta = 0.15$, is plotted in Figure 69. The optimum Θ-value for this case is $\Theta = 1.7$. This optimum is also evident in Figure 70 which illustrates, for the present combination of parameters, the relation between Θ and the amplitude decay.

The relation between δ and the numerical damping for a free, physically nondamped vibration when assuming the parametric combination $\alpha = 0.25$

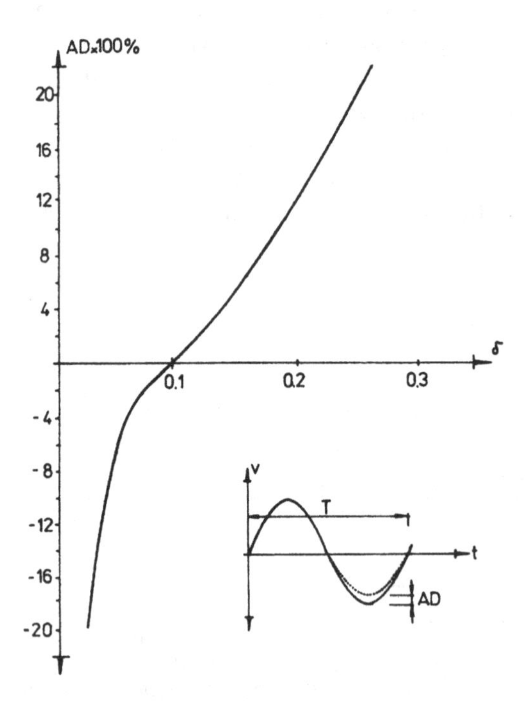

Fig. 68. Numerical damping as a function of the integration constant δ for the combination of parameters $\alpha = 1/6$ and $\Theta = 1.8$.

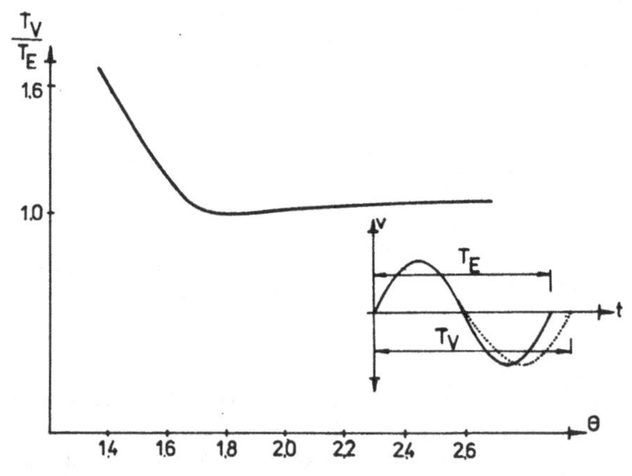

Fig. 69. Period elongation as a function of Θ for the combination of parameters $\alpha = 0.25$ and $\delta = 0.15$.

and $\Theta = 1.7$, is illustrated in Figure 71. The optimum zero of numerical damping is obtained for $\delta = 0.15$.

The present analysis determines some optimizing limits for individual parameters of the present mixed schemes for direct time integration, when applying the variable multigrid simulations of nonlinear discretization in space and time.

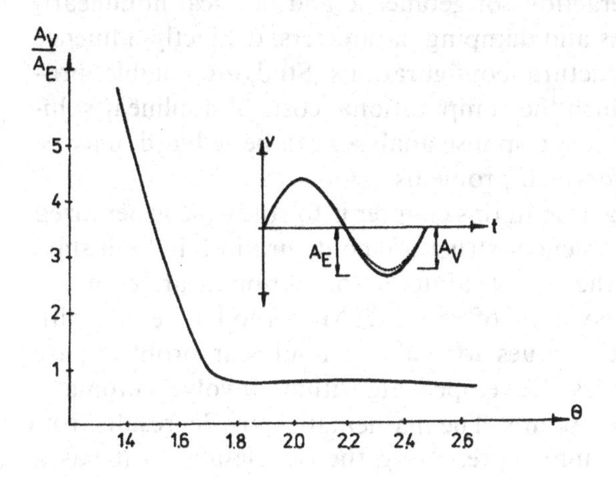

Fig. 70. Amplitude decay as a function of Θ for the combination of parameters $\alpha = 0.25$ and $\delta = 0.15$.

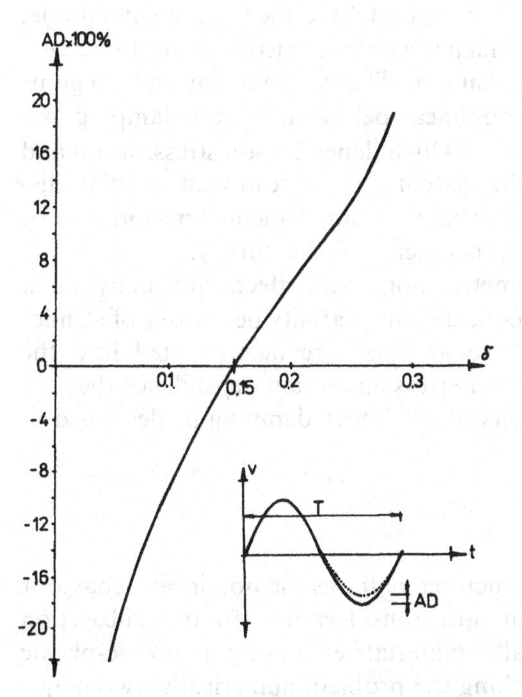

Fig. 71. Numerical damping as a function of the integration constant δ for the combination of parameters $\alpha = 0.25$ and $\Theta = 1.7$.

3.9. Nonlinear Interactions

An ultimate strength and strain analyses in critical, postcritical and resonance parametric vibrations determine the dynamic limit states of structural simulations. High stress and strain levels predominating all static influences, require the realization of nonlinear analyses of generalized spatial dynamic behaviour. The interactions of geometric and physical nonlinearities, corresponding to stiffness and damping parameters, distinctly influence the limit state response of structural configurations. Studying reliable, problem-oriented methods by which the computational costs of nonlinear solutions and particularly of dynamic response analyses can be reduced, may be of considerable importance for such problems.

The purpose of the investigation in this chapter is to study the generalized nonlinear spatial behaviour of slender structural configurations in limit state zones of vibration regarding the complex interaction of nonlinear geometric and material effects. The algorithm of the FETM-method together with aforementioned numerical techniques for solving nonlinear problems are adopted for numerical analyses. Developed algorithms involve automatic detection of possible instability points. The mathematical model results from a discretization of the continuum representing the simulation — it has a spectrum with minimum frequency corresponding to the fundamental mode, and a maximum cut-off frequency which is a characteristic of the discretized model rather than of the original continuum. The system in limit state regions may also have varying degrees of nonlinear behaviour — the damping and stiffness matrices may have coefficients which depend upon stress, strain and loading history. The motions of the system are not restricted to small displacements from the position of static equilibrium. Time integration is to be analysed with respect to stability, convergence and accuracy.

Two typical physical and geometric nonlinear effects are analysed in following subsections — the elastic-plastic and stability behaviour of slender structural configurations. Both these analyses are incorporated into the algorithm of the FETM-method. Over stress and strain dependence they are interactively coupled with the effects of nonlinear damping as described in Section 2.5.

a) *Elastic-Plastic Analysis*

An ultimate strength analysis of structures requires the nonlinear behaviour of structure material to be taken into consideration. In this subsection attention will be confined to metallic materials exhibiting an elastic-plastic type of behaviour. Before approaching the problem numerically, two major decisions have to be made. The first is the selection of a mathematical model (plasticity theory) by which the macroscopic behaviour of the material is described. Two major plasticity theories that can be applied are the *deformation theory* [31, 55] and the *flow theory*. For the sake of mathematical consistency and physical appropriateness the flow theory will be adopted here [38, 65].

The second important decision is related to the numerical technique itself and concerns the incorporation of the elastic-plastic effects into the FETM-method implementation of studied structure. Two approaches are readily available; they are the so-called "*initial strain*" method and the "*tangent stiffness*" method [3, 43, 51]. Although the initial strain method appears to be the simpler one, it is less reliable and breaks down for elastic-perfectly plastic materials. Furthermore, when plasticity is combined with large displacements, the incremental transfer matrix has to be recalculated anyway during load incrementation. For these reasons, the tangent stiffness method was chosen to be used here.

The plastic behaviour of a material is characterized by the following items:

— an *initial yield criterion* defining the dynamic elastic limit of the material,
— a *flow rule* relating the dynamic plastic strain increments to the stresses and stress increments,
— a *hardening rule* specifying the conditions for the subsequent dynamic yielding from a plastic state.

The following considerations describe the mathematical modelling of these items used in the present discrete analysis.

Since most experimental works on ductile metals support the von *Mises yield criterion*, the present formulation is based on this criterion. At initial yielding the expression for the von Mises criterion reads

$$f = \bar{\sigma} - \sigma_Y = 0, \tag{3.64}$$

where σ_Y is the uniaxial yield stress and

$$\bar{\sigma} = \sqrt{\frac{3}{2} s_{ij} s_{ij}}, \tag{3.65}$$

is the equivalent stress, where s_{ij} denotes the deviatoric stresses defined as

$$s_{ij} = \sigma_{ij} - \frac{1}{3} \delta_{ij} \sigma_{kk}, \tag{3.66}$$

and δ_{ij} is the Kronecker delta.

The first assumption made in the flow theory is that the total strain ε_{ij} may be decomposed into a *recoverable elastic part* ε^e_{ij} and an *irrecoverable plastic part* ε^p_{ij}

$$\varepsilon_{ij} = \varepsilon^e_{ij} + \varepsilon^p_{ij}. \tag{3.67}$$

In addition it is assumed that the loading function f exists at every stage of plastic deformation. For hardening materials f depends on the *stress state* σ_{ij}, on the *plastic strains* ε_{ij}^{p} and on a *parameter of hardening* γ

$$f = f(\sigma_{ij}, \varepsilon_{ij}^{p}, \gamma). \tag{3.68}$$

The material state may be

$$\begin{aligned} f &< 0 \quad \text{elastic}, \\ f &= 0 \quad \text{plastic}, \\ f &> 0 \quad \text{inadmissible}, \end{aligned} \tag{3.69}$$

for which three different loading conditions can be defined

$$\begin{aligned} &\frac{\partial \sigma_{ij}}{\partial f} \mathrm{d}\sigma_{ij} < 0, \quad f = 0 \quad \text{(unloading)}, \\ &\frac{\partial f}{\partial \sigma_{ij}} \mathrm{d}\sigma_{ij} = 0, \quad f = 0 \quad \text{(neutral loading)}, \\ &\frac{\partial f}{\partial \sigma_{ij}} \mathrm{d}\sigma_{ij} > 0, \quad f = 0 \quad \text{(loading)}. \end{aligned} \tag{3.70}$$

For a more detailed description of the flow theory the fundamental works [31] and [55] are recommended.

By comparing several hardening rules with experimental results in [38], it was proved that the *mechanical sublayer technique* is the best one available to represent load reversal. Therefore, this hardening rule is chosen in the present analysis. The *strain hardening phenomenon* may be interpreted as elastic-perfectly plastic behaviour in separate crystals which initially yield due to varying crystalline orientations and properties. The resulting stresses are found as a weighted sum of the sublayer stresses as follows

$$\sigma = \sum_{i=1}^{n} \sigma_i t_i, \tag{3.71}$$

where n is the number of sublayers at the point under consideration. The weight t_i denotes the relative thickness of each sublayer. These weights must satisfy the relations

$$\sum_{i=1}^{n} t_i = 1. \tag{3.72}$$

The incremental version of equation (3.71) reads

$$\mathrm{d}\sigma = \sum_{i=1}^{n} \mathrm{d}\sigma_i t_i. \tag{3.73}$$

The weights t_i are found by a fitting for n points of the corresponding stress-strain curve. Considering Figure 72, the weight of sublayer member i is given by the formula

$$t_i = \frac{E_i - E_{i+1}}{E}, \qquad i = 1, 2, 3, \ldots, n, \tag{3.74}$$

where E_{n+1} is set equal to zero. The yield stresses σ_{Y_i} of the individual sublayers become

$$\sigma_{Y_i} = E\varepsilon_{Y_i}, \tag{3.75}$$

where σ_{Y_i} is the strain at first yield of sublayer number i. The present plastic analyses are valid for infinitesimally small increments of stresses and strains. The numerical techniques described in Sections 1.7, 1.8 and 1.9 imply the finite increments. Attention must then be focused on special corrections in order to perform the dynamic stress calculations because of the implementation of the finite increment procedure into multigrid discretization schemes of the FETM-method.

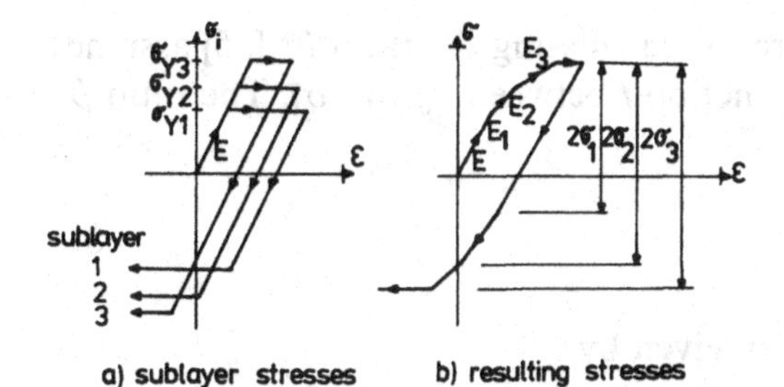

Fig. 72. Mechanical sublayer technique.

The first modification concerns the transition from an elastic to a plastic state at a stress point. The problem is illustrated in Figure 73. Prior to the increment the stress state denoted by σ_0 is in the elastic region. According to equation (3.70) the value of the function f is negative

$$f(\sigma_0) < 0. \tag{3.76}$$

The subsequent load increment causes a finite dynamic strain increment $\Delta\varepsilon$.

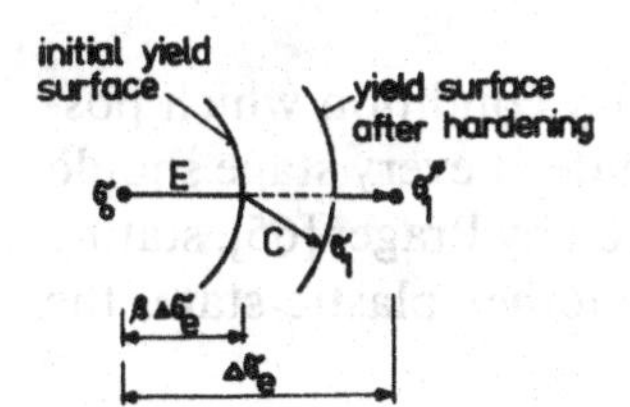

Fig. 73. Transition from elastic into plastic dynamic state.

The corresponding dynamic stress increment $\Delta\sigma_e$ is related to $\Delta\varepsilon$ by the elastic constitutive matrix $\mathbf{E}$

$$\Delta\sigma_e = \mathbf{E}\Delta\varepsilon. \tag{3.77}$$

The new stress state

$$\sigma_1^* = \sigma_0 + \Delta\sigma_0. \tag{3.78}$$

lies outside the yield surface so that

$$f(\sigma_1^*) > 0. \tag{3.79}$$

According to equations (3.70) the stress state σ_1^* is now in the inadmissible state. To be consistent with the plasticity theory the part of the stress increment that is created after yield should be based on the elastic—plastic constitutive relation. The problem is how to determine the factor β^* which defines the proportion of $\Delta\varepsilon$ associated with elastic material behaviour. The factor β^* should be calculated so that the yield condition

$$f(\sigma_0 + \beta^*\Delta\sigma_e) = 0 \tag{3.80}$$

is satisfied. The simplest procedure for calculating the ratio β^* [55] assumes a linear variation of the loading function f between σ_0 and σ_1^*. The ratio β^* becomes

$$\beta^* = -\frac{f(\sigma_0)}{f(\sigma_1^*) - f(\sigma_0)}. \tag{3.81}$$

The corrected stress state σ_1 is now given by

$$\sigma_1 = \sigma_0 + \Delta\sigma, \tag{3.82}$$

where

$$\Delta\sigma = \beta^*\Delta\sigma_e + (1 + \beta^*)\mathbf{C}_F\Delta\varepsilon, \tag{3.83}$$

with $\mathbf{C}_F$ as the elastic-plastic constitutive matrix based on the stress state $\sigma_0 + \beta^*\Delta\sigma_e$.

The incorporation of this simple modification into the present discrete analyses restricts the solution to be valid only for relatively small load increments.

The second modification deals with the consistency condition which postulates that, in case of plastic behaviour, the stress state at every stage should lie on the yield surface. From the consistency condition by Prager [65], stating that loading from one plastic state must lead to another plastic state, the following differential equation emerges

$$df = \frac{\partial f}{\partial \sigma_{ij}} d\sigma_{ij} + \frac{\partial f}{\partial \varepsilon_{ij}^{p}} d\varepsilon_{ij}^{p} = 0 . \tag{3.84}$$

Since only the first variation of the loading function is retained in equation (3.84), the mathematical formulation of the condition is valid only for infinitesimal increments. As a consequence, the finite increment technique will lead to a stress state which lies outside the yield surface defined by the plastic strains at the end of the increment. This problem is illustrated in Figure 74. The axes σ_{I} and σ_{II} denote principal stresses. The initial stress state σ_0 before the increment is assumed to lie on the yield surface so that the following relation holds

$$f(\sigma_0, \bar{\varepsilon}_0^{p}) = \bar{\sigma}(\sigma_0) - H(\bar{\varepsilon}_0^{p}) = 0. \tag{3.85}$$

In this equation $\bar{\varepsilon}_0^{p}$ is the *accumulated effective plastic strain* prior to the increment and the function H is found by specializing equation (3.85) to be valid for uniaxial stress. This is illustrated in Figure 75, which shows a typical uniaxial stress-strain curve. The stress state σ_1^* at the end of the increment lies outside the yield surface and results in the following inequality

$$f(\sigma_1^*, \bar{\varepsilon}_1^{p}) = \bar{\sigma}(\sigma_1^*) - H(\bar{\varepsilon}_1^{p}) \neq 0, \tag{3.86}$$

where $\bar{\varepsilon}_1^{p}$ is the accumulated effective plastic strain at the end of the increment.

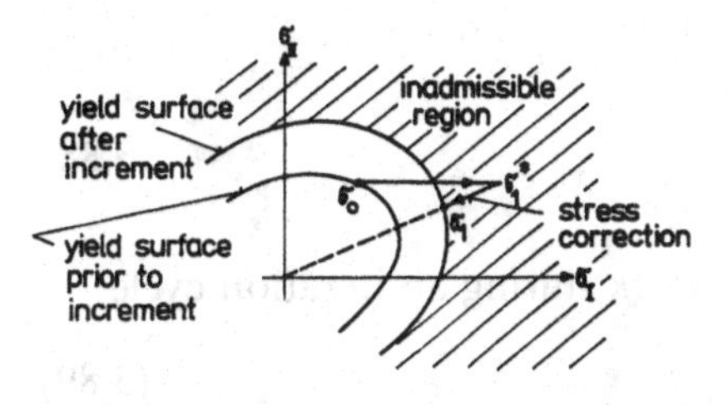

Fig. 74. Stress correction according to the consistency condition.

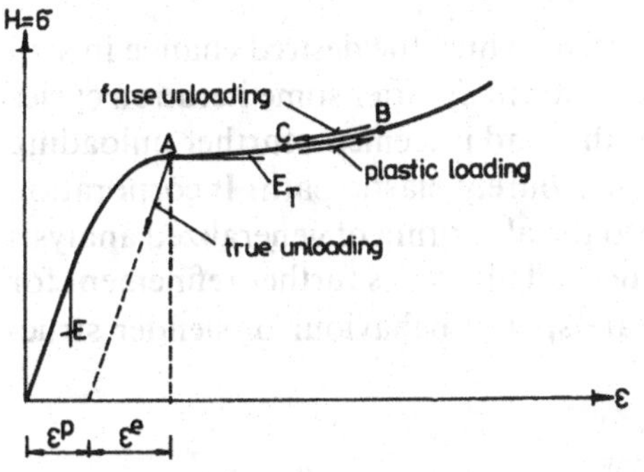

Fig. 75. Uniaxial test curve.

To prevent drift-off from the true solution, the stress state σ_1^* should be brought back to the yield surface after each increment in order to satisfy the second of equations (3.69). The stress components in σ_1^* are simply scaled proportionally back to the yield surface. The new stress state is given by expression

$$\sigma_1 = \frac{H(\bar{\varepsilon}_1^p)}{\bar{\sigma}(\sigma_1^*)}\sigma_1^*. \tag{3.87}$$

As is demonstrated in Figure 74, the corresponding stress vector $\sigma_1 - \sigma_1^*$ is directed towards the centre of the yield surface. The scaling technique is chosen for its simple mathematical form. However, special care has to be taken with so-called "false" unloading [8], which may occur during equilibrium iterations. The problem arises for a point which behaves plastically during load incrementation. Figure 75 illustrates the phenomenon for a uniaxial stress. The dynamic stress state prior to the increment is given by point A. When the incremental load is applied and described, correction of the stress state according to the consistency condition is applied and the new stress state is obtained. Subsequent equilibrium iteration may now result in unloading along B—C. This unloading is due to the numerical incrementation technique applied and has no physical relevance. The real deformation follows the direct route between A and G. Therefore, the elastic-plastic constitutive equations should be retained during equilibrium iteration between B and C, instead of assuming an elastic unloading from B. In case of such *false unloading* the effective plastic strain $\bar{\varepsilon}^p$ becomes smaller, taking account of the following expression

$$\mathrm{d}\bar{\varepsilon}^p = \frac{\Delta W^p}{|\Delta W^p|}\sqrt{\frac{2}{3}\mathrm{d}\varepsilon_{ij}^p\,\mathrm{d}\varepsilon_{ij}^p}, \tag{3.88}$$

in which $\mathrm{d}W^p$ is the increment in plastic work during an iteration cycle

$$\mathrm{d}W^p = \sigma_{ij}\Delta\varepsilon_{ij}^p. \tag{3.89}$$

For false unloading ΔW^p becomes negative. Thus, the desired change in sign for $\mathrm{d}\bar{\varepsilon}^p$ is obtained. If the effective plastic strain $\bar{\varepsilon}^p$, after some iteration cycles becomes less than the value $\bar{\varepsilon}_A^p$ prior to the load increment, further unloading during iteration is assumed to follow the purely elastic path. Incorporation of described elastic-plastic concepts into the algorithm of generalized analysis of motion on the basis of the FETM-method allows its further refinement for solution of special problems of ultimate spatial behaviour of slender structural simulations.

b) Stability and Post-Buckling Analysis

A thorough understanding of geometric nonlinear behaviour likely to be sustained by thin shell structures is essential for establishing reliable solution algorithms for stability and post-buckling analyses. The fundamentals of snap-through and post-buckling behaviour and its important effects on initial buckling and nonlinear analysis, have been put on a sound theoretical foundation, for example in reference [78].

The tracing of the load-deflection curve for a stability problem of the limit point type (Figure 76a) represents a formidable task. The major difficulty is associated with the singular incremental stiffness matrix that exists at the extremum points, indicating that an infinite increase in displacement takes place without any change of loading. The descending or unstable post-buckling path is characterized by an indefinite stiffness matrix.

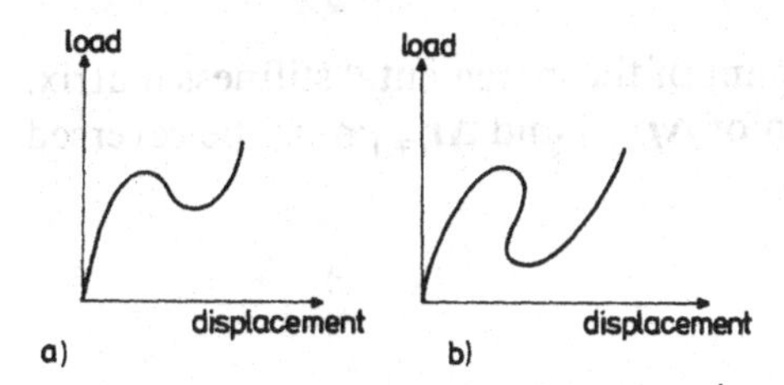

Fig. 76. Typical limit point phenomena in thin shell structures: a) standard snap-through behaviour, b) load-displacement curve with horizontal and vertical tangents.

Several schemes for tracing the entire deformation path of snap-through problems have been suggested. The common feature of these methods is to introduce modifications or constraints that render the stiffness matrix positive definite in the post-buckling range [77, 121]. Another popular method for handling *snap-through behaviour* consists of incrementing displacements rather than loads. This approach was proposed and applied in references [46, 50] or [62]. The basic idea behind displacement incrementation is that the incremental stiffness matrix ceases to be indefinite when the diplacement of some nodal point is prescribed. Using prescribed displacement increments may enable a tracing of unstable post-buckling branches but the method would fail in the case of load-displacement curves having vertical tangents (Figure 76b). Another drawback is the difficulty of handling nonconservative loading during the equilibrium iteration.

Rather than introducing modifications to keep the incremental stiffness matrix positive definite, one should adopt an equation solver that is capable of handling indefinite matrices. This can be accomplished by the general symmetric decomposition

$$\mathbf{K}_I = \mathbf{L}\mathbf{D}\mathbf{L}^T, \tag{3.90}$$

where $\mathbf{L}$ is a lower triangular matrix with unit diagonal elements and $\mathbf{D}$ is a diagonal matrix containing the pivots of the elimination process. With the aid of equation (3.90) it is possible to extent the technique of incrementing loads to the entire nonlinear regime. Occurrence of instability points may be detected by checking the sign of the determinant of the incremental stiffness matrix. Thus, equation (3.90) yields

$$\det(\mathbf{K}_I) = \det(\mathbf{D}) = d_1, d_2, d_3, \ldots d_M, \tag{3.91}$$

where M is the dimension of the stiffness matrix and d_i ($i = 1, 2, 3, \cdots, M$) are the elements of diagonal matrix $\mathbf{D}$. A procedure must be developed by which the automatic load incrementation techniques, as described in Section 1.9, may be adopted to deal with extremum points of the load-displacement curve. The alternative method is illustrated in Figure 77 and the basic steps involved are summarized below:

1. Apply a load increment $\Delta \boldsymbol{p}_{i+1}$ and compute the corresponding displacement vector $\Delta \boldsymbol{r}_{i+1}$.

2. Compute the sign of the determinant of the incremental stiffness matrix. If the determinant is negative, the sign of $\Delta \boldsymbol{p}_{i+1}$ and $\Delta \boldsymbol{r}_{i+1}$ must be reversed (see Figure 77).

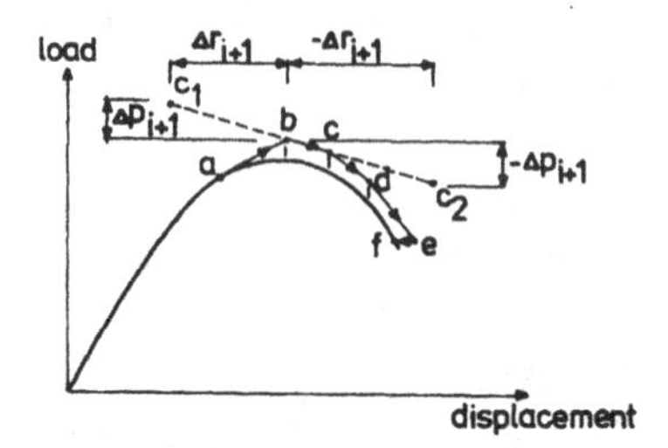

Fig. 77. Passing an extremum point.

3. In order to avoid excessive drifting of the solution in the vicinity of the extremum points, the load and displacement increments should be scaled according to the condition

$$|\Delta \boldsymbol{r}_{i+1}| \leqslant \varrho, \tag{3.92}$$

where ϱ is the prescribed value of the displacement norm.

4. One additional load step is carried out without the equilibrium iteration to make sure that the peak has been passed.

The above procedure for finding the extremum points only applies to the incremental part of the solution technique.

Automatic control of the load step size is based on the number of iterations required for the convergence in the subsequent steps. A proper method for the step size control was presented in Section 1.9. A more sophisticated concept is based on the *theorem of contractive mapping* (see reference [71]), which states that if the mapping $\mathbf{G}$ is contractive, then the sequence

$$\boldsymbol{x}^{k+1} = \mathbf{G}\boldsymbol{x}^k \tag{3.93}$$

converges towards a unique solution, i.e. $\boldsymbol{x}^* = \mathbf{G}\boldsymbol{x}$ and that

$$|\boldsymbol{x}^k - \boldsymbol{x}^*| \leqslant \alpha|\boldsymbol{x}^k - \boldsymbol{x}^{k-1}|/(1-\alpha). \tag{3.94}$$

The latter statement means that the norm of the error at any iterative step is bounded by the norm of the last correction. The contractive coefficient α depends on the step size. The convergence is assured for the values $\alpha \leqslant 1$. On each load level the value of α can be computed via

$$\alpha = \max(\beta^k), \tag{3.95}$$

where $\beta^k = |\boldsymbol{x}^{k+1} - \boldsymbol{x}^k|/|\boldsymbol{x}^k - \boldsymbol{x}^{k-1}|$ and k is the iteration number. The step size $\Delta\lambda$ is chosen so that

$$(\Delta\lambda)^{i+1}/(\Delta\lambda)^i = f, \tag{3.96}$$

where i denotes the load step number. If α, as computed from equation (3.95), is small at one load step, the step size should be increased in the next step ($f > 1$). This parameter may be modified automatically, rather than the step size directly, if the number of iterations per load step tends to fall outside the desired range or if nonconvergence occurs.

The selection of a proper time-integration method is another critical factor in the solution, especially of the snap-through and post-buckling resonance response of slender structures. The choice of the time discretization is problem-dependent and hence a variety of methods are available for the purposes of the analysis (see references [70] or [91]). One area of the research in time-integration methods is the development of schemes which permit different time integrators to be used simultaneously in different parts of structure. *Operator splitting methods* [120] are efficient for solving the present problem. The Newmark—Wilson mixed scheme explained above is combined with a splitting of transfer matrix such that each stage is integrated with the same size of the time step. Hence, splitting in space (element-wise) and splitting in time are allowed.

For advanced nonlinear problems it is difficult to prove stability or convergence. A common practice is to apply stability criteria derived for special cases and to accept the results if they appear to be stable. The use of operator splitting can sometimes simplify the development of stable methods.

3.10. Structural Synthesis and Optimization

The requirements for the utilization of the concepts of structural synthesis and optimization has become apparent, particularly with the development of slender thinwalled beams and thin shell structures in aviation and bridge engineering, where weight minimization or fully stressed design are of utmost importance. The synthesis of discrete approaches present in foregoing sections of this book, has provided all the necessary tools for the performance of optimization procedures in slender structures.

Structural optimization is accepted as the selection of design parameters which will allow the *minimum weight* or *fully stressed design* of structure. The selection of design parameters is subjected to the following types of constraints:

— *geometric constraints* — minimum and maximum areas and rigidities of structural members,
— *stress constraints* — maximum allowable stresses,
— *displacement constraints* — minimum and maximum deformations,
— *resonance and stability constraints.*

Such constraints form the *regional constraints*, and they are applied to all load conditions for which the structure is designed. They are represented by *constraint hypersurfaces.* Since the stresses and deformations as well as the stability and resonance limits are, in general, nonlinear functions of the design variables, the constraint hypersurfaces are also nonlinear functions of such variables. If the design point lies in the space above the constraint surfaces, the characteristic stress or displacement in the regional constraint will lie within the specified limits. The point where the constant-weight hypersurface touches the constraint hypersurfaces is the point for the minimum-weight design.

When constraints are imposed only on the stresses, an iterative procedure can be used to redesign the structure so that each element reaches limiting stress under at least one of the load conditions. Such design is described as a fully stressed design. The design variables for fully stressed design converge to a vertex of the n hypersurfaces representing n constraints on the stresses. The paper [68] gives an excellent account of the mathematical formulation of both the fully stressed and minimum-weight design procedures.

The total number of design variables, particularly in large systems, is often large. In many synthesis methods the solution efficiency is highly dependent on the number of variables optimized simultaneously and it is desired to reduce this number. While the optimal design is usually improved by considering geometric variables, the objective functions of a minimum-weight or fully stressed design are often not sensitive to changes in these variables near the optimum. The result might be slow convergence and further computational effort needed to reach the optimum.

The general optimization problem discussed in this section can be stated as follows: find the geometric design variables Y and the cross-sectional design variables X or Z (as other types of variables) such that

$$W = f(X,\ Y,\ Z) \rightarrow \min \qquad \text{(objective function)}, \tag{3.97}$$

$$\left.\begin{array}{l} X^L \leqslant X \leqslant X^U \\ Y^L \leqslant Y \leqslant Y^U \\ Z^L \leqslant Z \leqslant Z^U \end{array}\right\} \qquad \text{(geometric constraints)}, \tag{3.98}$$

$$\boldsymbol{\sigma}^L \leqslant \boldsymbol{\sigma} \leqslant \boldsymbol{\sigma}^U \qquad \text{(stress constraints)}, \tag{3.99}$$

$$\boldsymbol{r}^L \leqslant \boldsymbol{r} \leqslant \boldsymbol{r}^U \qquad \text{(displacement constraints)}, \tag{3.100}$$

$$\boldsymbol{\gamma}^L \leqslant \boldsymbol{\gamma} \leqslant \boldsymbol{\gamma}^U \qquad \text{(resonance and stability constraints)}, \tag{3.101}$$

in which L and U are superscripts denoting the lower and upper bounds, respectively. Symbols $\boldsymbol{\sigma}$, $\boldsymbol{r}$ and $\boldsymbol{\gamma}$ are the vectors of stresses, displacements and stability or resonance limits which are implicit functions of design variables and may be calculated using the analysis in foregoing sections.

The straightforward optimization approach is to optimize simultaneously all the variables by one of the available nonlinear programming methods. A possible two-level solution procedure of the optimization problem is as follows:

1. Assume an initial structural geometry.
2. Optimize the cross-sectional variables and the forces for a given geometry by satisfying equations (3.97—3.101).
3. Modify the geometric variables (location of stiffeners, ribs, supports, depths, etc.).
4. Repeat steps 2 and 3 until the optimal geometry is obtained.

The number of design variables is reduced by expressing all the geometric variables in terms of a small number of independent ones. Design variable linking is often necessary due to such considerations as functional requirements, fabrication limitations, etc. Another possibility to reduce the number of candidate geometries is to use a coarse grid in the space of geometric variables, so that only a small number of X, Y or Z values is considered. This is justified in many cases where the objective function (minimum-weight design, fully stressed design, etc.) is not sensitive to changes in the geometric variables near the optimum. To optimize the X, Y or Z variables at this step, one of the known unconstrained minimization techniques [22, 45, 73, 96] can be used.

Evolution strategies appear as most effective techniques for nonlinear optimization analyses. Evolution strategies apply the principles of biologic evolution for the optimization of technical problems. The principle of such methods may be written as

$$\boldsymbol{x}^{(k+1)} = \begin{cases} \boldsymbol{x}^{(k)} + \boldsymbol{p}^{(k)}, & \text{for} \quad z(\boldsymbol{x}^{(k)} + \boldsymbol{p}^{(k)}) < z(\boldsymbol{x}^{(k)}), \\ \boldsymbol{x}^{(k)}, & \text{for} \quad z(\boldsymbol{x}^{(k)} + \boldsymbol{p}^{(k)}) \geqslant z(\boldsymbol{x}^{(k)}), \end{cases} \tag{3.102}$$

where $\boldsymbol{x}^{(k+1)}$ is the optimization vector for analysed step $k+1$, $\boldsymbol{x}^{(k)}$ is the corresponding optimization vector of design variables for the step k, $\boldsymbol{p}^{(k)}$ is the random vector for the step k and z is the value of objective function.

The laws of the evolution in biology specify the selection of better individuals (parents or ancestors) and their installation as parents of the new generation. The worser of both are eliminated.

The so-called (1 + 1)-evolution strategy [69, 74] adopts the analogy with biologic evolution for the mutation of the decisive vector of design variables of structural configurations. The checking of obtained values of the objective function determines the selection of the better approximating design parameters and their utilization as starting values for the next iteration step. This concept is defined by the following set of operations:

1. Applying the allowable values of the vector $\boldsymbol{x}^{(k)}$ and of the random vector $\boldsymbol{p}^{(k)}$ there will be specified the new optimization vector $\boldsymbol{x}^{(k+1)}$ as

$$\boldsymbol{x}^{(k+1)} = \boldsymbol{x}^{(k)} + \boldsymbol{p}^{(k)}. \tag{3.103}$$

Thus, the random vector $\boldsymbol{p}^{(k)}$ performs the function of mutation.

2. If the optimization vector $\boldsymbol{x}^{(k+1)}$ is allowable, there may be calculated the value of objective function $z(\boldsymbol{x}^{(k+1)})$, with subsequent selective part of the solution. The new optimization vector $\boldsymbol{x}^{(k+1)}$ is accepted only if it has better approximation to desired optimum as in all foregoing cases. The iteration scheme of optimization procedure is given by equation (3.102). For the derivation of the random vector $\boldsymbol{p}^{(k)}$ the density and deviation of each of its components must be specified as shown in references [69, 74, 96].

The automatic control mechanisms for present nonlinear optimization analyses may be adopted in accordance with achievements in references [69, 74, 96].

Chapter 4
Applications

4.1. Specified Topics

The real applications of the given theoretical considerations are presented in this chapter. Attention is paid to the solution of typical slender structural configurations in linear and nonlinear limit regions of their static and dynamic behaviour, using the numerical and experimental approaches.

First of all, the problems of spatial, torsional and distortional behaviour of thinwalled box beams are studied. Subsequently, the resonance analyses and harmonic vibrations of corrugated and sandwich panels, shells and curved thinwalled systems are dealt with. Some problems of elastic-plastic, stability, post-buckling and aeroelastic demeanour of the above structures are also solved. Such problems are typical for advanced nonlinear ultimate and limit state regions of structural response. Finally, using the synthesis approach, the structural optimization of subtle shell bridge systems is performed.

Explanations of individual studied cases are concise; they are concerned only with physical parameters and the obtained results without closer discussions on the discretization schemes in space and time as well as on the details of numerical solutions. However, all examples are provided with reference citations containing further details of the present solutions.

Present numerical and experimental results come to the fore especially when checking or verifying the computer programs developed on the basis of techniques described in this book.

4.2. Thinwalled Beams

a) Thinwalled Box Beam — Natural Spatial Vibration

Theoretical analyses in Section 2.4 are numerically and experimentally verified on a thinwalled, antisymmetric steel box model with cross-section, elevation and physical parameters as shown in Figure 78. The frequency spectrum of the spatial natural vibration is studied regarding the coupling of bending, torsional and distortional oscillations of the studied model. For a registration of all these kinds of vibrations, including the distortional behaviour, a distinctly thinwalled box model was designed having low distortional rigidity of the cross-section.

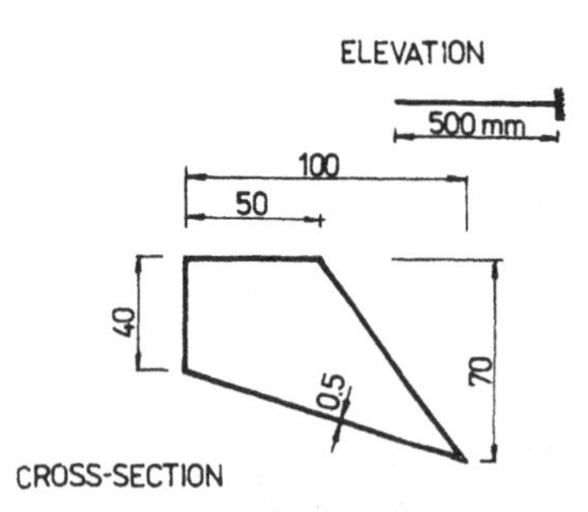

Fig. 78. Cross-section and elevation of the studied box structure.

For the numerical solution of the present example, the terms A-10 and A-11 of the CTM were applied.

The calculated and experimentally verified values of the spectrum of natural frequencies for the spatial vibration of the studied model are listed in Table IV. The natural frequencies of principal coupled kinds of vibration in bending, torsional and distortional actions of the studied model are presented.

Further details of the performed analysis are explained in reference [83].

b) *Torsional Time Response*

A thinwalled box bridge with the cross-section and elevation as illustrated in Figure 79 is subjected to the action of midspan time variable torques as shown in Figure 80. The corresponding time response of the midspan angle of torsion for all three studied torque impulses and for various values of the torsion-bending characteristics $k = \lambda_1^{(s)}$ (see term A-7 of the CTM) is illustrated in Figure 81. For the calculation the combination of the TMM with a Wilson direct time-integration scheme was applied using the term A-7 of the CTM.

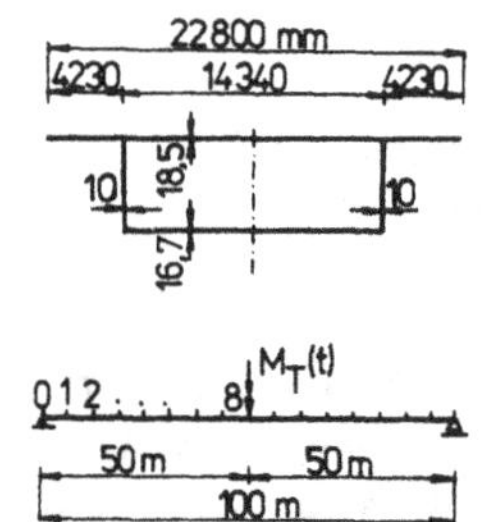

Fig. 79. Cross-sectional and elevation scheme of the thinwalled box bridge under consideration.

Table IV. Calculated natural frequencies of separate and coupled vibrations of the investigated thinwalled box model together with experimental verifications

Frequency spectrum	Bending in direction y [Hz]	Bending in direction x [Hz]	Pure torsion [Hz]	Torsion-bending [Hz]	Pure distortion [Hz]	Distortion-bending [Hz]	Space analysis [Hz]	Refined space analysis [Hz]	Experimental values [83] [Hz]
f_{1B_y}	195.231						181.412	173.122	185
f_{1T}			399.334	406.931			416.772	432.712	393.7
f_{1B_x}		458.930					461.038	458.293	448.3
f_{2T}			1148.113	1264.754			1260.355	1249.327	1240
f_{1D}					1275.436	1382.275	1398.835	1382.327	1330
f_{2B_y}	1198.343						1642.341	1808.351	1780
f_{3T}			1829.816	2312.841			2621.711	2687.831	2364
f_{2B_x}		2558.125					2683.273	2680.348	2637
f_{4T}			2458.412						
f_{3B_y}	3181.243						3747.352	3732.439	3820
f_{5T}			2943.700						
f_{6T}			3421.129						
f_{4B_y}	5942.277						4503.123	4491.141	4350

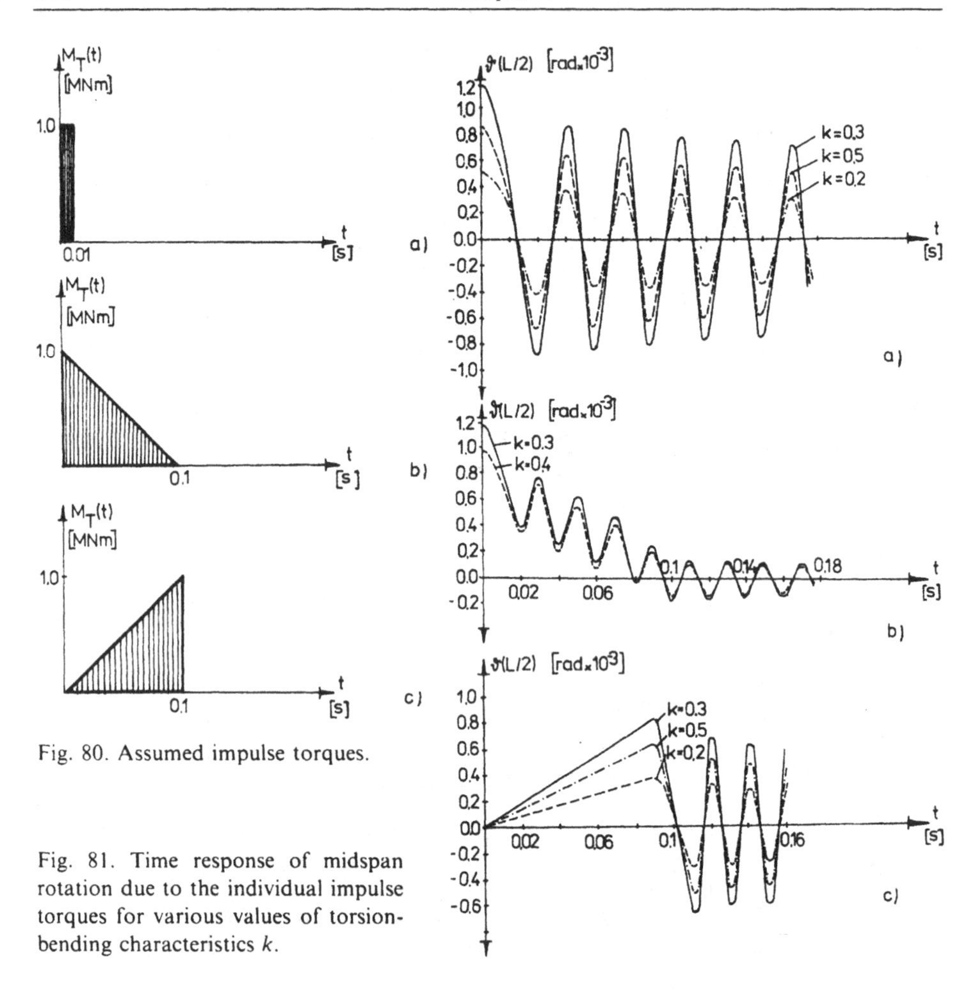

Fig. 80. Assumed impulse torques.

Fig. 81. Time response of midspan rotation due to the individual impulse torques for various values of torsion-bending characteristics k.

Further details of the performed numerical analysis are explained in reference [108].

c) Distortional Behaviour

For the optimization of locations and rigidities of transverse stiffeners in thinwalled beams, the problem of distortional behaviour is coming distinctly to the fore. The same thinwalled box beam as in foregoing example is studied. The beam is subjected to the action of an antisymmetrically positioned and continuously distributed line load $q_0 = 20 \text{ kN m}^{-1}$ which is located over one of vertical webs of the cross-section. The studied thinwalled beam is provided with a variable number of transverse stiffeners having the distortional rigidity

J_{RV}. The distortional rigidity of the cross-section is denoted by J_D. The influence of the variability of the ratio $\varrho = J_{RV}/J_D$ and of the number of stiffeners n on distortional behaviour of the structure is studied.

The diagrams of distortional bimoments in individual discrete points of the studied box beam, in accordance with the simulation scheme in Figure 79, for various parameters ϱ and n are illustrated in Figure 82. The diagrams of the corresponding angles of distortion are depicted in Figure 83. The diagrams of the angle of distortion midspan of the studied bridge (node No. 8), for various parameters ϱ and n, are plotted in Figure 84.

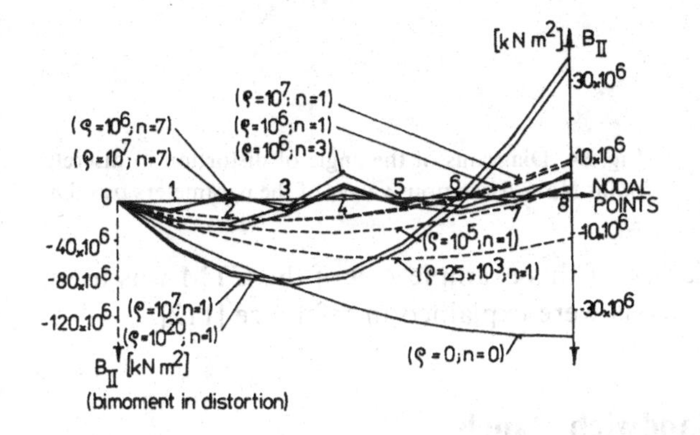

Fig. 82. Diagrams of distortional bimoments in individual discrete points of the studied box bridge for various values of parameters ϱ and n.

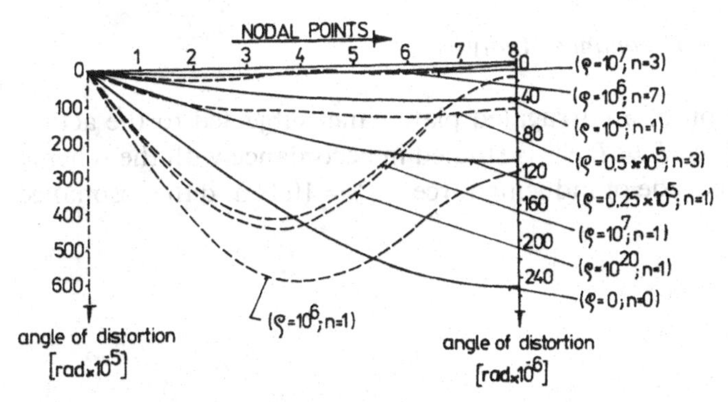

Fig. 83. Diagrams of the angles of distortion in individual discrete points of the studied box bridge for various values of the parameters ϱ and n.

The results allow to determine the optimum values of the parameters ϱ and n which must be satisfied in order to obtain the sufficient stiffening of the box cross-section by transverse diaphragms or stiffeners. Figure 83 shows that for one stiffener positioned midspan of the studied beam, the value $\varrho = 10^7$ simulates approximately the action of a rigid stiffener ($\varrho = 10^{20}$). Similar conclusions may be drawn also when analysing Figure 84. In analogy with aforementioned, the number of stiffeners n influences the distortional behaviour of the thinwalled box beam depending on the parameter ϱ. The minimal values of the parameter ϱ with regard to the number of stiffeners can be derived from the present results.

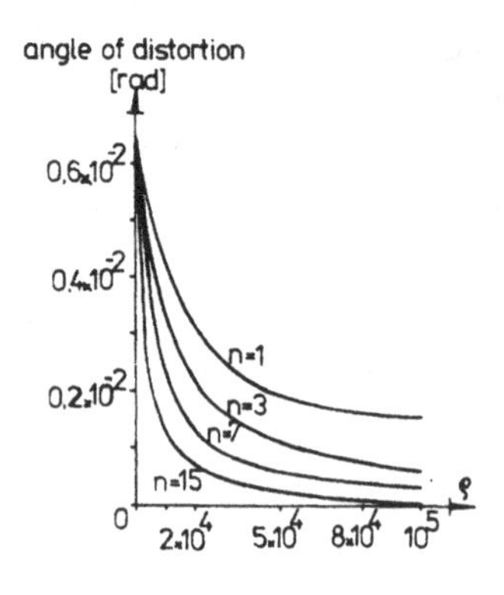

Fig. 84. Diagrams of the angle of distortion in discrete point No. 8 for various values of the parameters ϱ and n.

For the numerical evaluation of this example A-9 of the CTM was used. Further details of the analysis were explained in reference [112].

4.3. Corrugated and Sandwich Panels

a) Corrugated Plate — Resonance Analysis

The harmonic vibration of a corrugated plate panel subjected to the action of central midspan force $P = P_0 e^{i\Omega t}$ is studied in accordance with the scheme in Figure 85. The amplitude of inducing force is $P_0 = 10$ N and the resonance

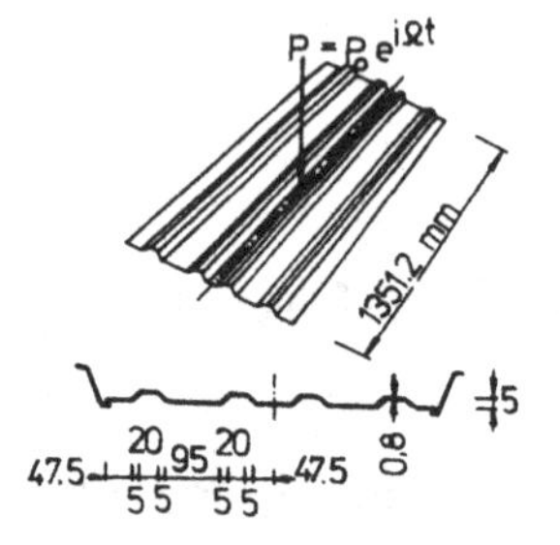

Fig. 85. Geometric and loading parameters of the studied thinwalled panel.

peaks were studied numerically and experimentally. The panel is simply supported at the shorter edges, whereas the longitudinal boundaries are elastically clamped. The stiffness characteristics of the longitudinal boundaries are given by the averaged constant value $c = 4 \times 10^4$ N m^{-1}. All physical parameters of the studied case are given in Figure 85. The calculated and experimentally found peaks of the resonance spectrum are listed in Table V.

Table V. Resonance peaks of the frequency spectrum for the studied corrugated plate panel

Resonance frequencies [Hz]	
Numerical calculation	Experiment
23.112	21.7
35.041	37.4
41.175	43.4
64.357	69.9

For the numerical calculation of this example expressions C-1 to C-11 of the CTM were used.

A further analysis of this example is given in reference [110].

b) Sandwich Panel — Parametric Study

Attention is now focused on a parametric study of the dynamic behaviour of the sandwich panel provided with longitudinal ribs as shown in Figure 86. The panel is subjected to the action of a central midspan force $P = P_0 e^{i\Omega t}$, with amplitude $P_0 = 200$ N. The problem is studied in the frequency interval $\Omega = 0$—1000 Hz. The short transverse edges of the panel are simply supported. The longitudinal edges of the panel are elastically clamped. The variable

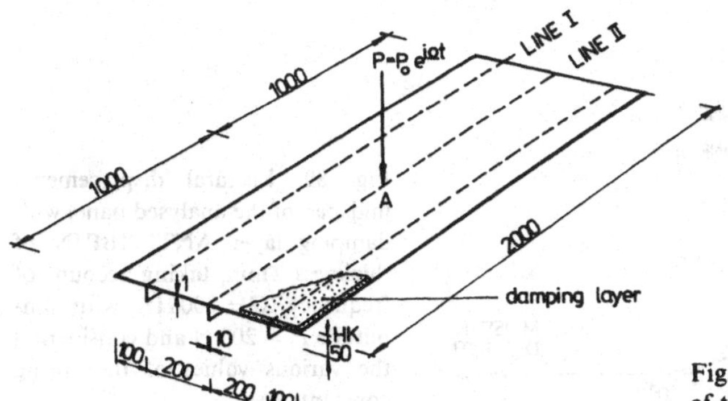

Fig. 86. Geometry and elevation of the analysed panel.

stiffness of spring constants c is assumed. The stiffened steel panel is provided with two configurations of damping layers as shown in Figure 87. As damping layer is used the material ANTIVIBRIN V 2100, with elasticity moduli $E = 64.8 \times 10^8\ \mathrm{N\,m^{-2}}$; $G = 21.6 \times 10^8\ \mathrm{N\,m^{-2}}$, with mass $\tau = 1.092 \times 10^3\ \mathrm{kg\,m^{-3}}$ and with the parameter of damping $\eta = 0.15$.

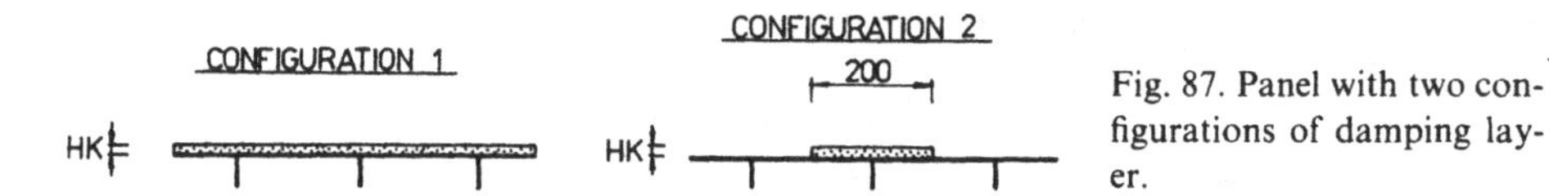

Fig. 87. Panel with two configurations of damping layer.

Flexural displacements in typical nodes of the panel simulation for various thicknesses of the damping layer in the first configuration, taking account of the inducing frequency $\Omega = 390$ Hz and spring constants $c = 50\ \mathrm{kN\,m^{-1}}$, are illustrated in Figure 88. Analogous flexural deformations, taking the thickness of the damping layer in the first configuration to be 2 mm and the inducing frequency $\Omega = 200$ Hz, for various values of the spring constants c, are as shown in Figure 89. The comparison of nodal flexural displacements, for variable thicknesses of the damping layers in the two configurations, is

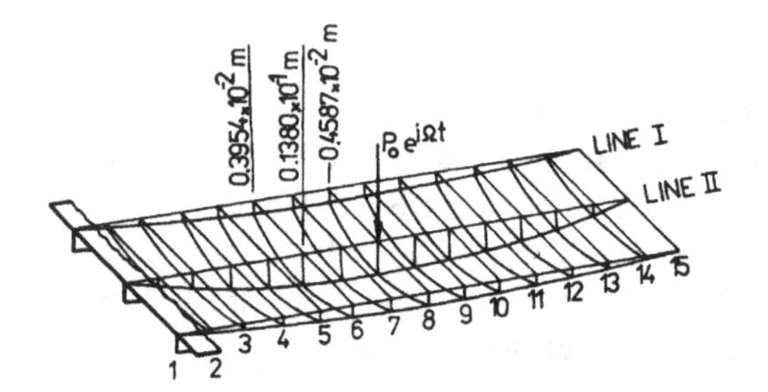

Fig. 88. The mode of basic panel at frequency $\Omega = 390$ Hz, with the amplitude of the inducing force $P_0 = 200$ N and with spring constants on longitudinal boundaries $c = 50\ \mathrm{MN\,m^{-1}}$.

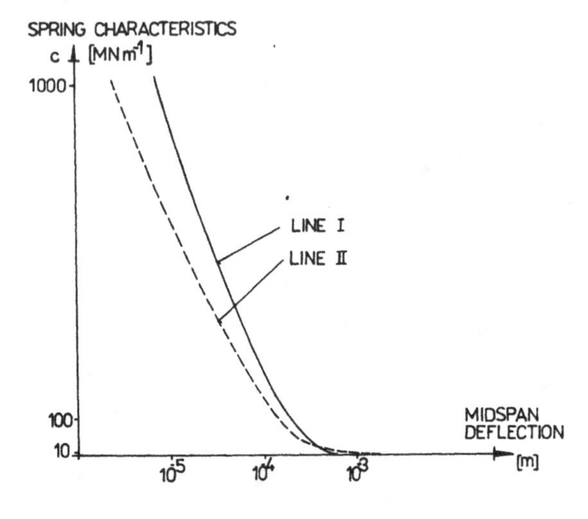

Fig. 89. Flexural displacements midspan of the analysed panel with damping layer ANTIVIBRIN of thickness 2 mm, taking account of frequency $\Omega = 200$ Hz, with amplitude $P_0 = 200$ N and considering the various values of the spring constants c.

plotted in Figure 90. Maximum damping is obtained when applying the second of the assumed configurations taking the thickness of the damping layer to be 6 mm.

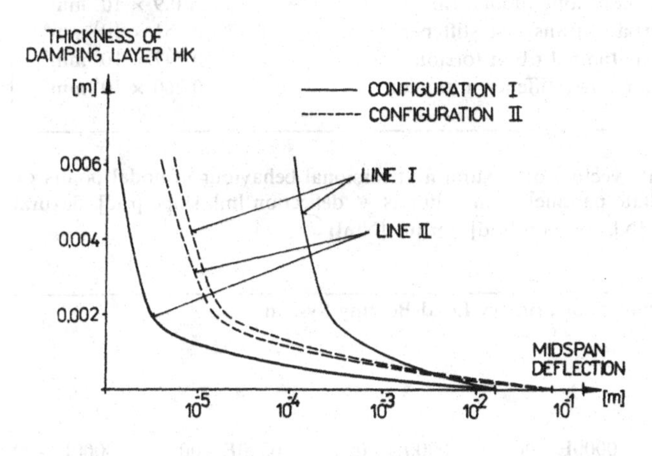

Fig. 90. Flexural displacements midspan of the studied panel regarding the indicated thickness and location of damping layer ANTIVIBRIN at frequency $\Omega = 400$ Hz, amplitude of inducing force $P_0 = 200$ N and spring constant $c = 0$.

For the numerical solutions B-1 to B-13 and C-1 to C-11 of the CTM were used.

Further details of the analysis are presented in reference [111].

4.4. Shells

a) Shell Grillage — Static Analysis

The static spatial behaviour of a circular shell grillage as shown in Figure 91 is studied. The analysed simulation is provided with longitudinal ribs and transverse stiffeners and is loaded by its own dead weight. Volume unit weight loads are concentrated in the nodal points of assumed theoretical simulation as shown in Figure 91. All longitudinal members are simply supported on both ends of the studied simulation. All physical parameters, together with the parameters of the applied transverse and longitudinal discretization are as shown in Figure 91 and in Table VII, using the notation in Table VI. The calculated state vectors of spatial behaviour in the individual

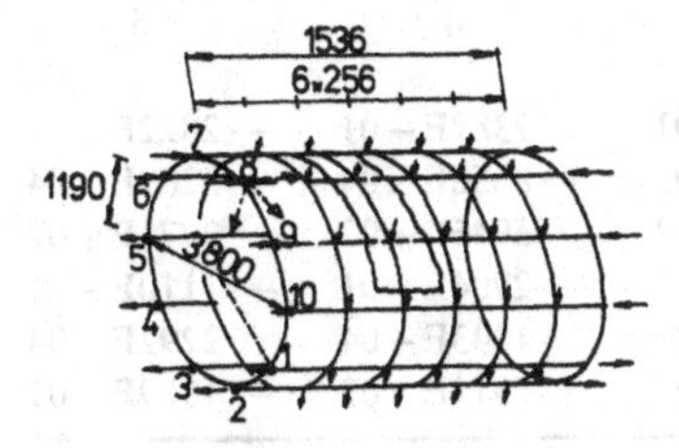

Fig. 91. Geometric scheme of the studied cylindrical shell.

Table VI. Physical parameters of the studied shell grillage

Parameter		Value
Cross-sectional area of longitudinal rib	—	3590 mm^2
Flexural moment of inertia of longitudinal rib	—	0.9×10^6 mm^4
Flexural moment of inertia of transverse stiffener	—	0.651×10^3 mm^4
Moment of inertia of longitudinal rib in torsion	—	0.747×10^4 mm^4
Moment of inertia of transverse stiffener in torsion	—	0.260×10^4 mm^4

Table VII. Calculated state vectors of flexural and torsional behaviour in nodal points of the studied shell grillage (State parameters in columns — deflection [m], slope [rad], flexural moment [Nm], lateral force [N], rotation [rad], torque [Nm])

State Parameters of the Members of Primary Load-Bearing System
1—10 Section 1

Point	1	2	3	4	5
	0000E+00	·0000E+00	·0000E+00	·0000E+00	·0000E+00
	−·6728E−04	−·1762E−03	−·2178E−03	−·1762E−03	−·6729E−04
	·0000E+00	·0000E+00	·0000E+00	·0000E+00	·0000E+00
	−·5546E−01	−·1452E+00	−·1795E+00	−·1452E+00	−·5547E−01
	·0000E+00	·0000E+00	·0000E+00	·0000E+00	·0000E+00
	·2984E−01	·2497E−01	−·1603E−02	−·1720E−01	−·3417E−01

State Parameters of the Members of Primary Load-Bearing System
1—10 Section 2

Point	1	2	3	4	5
	−·1640E−01	−·4295E−01	−·5309E−01	−·4295E−01	−·1641E−01
	−·5766E−04	−·1510E−03	−·1866E−03	−·1510E−03	−·5768E−04
	−·1419E+02	−·3716E+02	−·4593E+02	−·3716E+02	−·1419E+02
	−·3328E−01	−·8713E−01	−·1077E+00	−·8714E−01	−·3329E−01
	−·1270E−04	−·1062E−04	·6822E−06	·7319E−05	·1454E−04
	·1416E−01	·1737E−01	−·1471E−02	−·7311E−02	−·1969E−01

State Parameters of the Members of Primary Load-Bearing System
1—10 Section 3

Point	1	2	3	4	5
	−·2821E−01	−·7387E−01	−·9131E−01	−·7387E−01	−·2822E−01
	−·3268E−04	−·8556E−04	−·1058E−03	−·8556E−04	−·3269E−04
	−·2271E+02	−·5946E+02	−·7350E+02	−·5946E+02	−·2271E+02
	−·1109E−01	−·2904E−01	−·3590E−01	−·2904E−01	−·1110E−01
	−·1872E−04	−·1802E−04	·1308E−05	·1043E−04	·2292E−04
	·1416E−01	·1737E−01	−·1471E−02	−·7311E−02	−·1969E−01

nodal points are given in Table VII. Out-of-circle flexural displacements are negative, in-circle displacements are positive. The same convention is valid for other components of the calculated state vectors. The registered effects of antisymmetry in the transversal direction, are caused by the influences of a variable axial tension and compression in the individual longitudinal ribs of the shell grillage. The symmetry in the longitudinal direction remains undisturbed.

For the numerical evaluation the terms B-17 to B-19 of the CTM were used.

Further details of the analysis are presented in reference [103].

6	7	8	9	10
·0000E+00	·0000E+00	·0000E+00	·0000E+00	·0000E+00
·7085E−04	·2003E−03	·3150E−03	·2003E−03	·7085E−04
·0000E+00	·0000E+00	·0000E+00	·0000E+00	·0000E+00
·5831E−01	·1645E+00	·2570E+00	·1645E+00	·5831E−01
·0000E+00	·0000E+00	·0000E+00	·0000E+00	·0000E+00
−·3585E−01	−·3409E−01	·3086E−03	·3291E−01	·3302E−01

6	7	8	9	10
·1728E−01	·4886E−01	·7686E−01	·4886E−01	·1727E−01
·6075E−04	·1719E−03	·2708E−03	·1719E−03	·6075E−04
·1489E+02	·4186E+02	·6480E+02	·4186E+02	·1489E+02
·3571E−01	·1037E+00	·1743E+00	·1036E+00	·3571E−01
·1526E−04	·1451E−04	−·1313E−06	−·1401E−04	−·1405E−04
−·2076E−01	−·1823E−01	·3132E−03	·1638E−01	·1629E−01

6	7	8	9	10
·2972E−01	·8410E−01	·1325E+00	·8410E−01	·2972E−01
·3446E−04	·9762E−04	·1543E−03	·9762E−04	·3446E−04
·2391E+02	·6755E+02	·1060E+03	·6755E+02	·2390E+02
·1247E−02	·3843E−01	·7383E−01	·3842E−01	·1247E−01
·2409E−04	·2227E−04	−·2646E−06	−·2098E−04	−·2099E−04
−·2076E−01	−·1823E−01	·3132E−03	·1638E−01	·1629E−01

Table VII (Continued)

State Parameters of the Members of Primary Load-Bearing System
1—10 Section 4

Point	1	2	3	4	5
	−·3248E−01	−·8504E−01	−·1051E+00	−·8504E−01	−·3248E−01
	·6658E−09	−·4657E−09	−·7722E−09	−·9104E−09	−·1069E−08
	−·2555E+02	−·6690E+02	−·8269E+02	−·6690E+02	−·2556E+02
	·1109E−01	·2904E−01	·3590E−01	·2904E−01	1109E−01
	−·2475E−04	−·2541E−04	·1934E−05	·1354E−04	·3130E−04
	−·1623E−01	·6114E−02	−·5262E−03	·1243E−01	·6270E−02

State Parameters of the Members of Primary Load-Bearing System
1—10 Section 5

Point	1	2	3	4	5
	−·2821E−01	−·7387E−01	−·9131E−01	−·7387E−01	−·2822E−01
	·3268E−04	·8556E−04	·1058E−03	·8556E−04	·3268E−04
	−·2271E+02	−·5946E+02	−·7350E+02	−·5946E+02	−·2272E+02
	·3328E−01	·8714E−01	·1077E+00	·8714E−01	·3328E−01
	−·1784E−04	−·2801E−04	·2158E−05	·8254E−05	·2863E−04
	−·1623E−01	·6114E−02	−·5262E−03	·1243E−01	·6270E−02

State Parameters of the Members of Primary Load-Bearing System
1—10 Section 6

Point	1	2	3	4	5
	−·1640E−01	−·4295E−01	−·5309E−01	−·4295E−01	−·1641E−01
	·5766E−04	·1510E−03	·1866E−03	·1510E−03	·5768E−04
	−·1419E+02	−·3716E+02	−·4593E+02	−·3716E+02	−·1420E+02
	·3328E−01	·8714E−01	·1077E+00	·8714E−01	·3328E−01
	−·1094E−04	−·3061E−04	·2382E−05	·2965E−05	·2597E−04
	−·1623E−01	·6114E−02	−·5262E−03	·1243E−01	·6270E−02

State Parameters of the Members of Primary Load-Bearing System
1—10 Section 7

Point	1	2	3	4	5
	−·4116E−06	−·7283E−06	−·3148E−06	−·1103E−05	−·6647E−06
	·6727E−04	·1762E−03	·2178E−03	·1762E−03	·6729E−04
	1783E−03	·3357E−03	·7172E−03	·9918E−03	·2298E−03
	·5546E−01	·1452E+00	·1795E+00	·1452E+00	·5548E−01
	·1426E−05	−·3744E−04	·8406E−06	−·6949E−05	·2240E−04
	−·2905E−01	·1604E−01	·3623E−02	·2330E−01	·8367E−02

6	7	8	9	10
·3422E−01	·9685E−01	·1526E+00	·9684E−01	·3422E−01
·4948E−09	·8404E−09	·6685E−09	·4511E−09	−·3865E−09
·2693E+02	·7621E+02	·1202E+03	·7621E+02	·2692E+02
−·1109E−01	−·2905E−01	−·3587E−01	−·2905E−01	−·1109E−01
·3292E−04	·3003E−04	−·3979E−06	−·2794E−04	−·2792E−04
·5831E−02	·1162E−01	·1991E−03	−·1635E−01	−·1685E−01

6	7	8	9	10
·2972E−01	·8410E−01	·1325E+00	·8410E−01	·2972E−01
−·3446E−04	−·9762E−04	−·1543E−04	−·9762E−04	−·3446E−04
·2391E+02	·6755E+02	·1060E+03	·6755E+02	·2390E+02
−·3466E−01	−·9652E−01	−·1456E+00	−·9652E−01	−·3466E−01
·3044E−04	·2508E−04	−·4826E−06	−·2099E−04	−·2075E−04
·5831E−02	·1162E−01	·1991E−03	−·1635E−01	−·1685E−01

6	7	8	9	10
·1728E−01	·4886E−01	·7686E−01	·4886E−01	·1727E−01
−·6075E−04	−·1719E−03	−·2708E−03	−·1719E−03	−·6075E−04
·1489E+02	·4186E+02	·6480E+02	·4186E+02	·1489E+02
−·3571E−01	−·1037E+00	−·1743E+00	−·1036E+00	−·3571E−01
·2796E−04	·2014E−04	−·5673E−06	−·1403E−04	−·1357E−04
·5831E−02	·1162E−01	·1991E−03	−·1635E−01	−·1685E−01

6	7	8	9	10
·6915E−06	·4498E−06	·7548E−06	·6938E−06	−·7560E−06
−·7085E−04	−·2003E−03	−·3150E−03	−·2003E−03	−·7085E−04
−·3061E−03	−·6256E−03	−·2289E−03	−·2747E−03	·8078E−03
−·5831E−01	−·1645E+00	−·2570E+00	−·1645E+00	−·5830E−01
·2567E−04	·1184E−04	−·2774E−06	−·4321E−07	·1072E−05
·5383E−02	·1951E−01	−·6813E−03	−·3287E−01	−·3442E−01

b) Shell Panel — Linear and Nonlinear Resonance

Due to the variable degrees of nonlinearity in various regions of thin shell structure, defined over the magnitudes of external loads and local displacements as well as over the interactions of geometric and physical nonlinearities, the spectral variability of nonlinear peaks occurs in the corresponding resonance curves of large amplitude vibrations. Depending on the degree and type of nonlinearity, the values of nonlinear resonance frequencies are variable compared with their linear equivalents. For high load levels or for advanced plastifications of a structure, the resonance spectrum smears up completely, e.g. the resonance curves lose their local peaks and physical significance. Such extremal cases are not dealt with in the present example, because for such situations the structure is accepted to be in the state of collapse.

The resonance response for the linear and nonlinear flexural vibration of a steel corrugated shell panel, as shown in Figure 92, is studied. The panel is subjected to the action of harmonically variable transverse force $P = P_0 e^{i\Omega t}$, with amplitude $P_0 = 10$ kN and with a variable forcing circular frequency Ω. The amplitude P_0 has the value which initiates a distinct nonlinear behaviour of the studied panel. The spectral shift of the first nonlinear peak, compared with its linear equivalent for the flexural behaviour of the midspan point A located directly under the force P, is shown in Figure 93. Note the distinct decrease of the first linear resonance frequency $\Omega = 1260$ Hz compared with

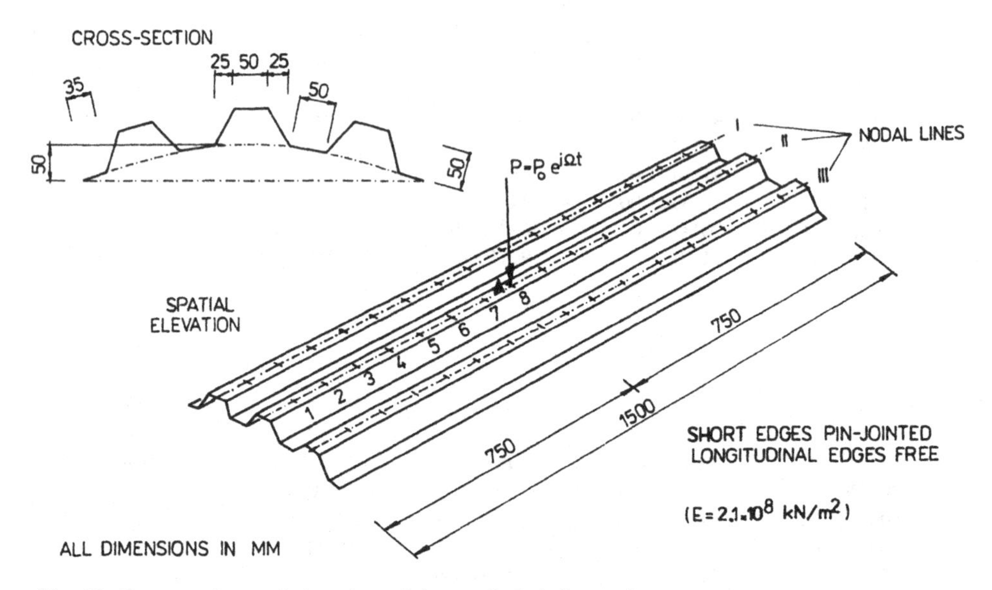

Fig. 92. Cross-section and elevation of the studied shell panel.

its nonlinear equivalent $\Omega = 479$ Hz . Such a relation has the validity for the assumed value of inducing force. The typical nonlinear inclination of the resonance peak in the direction of lower frequencies, which corresponds to the assumed load level and degree of nonlinearity, can be registered in Figure 93. The geometric shape of the inclined resonance peak specifies several solutions (roots) of the resonance problem in nonlinear zones of vibration. The present problem was analysed numerically; for deformations corresponding to the set of pseudo-forces for instantaneous incremental configurations of the shell panel, the resonance peaks of the nonlinear vibration were evaluated taking into account the assumed convergence criteria. In nonlinear dynamic analyses, however, individual particles or zones of thin shell skin are characterized by their own resonance curves or peaks, depending on the degree of nonlinearity for given micro- or macroelements of the applied simulation mesh. Local resonance zones occur with accumulated resonance peaks for neighbouring regions of the used discrete simulation. Resonance peaks of local structural regions are interactively coupled. The application of microdiscretization simulations in local resonance zones of a thin shell skin with a dynamically variable mesh is inevitable in order to account for the local displacements, stresses and variable degrees of nonlinearity. The automatic variability of the load steps is significant with regard to the economy of the computer calculation.

For the numerical analysis the terms F-1 to F-12 of the CTM were used. Further details of the analysis are presented in reference [93].

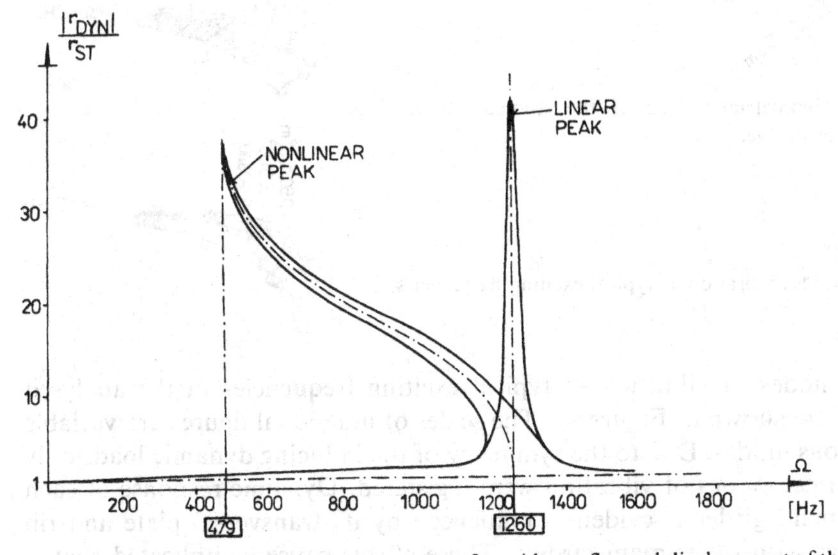

Fig. 93. Linear and nonlinear resonance curves for midspan flexural displacement of the studied shell panel.

4.5. Curved System

Curved Bridge with Several Main Girders — Resonance Mode Analysis

The spatial dynamic behaviour is studied of a curved plate-girder steel bridge, provided with longitudinal stiffeners of upper roadway plate, which is subjected to the action of harmonically variable exciting forces. The cross-sectional and longitudinal scheme of the plate-girder bridge with three main girders is shown in Figure 94. The bridge is subjected to the action of two concentrated dynamic forces $P_D = P_0 e^{i\Omega t}$, with constant amplitude $P_0 = 1.0$ MN and with variable inducing frequency Ω. The positions of both inducing forces are depicted in Figure 94. The harmonic vibration of the structure is studied at various inducing frequencies in the interval $\Omega = 0$—30 Hz. The applied scheme of discretization is as shown in Figure 94.

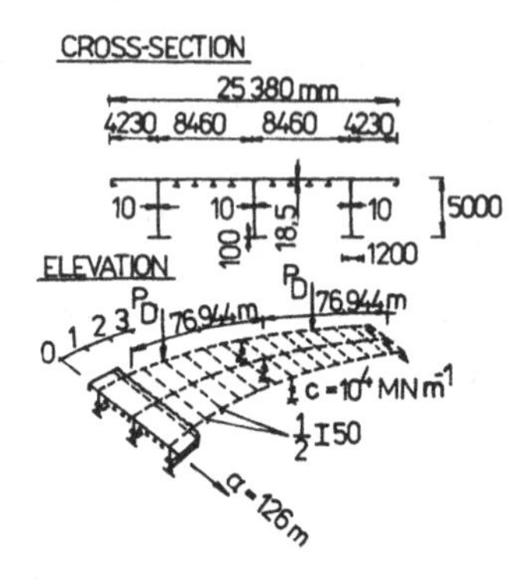

Fig. 94. Geometric scheme of the studied curved plate-girder bridge.

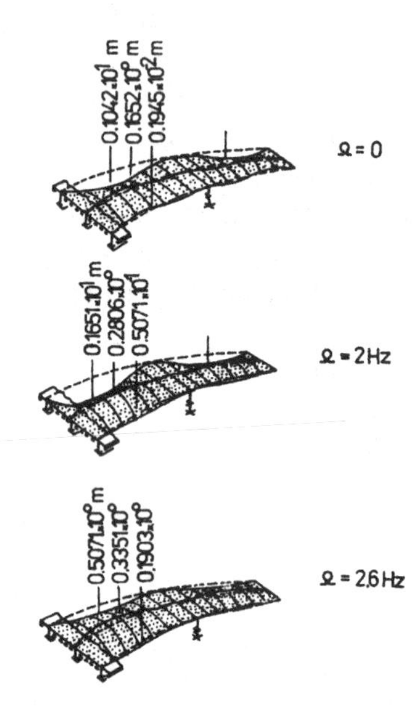

Fig. 95. Modes vibration at typical exciting frequencies.

The modes of vibration at typical exciting frequencies in the analysed interval are shown in Figure 95. The scales of individual figures are variable for various modes. Due to the symmetry of the inducing dynamic loads only symmetrical modes of vibration were registered. Dynamic response of each curved main girder is evidently influenced by its transversal plate and rib couplings with other main girders. These effects cause complicated spatial strain and stress states of this type of structures.

For the calculation the terms B-7, B-8 and B-9 as well as C-1 to C-11 of the CTM were used.

Further details of the analysis are presented in reference [87].

4.6. Elastic-Plastic Behaviour

Thinwalled Box Girder — Elastic-Plastic Analysis

The elastic-plastic behaviour is analysed of a thinwalled box beam subjected to the action of antisymmetric midspan load in accordance with the scheme in Figure 96. A transverse stiffener having parameter $\varrho = 500$ is assumed to be located midspan of studied beam. The parameter ϱ again gives the ratio of the frame rigidity of the transverse stiffener and the frame distortional rigidity of the nonstiffened cross-section. The end diaphragms were assumed to be infinitely rigid in their plane and nonrigid out the plane. The application of the midspan stiffener dominated the distribution of distortion-bending plastic deformation zones. The flexural, torsion-bending and distortion-bending elastic-plastic behaviour of the box beam in its spatial comprehension have been treated theoretically (see Section 2.4).

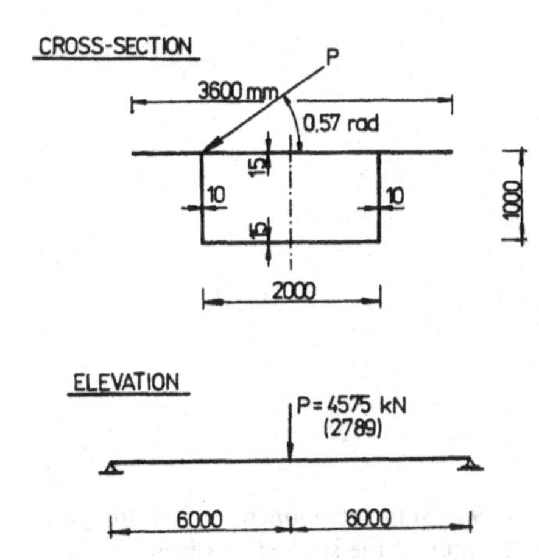

Fig. 96. Cross-sectional and loading scheme of the studied box beam.

The distribution of plastic areas together with the diagrams of normal and shear stresses in three characteristic sections of the studied box beam are shown in Figure 97. The distribution of the plastic zones was followed until the collapse of the structure. The calculated force components of state vectors in typical elastic-plastic zones are plotted in Figure 98. The dotted lines and values in brackets determine the limit elastic state whereas the solid lines and single values express the plastic state of the studied structure. The first four components express the corresponding bending moments and lateral forces in the direction of both principal axes, whereas the following four components determine the force state components of torsion-bending and distortion-bending behaviour of the studied box beam.

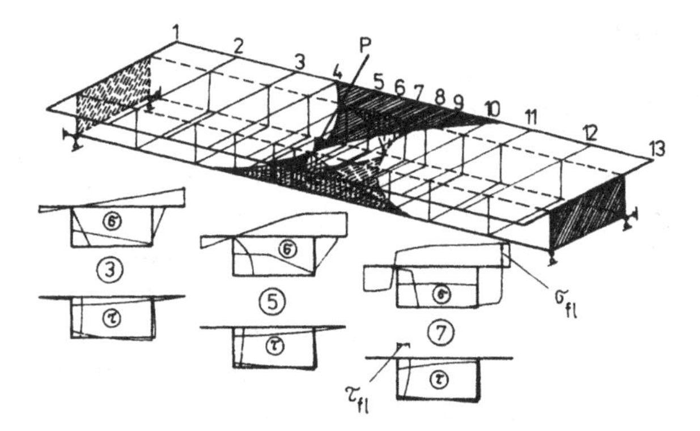

Fig. 97. Distribution of plastic zones in the box beam (σ_{fl}) — normal yield stress, τ_{fl} — shear yield stress).

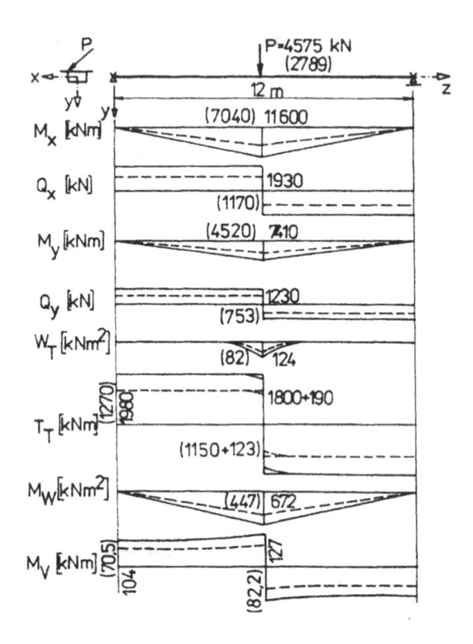

Fig. 98. State components of elastic-plastic behaviour of the studied box beam.

For the solution of the present example the terms A-1 to A-11 of the CTM were used.

Further details of the analysis are given in reference [101].

4.7. Nonlinear Interrelations

Shell Panel — Physical and Geometrical Nonlinear Interactions

The nonlinear resonance response is studied of a longitudinally stiffened thinwalled steel shell panel, shown in Figure 99. The analysed panel is subjected to the action of a harmonically variable central force positioned in midspan point A. The amplitude of the inducing force $P_{D,i}$ is continually variable in the interval $\alpha_i = 1, \dots 2$, with the ratio $\alpha_i = P_{D,i}/P_{D,ref}$, where $P_{D,ref} = 5$ kN is the corresponding reference value. The attention is focused on the analysis of temporary phenomena, arising in resonance zones of vibration taking account of the mutual interaction of geometric (stiffness parameters) and physical (influences of damping) nonlinearities. The degree of nonlinearity is increased or reduced by varying the values α_i.

Figure 100 shows the linear and nonlinear resonance curves of the flexural displacement in point A for the value $\alpha_i = 1.8$. Line ① indicates the resonance curve of the linear vibration. Line ② illustrates the diagram of the corresponding resonance curve taking into account geometric nonlinearities. Note the characteristic nonlinear inclination of the resonance curve directed towards lower frequencies. The inclination becomes more pronounced with increasing values of the parameter α_i (see Figure 101). Line ③ in Figure 100 describes the graph of the analysed resonance curve taking into account the interaction of geometric and physical nonlinearities. The influence of linear and nonlinear damping on the nonlinear resonance flexural deflection of point A is shown in Figure 102. In this figure the region denoted by "a" is

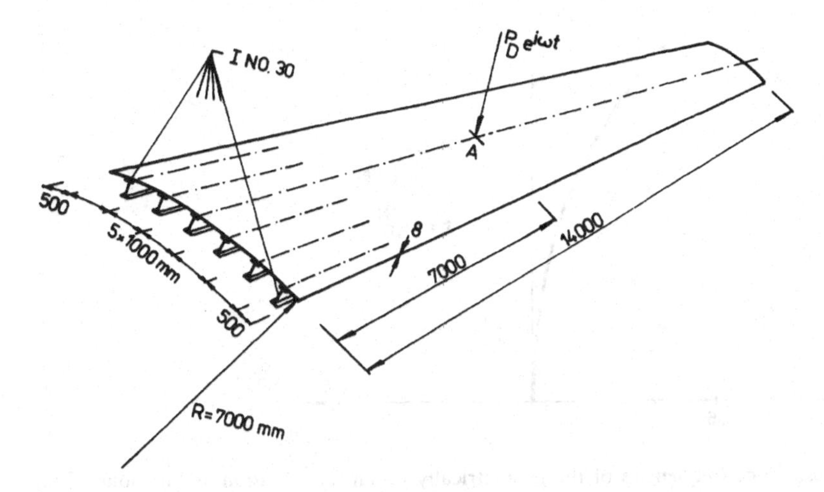

Fig. 99. Cross-section and elevation of the studied stiffened shell panel.

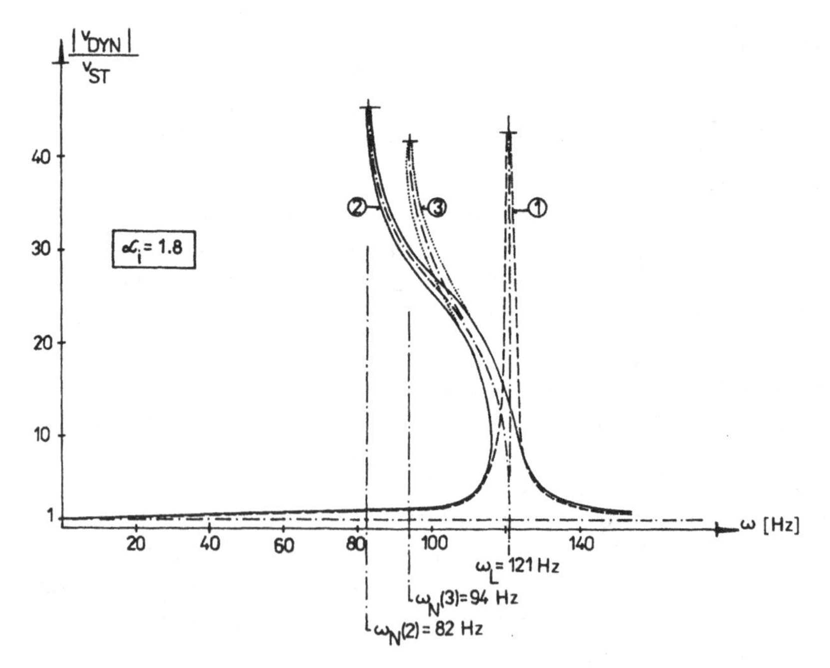

Fig. 100. Linear and nonlinear resonance curves for the midspan flexural deflection of the studied shell panel: ① — resonance curve of linear vibration, ② — resonance curve with regard to nonlinear stiffness parameters, ③ — resonance curve with regard to nonlinear stiffness and damping parameters.

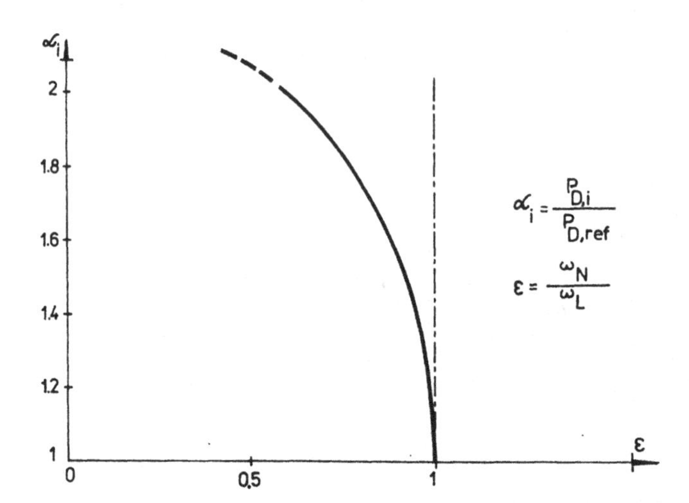

Fig. 101. Resonance frequencies of the geometrically nonlinear vibration as functions of the loading level.

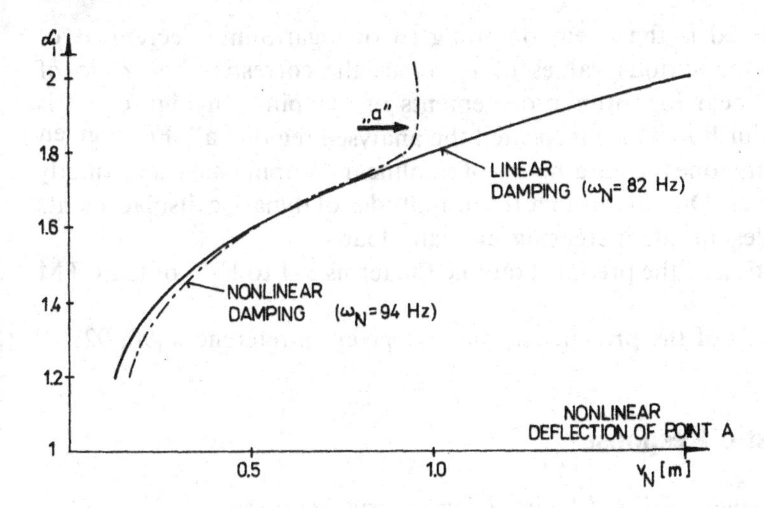

Fig. 102. Influence of linear and nonlinear damping on nonlinear resonance deflection in point A.

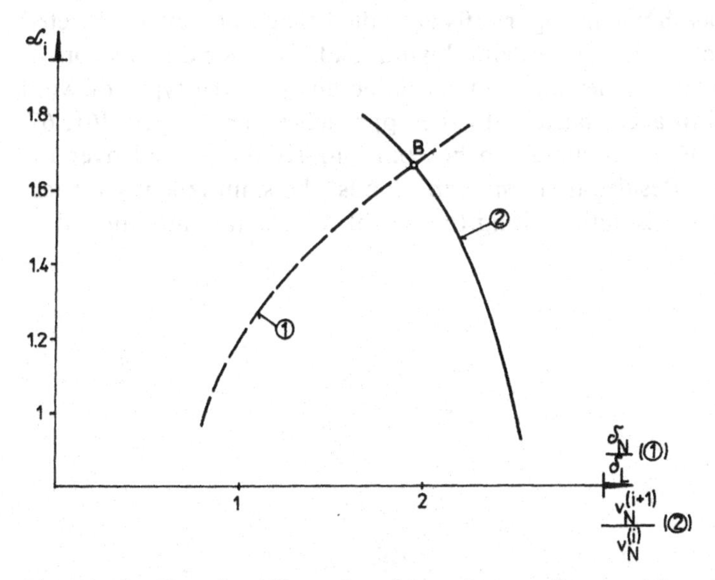

Fig. 103. Nondimensional illustration of the ratio on nonlinear and linear logarithmic decrement of damping (curve ①) and of geometrically nonlinear deflection at the point A regarding the linear damping (curve ②), versus variable values of the parameter a_i.

important, with constant values of the flexural deflection in point A despite the increasing dynamic loads. This phenomenon is called the *paradox of interaction of the geometric and physical nonlinear vibration*. In zone "a" an interactive overcovering of geometric and physical nonlinear displacements occurs. Considered is the linear damping (with logarithmic decrement of damping 0.03) for various values of α_i versus the corresponding ratio of nonlinear and linear logarithmic decrements of damping. In Figure 103 is specified the point B in which is located the analysed region "a". For a given stress state in this zone the parameters of nonlinear damping have a distinctly growing character. Due to this fact the amplitudes of dynamic displacements remain stable despite the increasing inducing loads.

For the solution of the present example the terms F-1 to F-12 of the CTM were used.

Further details of the present analysis are given in references [90, 92].

4.8. Aeroelastic Response

Thinwalled Chimney — Wind Induced Parametric Response

The time-dependent response of a thinwalled steel chimney with elevation and cross-section as shown in Figure 104 is studied. The chimney is subjected to the action of time variable horizontal wind loads. The wind loads consist of stationary constant values and time variable gusts. Three types of wind gusts are assumed in accordance with the approaches in references [61, 67, 72]. The wind loads are assumed to be continuously distributed over the height of structure. Besides dynamic wind loads, the simultaneous vertical static dead-weight loads acting along the height of structure are applied.

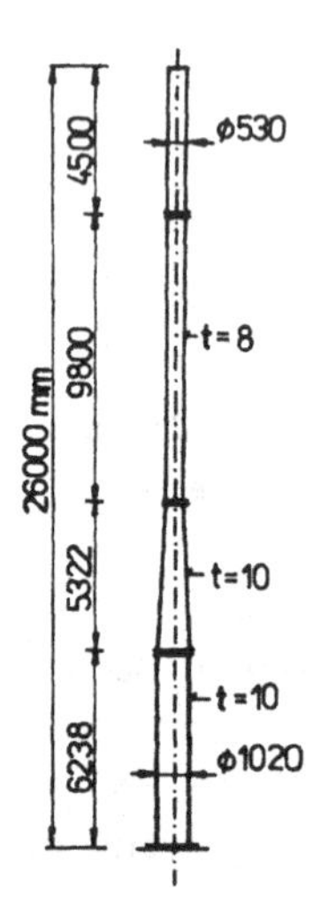

Fig. 104. Structural scheme of the studied chimney stack.

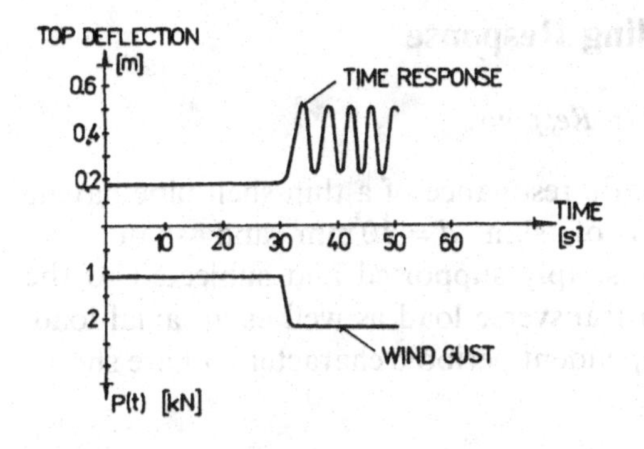

Fig. 105. Input signal (loading parameters according to Rausch) and structural response (amplitudes of dynamic displacements on top of the chimney).

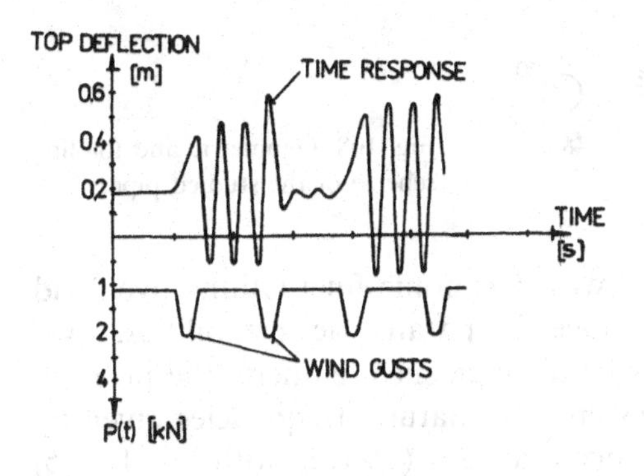

Fig. 106. Input signal (loading parameters according to Schlaich) and structural response (amplitudes of dynamic displacements on top of the chimney).

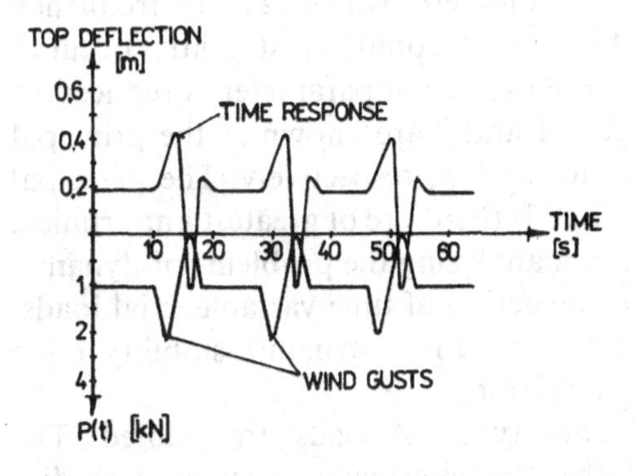

Fig. 107. Input signal (loading parameters according to Petersen) and structural response (amplitudes of dynamic displacements on top of the chimney).

The assumed input signals (stationary and time variable action of wind) and the calculated output singnals (flexural deflection on the top of the chimney) for all three studied cases are shown in Figures 105, 106 and 107.

For the calculation the terms A-1 to A-11 of the CTM were used.

Further details of the present analyses as well as other problems concerning the aeroelastic response of slender structural configurations when applying transfer matrix or FETM-methods, are given in references [82, 84, 85, 97, 107].

4.9. Stability and Post-Buckling Response

a) Shell Pipe — Parametric Stability Response

The dynamic stability and parametric resonance of a thin shell pipe having diameter $\Phi = 508$ mm, thickness of skin $d = 10$ mm and span $L =$ 10 000 mm is studied. The pipe is simply supported and subjected to the simultaneous action of a midspan transverse load as well as an axial load. Both types of loads have a time-dependent periodic character and are shown in Figure 108.

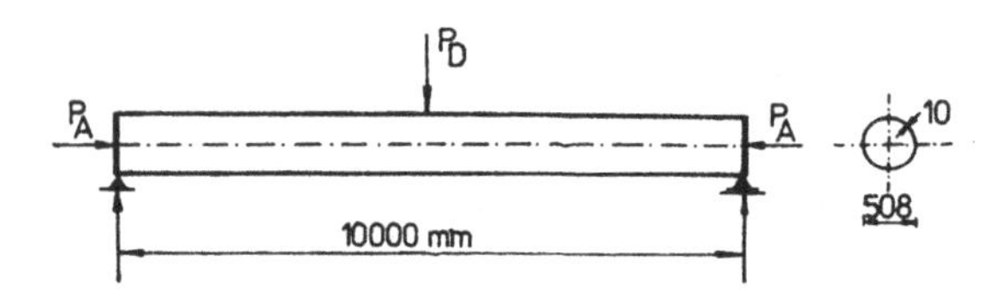

Fig. 108. Geometric and loading scheme of the studied pipe.

The action of the transverse load with harmonic fluctuations over and above a constant mean value can induce the parametric instabilities. Two types of parametric instabilities may be distinguished, namely: the primary and secondary instabilities. For a system with natural frequencies given by $\Omega_n = 1, 2, 3, \ldots$, primary instabilities occur at $\Omega = (2/k)\,\Omega_n$, with $k = 1, 3, 5, \ldots$, as the excitation parameter is reduced to zero, where Ω is the frequency of a periodically varying parameter. Likewise, secondary instabilities occur at $\Omega = (2/k)\,\Omega_n$, with $k = 2, 4, 6, \ldots$, as the excitation parameter is reduced to zero. Instabilities corresponding to $k = 1$ and 2 are known as the principal primary and the principal secondary instability, respectively. The principal primary instability is the most critical and is therefore of greatest importance.

Similar considerations are valid when analysing the problems of dynamic stability of pipe systems subjected to the action of time-variable axial loads. Both effects, the parametric resonance as well as dynamic stability must therefore be considered in mutual interaction.

Three various combinations of both types of loads are studied. The magnitudes and time variability of the dynamic loads, together with diagrams of the corresponding time response for the flexural deflection midspan of the studied pipe are shown in Figure 109.

In order to investigate the parametric response and dynamic stability of the studied pipe, it is interesting to analyse the results of Figure 109a. The obtained results illustrate that the primary parametric response of the pipe is distinctly influenced by increasing values of the time periodic axial forces. For the analysed region 10^2—10^8 N of magnitudes of the axial forces, an evident phase shift occurs for the start of the periodic time response midspan of the studied pipe. Such results indicate the possibilities for control of dynamic response of pipes in advanced nonlinear regions of their parametric resonance and dynamic stability.

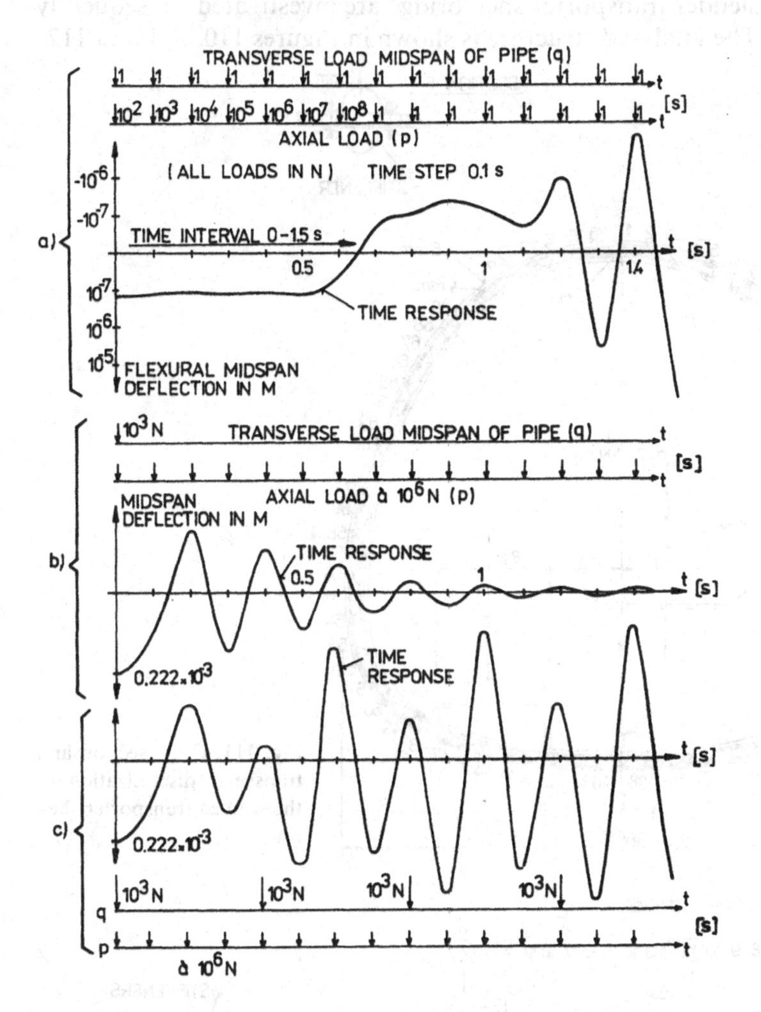

Fig. 109. Patterns of magnitudes and time variability of dynamic loads and diagrams of corresponding time response for midspan flexural deflection for the three studied cases.

The Wilson direct integration technique incorporated into the terms F-1 to F-12 of the CTM was used for solving the present case.

Further information on the performed numerical analysis may be obtained in reference [98].

b) Transporter Shell Bridge — Stability and Post-Buckling Response

As an illustrative application of the present theoretical analyses for the solution of real structural configurations, special problems of the spatial behaviour of a slender transporter shell bridge are investigated subsequently in this chapter. The analysed structure is shown in Figures 110, 111 and 112.

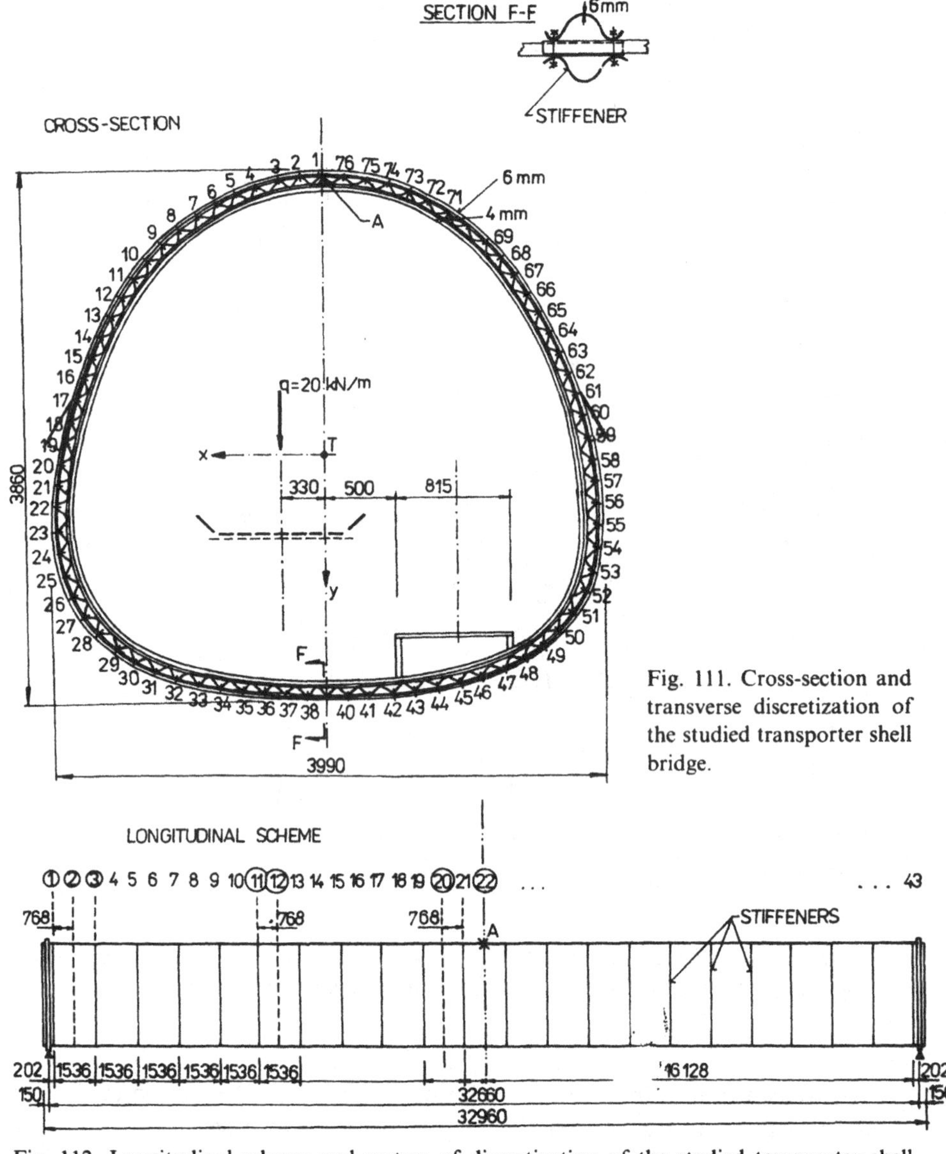

Fig. 111. Cross-section and transverse discretization of the studied transporter shell bridge.

Fig. 112. Longitudinal scheme and system of discretization of the studied transporter shell bridge.

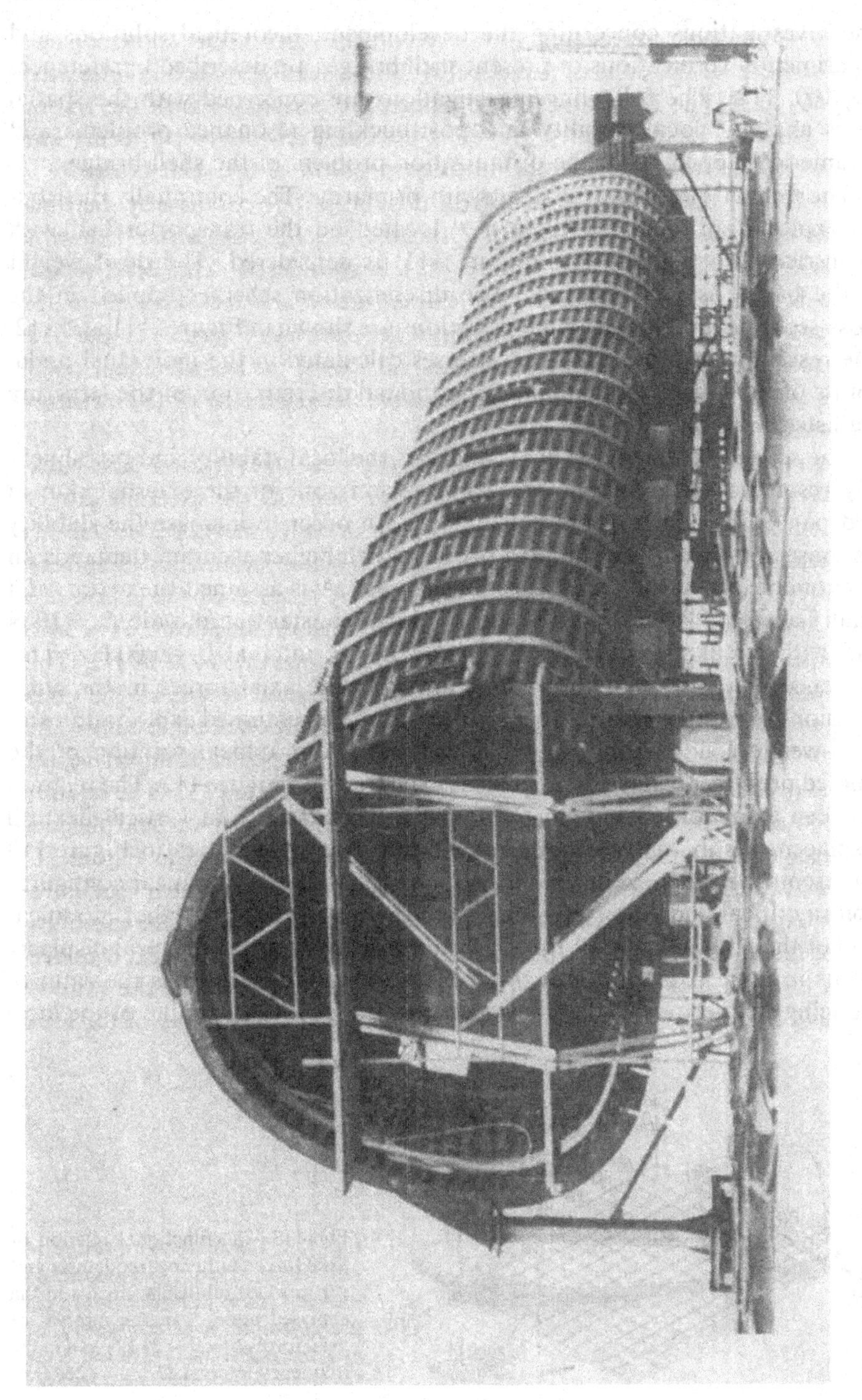

Fig. 110. Photo of the studied experimental prototype of transporter shell bridge.

The investigations concerning the development, theoretical solutions and experimental verifications of present shell bridges are described in references [99, 100, 112]. The following investigations are concerned with the spatial stress analysis, local stability and post-buckling resonance problems and a numerical approach to the optimization problem of the shell bridge.

The first of these items is dealt with primarily. The continually distributed exploitation load $q = 20\,\mathrm{kN\,m^{-1}}$ located on the transporter belt with eccentricity $e = 350\,\mathrm{mm}$ (see Figure 111) is considered. The dead weight of the bridge is $8.936\,\mathrm{kN\,m^{-1}}$. The discretization schemes applied in the cross-sectional and longitudinal directions are shown in Figures 111 and 112. The resulting absolute values of stresses calculated in the individual nodal points of the cross-sectional and longitudinal discretization of the structure are listed in Table VIII.

The second of the above items analyses the local stability and post-buckling resonance behaviour in the compression zone of the external skin in midspan point A in Figures 111 and 112. In order to increase the stability response and thereby obtaining a problem with higher accuracy demands on the nonlinear solution techniques, in the point A is assumed the action of a small harmonic lateral force $P = P_0 e^{i\Omega t}$, with constant amplitude $P_0 = 10\,\mathrm{N}$ and with circular frequency Ω variable in the interval 0—8000 Hz. This influence is analysed in the interaction with the axial forces in the compression zone of the external skin due to the aforementioned exploitation and dead-weight loads acting on the structure. The graphical solution of the studied problem of stability response is illustrated in Figure 113. The relation between axial forces, inducing resonance frequencies and vertical flexural displacements in analysed node A is shown. The shaded area in Figure 113 represents the sought zone of instability with a distinctly nonlinear configuration in critical and resonance regions of the diagram. The further modification of the relation of axial forces, frequencies and vertical flexural displacements in node A is plotted in Figure 114. The parameter α_i is the ratio of inducing frequency and the first resonance frequency. The value of the local

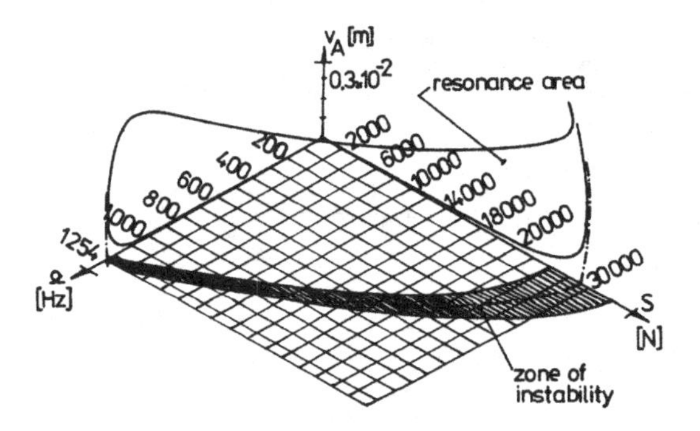

Fig. 113. The functional relation of axial forces, inducing frequencies and vertical flexural displacements in the analysed node A in the interval of axial forces $S = 0 - 31\,\mathrm{kN}$ and in the frequency interval $\Omega = 0 - 1260\,\mathrm{Hz}$.

Table VIII. Resulting spatial stresses in indicated nodal points of the studied experimental prototype of the transporter shell bridge (all stresses in MPa)

Transverse discretization	Longitudinal discretization						
	1	2	3	11	12	20	22
1	6.08	12.06	21.55	83.44	88.65	113.61	114.65
2	10.54	14.94	23.20	84.15	89.33	113.99	115.02
3	15.01	18.41	25.24	84.26	89.36	113.49	114.50
4	19.38	22.10	27.50	83.60	88.57	111.86	112.83
5	23.75	25.98	30.01	82.42	87.22	109.41	110.33
6	27.75	29.61	32.43	80.42	84.97	105.86	106.54
7	31.79	33.36	35.05	78.24	82.54	101.64	102.43
8	35.51	36.81	37.41	75.21	79.23	96.61	97.33
9	39.03	40.12	39.72	71.79	75.51	90.98	91.62
10	42.23	43.11	41.77	67.48	70.84	84.08	84.63
11	45.19	45.90	43.73	63.08	66.05	76.90	77.35
12	47.87	48.43	45.52	58.48	61.07	69.34	69.69
13	50.27	50.71	47.18	54.13	56.33	61.89	62.12
14	52.43	52.68	48.61	49.65	51.44	53.94	54.05
15	54.03	54.25	49.61	43.97	45.35	44.59	44.56
16	55.44	55.58	50.51	39.15	40.14	35.65	35.45
17	56.72	56.60	51.20	34.86	35.46	26.50	26.06
18	57.27	57.30	51.68	31.43	31.70	17.46	16.61
19	57.69	57.70	51.94	29.23	29.27	8.40	6.15
20	57.74	57.74	51.97	28.98	28.99	6.70	3.44
21	57.41	57.43	51.76	30.53	30.71	14.45	13.35
22	56.75	56.81	51.35	33.81	34.27	24.11	23.59
23	55.77	55.89	50.73	38.24	39.99	33.89	33.66
24	54.40	54.61	49.88	43.32	44.71	43.46	43.42
25	52.66	52.99	48.83	49.36	51.32	53.81	53.93
26	50.55	50.99	47.44	54.63	56.15	62.74	62.99
27	48.15	48.71	45.75	58.79	61.66	70.02	70.39
28	45.47	46.13	43.79	62.21	65.17	76.21	76.68
29	42.46	43.25	41.60	65.58	68.88	82.30	82.87
30	39.27	40.20	39.31	68.46	72.43	87.33	87.97
31	35.88	36.95	36.79	70.22	74.21	90.91	91.61
32	32.31	33.54	34.19	71.56	73.94	93.63	94.38
33	28.69	30.09	31.56	72.18	73.10	95.25	96.04
34	24.97	26.61	29.02	72.85	74.10	96.77	97.59
35	21.17	23.11	26.60	73.57	75.61	98.28	99.13
36	17.31	19.66	24.28	73.81	76.35	99.13	100.00
37	13.46	16.35	22.18	73.72	77.12	99.44	100.33
38	9.55	13.24	20.25	73.13	77.95	99.20	100.09
39	6.08	10.94	18.99	72.81	78.25	99.12	100.03
40	2.60	9.34	18.09	71.12	78.20	98.49	99.40
41	1.31	8.89	17.63	71.26	78.20	97.67	98.59

Table VIII (Continued)

Transverse discretization	Longitudinal discretization						
	1	2	3	11	12	20	22
42	5.16	10.09	17.95	70.58	75.10	96.86	97.77
43	9.02	12.38	18.81	69.50	73.97	95.44	96.35
44	12.82	15.27	20.19	68.19	72.57	93.57	94.45
45	16.54	18.40	21.96	66.94	71.23	91.69	92.56
46	20.16	21.61	23.92	65.32	69.50	89.36	90.20
47	23.73	24.86	25.98	62.89	66.93	85.86	86.67
48	27.12	27.96	27.95	59.39	63.21	80.94	81.71
49	30.31	30.94	29.99	55.54	59.09	75.21	75.92
50	33.32	33.76	31.87	50.41	53.61	67.72	68.35
51	36.00	36.29	33.68	45.55	48.39	60.33	60.89
52	38.40	38.59	35.39	40.61	43.04	52.35	52.81
53	40.51	40.63	36.97	35.82	37.76	43.77	44.10
54	42.25	42.32	38.32	31.53	32.94	34.96	35.11
55	43.62	43.65	39.42	28.29	29.23	27.13	28.10
56	44.60	44.62	40.22	25.55	26.06	19.13	18.83
57	45.26	45.27	40.76	23.73	23.90	11.53	10.83
58	45.59	45.59	41.04	22.93	22.95	5.69	3.44
59	45.54	45.55	41.00	23.06	23.11	6.95	5.31
60	45.12	45.13	40.65	24.19	24.46	13.80	13.20
61	44.37	44.39	40.02	25.90	26.49	20.61	20.36
62	43.33	43.37	39.18	28.57	29.58	28.16	28.18
63	41.92	41.99	38.03	31.63	33.08	35.52	35.69
64	40.18	40.28	36.60	34.68	36.59	42.50	42.83
65	38.12	38.29	35.06	39.51	41.88	51.93	51.38
66	35.72	35.99	33.32	44.22	47.01	58.64	59.29
67	33.04	33.43	31.43	49.04	52.24	66.46	68.13
68	30.08	30.64	29.47	53.95	57.53	74.00	74.73
69	26.83	27.62	27.45	58.79	62.73	81.26	82.08
70	23.35	24.34	25.44	63.09	67.33	86.53	88.40
71	19.64	21.12	23.52	67.11	71.62	93.28	94.20
72	15.60	17.72	21.92	71.31	76.04	98.86	99.83
73	11.46	14.61	20.86	75.42	80.36	104.20	105.21
74	7.23	11.97	20.16	78.39	83.47	108.99	109.02
75	2.86	10.33	20.02	80.76	85.94	110.89	111.94
76	1.61	10.33	20.47	82.36	87.57	112.66	113.81

resonance frequency of the external skin in the compression zone midspan of two stiffeners is 1251 Hz. The parameter $\beta_i = S_i/S_{cr}$ is the ratio of axial force S_i versus critical force S_{cr}. The value of S_{cr} for the analysed zone is 29.835 kN. The parameter γ determines the ratio of dynamic and static flexural displacements in node A. In Figure 114 the influence of nonlinear effects on the post-buckling resonance behaviour of the external skin is shown. The resonance peak B and the corresponding line C express the dynamic behaviour covering the geometrically nonlinear effects in the studied case. The behaviour, when the interactions of all geometric and physical nonlinearities are taken into consideration, is depicted by line V in Figure 114. The figure visualizes furthermore the functional shape of the interacting resonance and critical curves in typical sections M and N of the presented diagram.

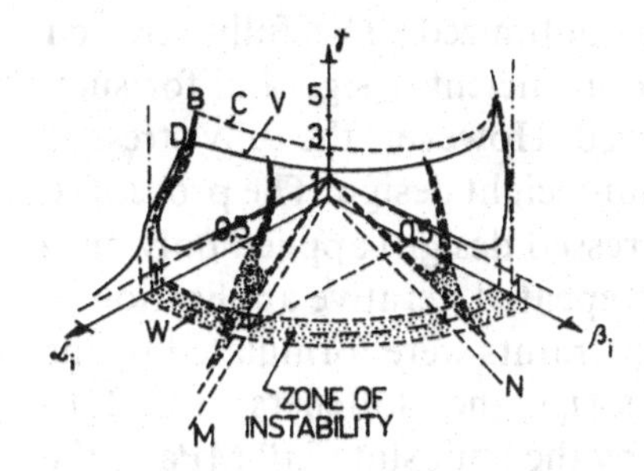

Fig. 114. Post-buckling resonance behaviour of the external skin in node A.

A detailed spatial shape of the studied resonance area in section M is shown in Figure 115. The nonlinear character of the resonance response in the post-buckling regions of vibration is noteworthy. For the region M the different shape of concave and convex zones of an interactive spatial resonance curve and of the projection of the separate resonance curve into plane $\Omega_M - v_A$ is typical, as shown in Figure 115. This effect is initiated by various nonlinear characteristics in the resonance and critical regions of spatial behaviour of the analysed shell bridge.

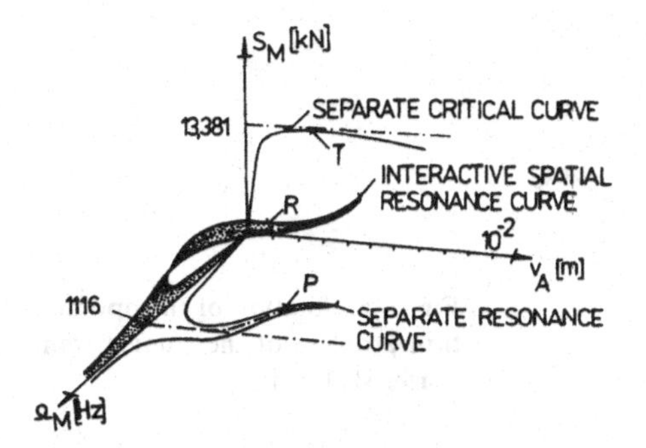

Fig. 115. Interactive spatial post-buckling resonance curve in node A (section M in Figure 114).

For the present numerical analysis all terms contained in the CTM were used.

Further details of the theoretical and numerical solutions of this case are given in references [91, 95].

4.10. Optimization

Transporter Shell Bridge — Structural Synthesis and Optimization

Taking account of the present theoretical and numerical analyses, the numerical approach to structural synthesis and optimization procedures for the transporter shell bridge studied in the previous section is performed. The variability of the thickness of the corrugated sheets of the external skin versus the thickness of the transversal stiffeners is optimized. The fully stressed design for some loading conditions led to an inefficient design and for such cases the minimum-weight design was preferred. However, the fully stressed design was not far removed from the minimum-weight design. The procedure used, was made up of an initial step (fully stressed design) applied only once at the start of the computation, followed by repeated iterative application of the minimum-weight concept. Geometric constraints were formulated by the minimum and maximum thickness of sheets (t_p) and stiffeners (t_v), 2 to 8 mm for both. Stress constraints were given by the limit state 210 MPa of the applied material. The maximum value of the displacement constraint was assumed to be $L/400$, where L is the span of the studied bridge. The typical constraint surfaces of the studied optimization problem in three dimensions are illustrated in Figure 116. The solution of this optimization problem is schematically depicted in Table IX.

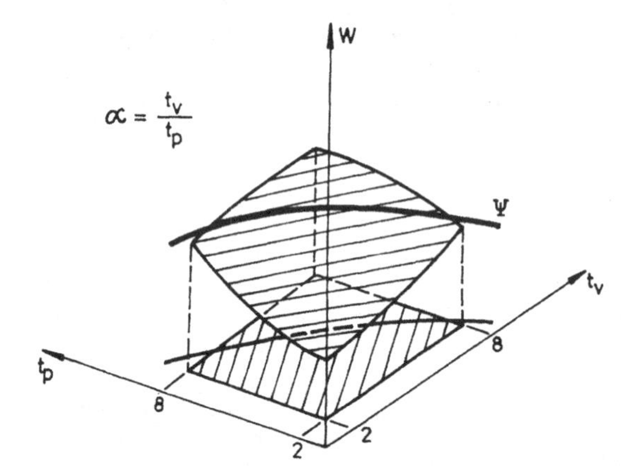

Fig. 116. Solution of the optimization problem of the studied transporter shell bridge.

The geometric shape of the shell bridge, optimized when considering the location of two interior transporter belts is ilustrated in Figure 117 and 118.

For the present numerical analysis all terms in the CTM were used.

Further details of the theoretical and numerical solutions of the present case are given in references [94, 96].

Table IX. Solution of optimization problem of the transporter shell bridge

α	Load parameter
1.5	1.000
2.0	0.938
2.5	0.892
3.0	0.858
3.5	0.790
4.0	0.683

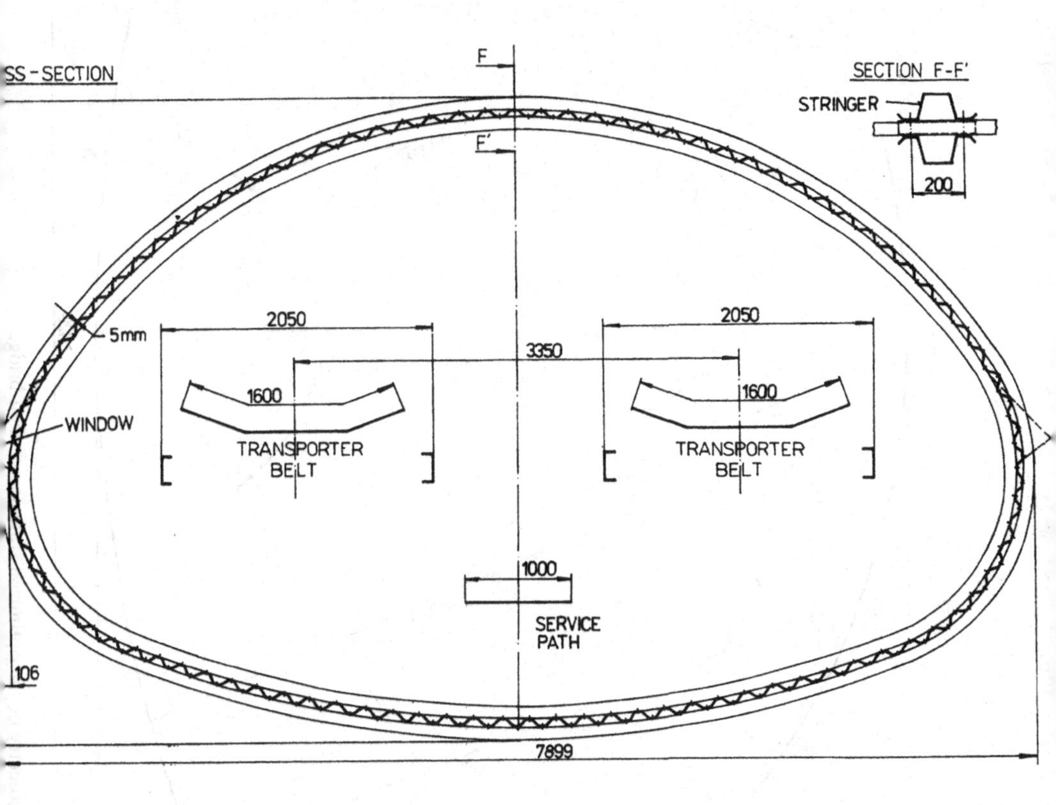

Fig. 117. Geometric shape of optimized transporter shell bridge with two transporter belts.

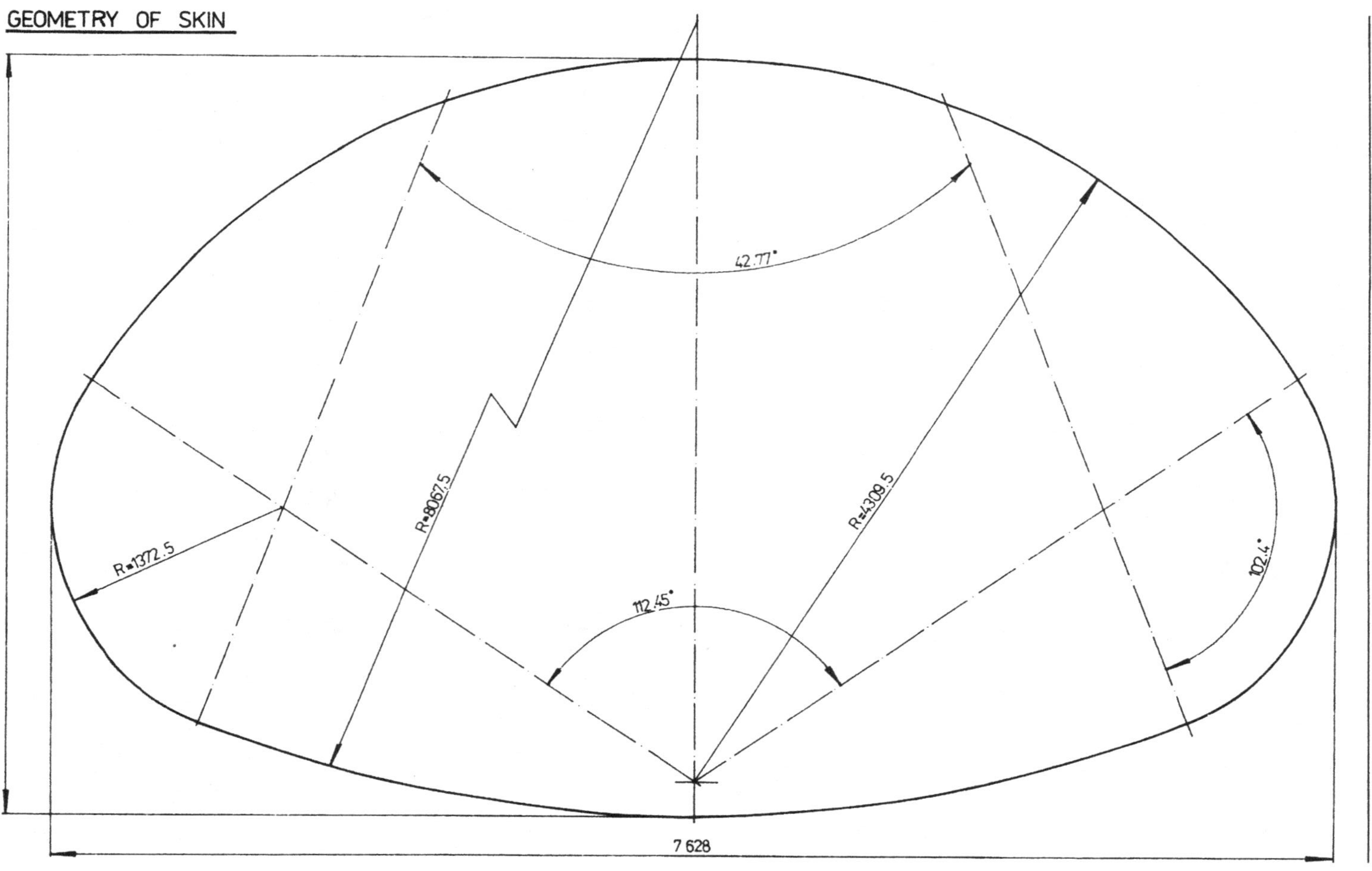

Fig. 118. Cross-sectional geometry of the optimized transporter shell bridge.

Chapter 5
Conclusions

If any predictions about possible future trends in the discrete analysis of structures are to be made, they must be based on the assumption that research and development in the area will be responsive to the needs of the engineering community. More efficient general purpose programs, derived in such a way that they are reasonably easy to use, are definitely needed.

Some new concepts were developed recently in the form of new ideas in discretization procedures and in the automation of numerical techniques. The prediction is simply that these are going to be developed further, thoroughly evaluated and included in commonly available software of computers or personal computers.

Further, the development of more efficient procedures for linear and nonlinear analyses continues to be needed. We are no longer satisfied with being able to predict whether or not a structure will endure a certain loading environment without damage, but often we also expect to be able to assess the damage in the case of an accidental overloading.

Because the efficiency of the available solution procedures is highly case-dependent, an optimally efficient computer program must contain a number of options and consequently requires an increasing degree of user sophistication. The quality of structural analysis would be greatly improved if more efficient solution procedures were derived and introduced into a single computer code, together with a way to select automatically from the available options the procedures that are best suited for each particular case.

The new advances are manifested further by the development of new material laws, new structural theories, sophisticated mathematical models, efficient optimization techniques and numerical algorithms as well as versatile and powerful software systems for structural analysis and design.

The driving force behind these activities has been, and continues to be, the need for realistic modelling, accurate analysis and optimized design of large complex structures subject to harsh environments. However, the rapid pace of the development is largely due to the opportunities provided by the extensive advances in computer and personal computer hardware and software technology as well as the growing interaction among a number of disciplines including applied mechanics, numerical analysis, software design and structural engineering.

This book, hopefully, is one step in direction of these future strivings.

CATALOGUE OF TRANSFER MATRICES

Contents

B-10 Plane and curved grillage — submatrices of transverse interaction
B-11 Plane grillage — in-plane behaviour — bending — transfer submatrix
B-12 Plane grillage — in-plane behaviour — bending, tension — transfer submatrix
B-13 Plane grillage — in-plane behaviour — bending, compression — transfer submatrix
B-14 Plane grillage — in-plane behaviour — transfer matrix
B-15 Plane grillage — in-plane behaviour — matrix of transverse interaction
B-16 Plane grillage — in-plane behaviour — submatrices of transverse interaction
B-17 Shell grillage — flexural behaviour — transfer matrix
B-18 Shell grillage — flexural behaviour — matrix of transverse interaction
B-19 Shell grillage — submatrices of transverse interaction
B-20 Three-dimensional grillage — transfer matrix
B-21 Three-dimensional grillage — matrix of transverse interaction
B-22 Three-dimensional grillage — submatrices of transverse interaction
B-23 Three-dimensional grillage — matrices of initial parameters
B-24 Three-dimensional grillage — matrices of boundary conditions

C. PLANE SUBSTRUCTURE

C-1 Flexural and torsional behaviour — transfer submatrix
C-2 Flexural and torsional behaviour — transfer matrix
C-3 Matrix of transverse interaction
C-4 Submatrix of transverse interaction P1
C-5 Submatrix of transverse interaction P2
C-6 Submatrix of transverse interaction P3 with substitutions
C-7 Incorporation of plane matrices into matrix of transverse interaction for shell systems
C-8 Plane substructure — nodal matrix
C-9 Plane substructure — nodal submatrices
C-10 Plane substructure — matrices of initial parameters
C-11 Plane substructure — matrices of boundary conditions

D. SHELL SUBSTRUCTURE

D-1 Spatial behaviour — transfer submatrix

E. CORRUGATED SHEET SUBSTRUCTURE

F. NONLINEAR MACROELEMENT

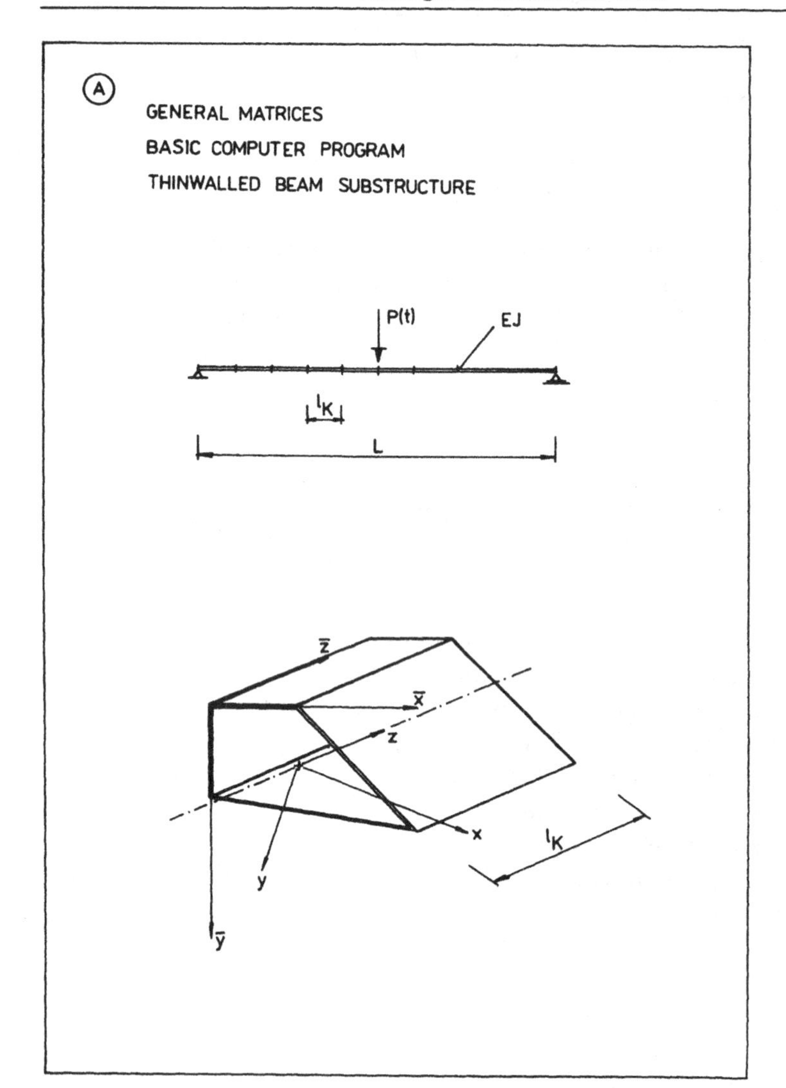
A
GENERAL MATRICES
BASIC COMPUTER PROGRAM
THINWALLED BEAM SUBSTRUCTURE
P(t)
EJ
l_K
L
$\bar{z}$
$\bar{x}$
z
x
y
$\bar{y}$
l_K

A-1

SIMPLY SUPPORTED BEAM – FLEXURAL BEHAVIOUR

TRANSFER MATRIX

BK =

1	l_K	$-\frac{l_K^2}{2EJ}$	$-\frac{l_K^3}{6EJ}$	
	1	$-\frac{l_K}{EJ}$	$-\frac{l_K^2}{2EJ}$	
		1	l_K	
			1	
				1

NODAL MATRICES

CK1=

1				
	1			
		1		
$-m\omega^2$			1	
				1

CK2=

1				
	1			
		1		
$-m\omega^2$			1	$\mp P$
				1

MATRIX OF INITIAL PARAMETERS

HL =

1		
	1	
		1

MATRIX OF BOUNDARY CONDITIONS

RP=

1				
		1		

A-2

```
      C      TIME RESPONSE OF SINGLE BEAM
      C      TRANSFER MATRIX VERSUS WILSON
             METHOD
      C      FLEXURAL VIBRATION — BASIC PROGRAM
      C      ------------------------------------------
      C      NUL     — IDENTIFICATION NUMBER
      C      E       — YOUNG MODULUS
      C      DE      — PARAMETER OF INTERNAL
                       DAMPING
      C      XL      — LENGTH OF ELEMENT
      C      ZX      — FLEXURAL RIGIDITY
      C      HM      — NODAL INERTIAL MASS
      C      NELEM   — NUMBER OF ELEMENTS
      C      NPOIN   — NUMBER OF NODES
      C      NLOAD   — NODE OF ACTING FORCE
      C      NOP     — INITIAL PARAMETERS (2, 4, 5)
      C      NPP     — BOUNDARY CONDITIONS (1, 3)
      C      NTIME   — NUMBER OF TIME STEPS
      C      STEP    — LENGTH OF TIME STEP
      C      PFOR    — VALUES OF INDUCING FORCE IN
                       TIME
      C      ------------------------------------------
0001         COMPLEX EE,EE1, JC, LKK, UN
0002         DIMENSION NOP(3), NPP(2), PFOR(200), PEF(100),
            +DISP(100), RYCH(100), ZRCH(100, DISTHE(100),
            +VZRCH(100), VRYCH(100),VDISP(100)
0003         READ 100,NUL
0004         READ 102,E,DE,XL,ZX
0005         READ 101,HM
0006         READ 100,NELEM,NPOIN,NLOAD
0007         READ 100,(NOP(I),I = 1,3)
0008         READ 100,(NPP(I),I = 1,2)
0009         DO 10 NNN = 1,NUL
0010         READ 100,NTIME
0011         READ 101,STEP
0012         READ 101,(PFOR(I),I = 1,NTIME)
0013         THE = 1.37
0014         A0 = 6./(THE*STEP)**2
0015         A1 = 3./(THE*STEP)
0016         A2 = 2.*A1
```

```
ØØ17        A3 = THE*STEP/2.
ØØ18        A4 = AØ/THE
ØØ19        A5 = - A2/THE
ØØ2Ø        A6 = 1. - 3./THE
ØØ21        A7 = STEP/2
ØØ22        A8 = STEP*STEP/6.
ØØ23        D = (DE/3.14)*E
ØØ24        HM1 = HM*AØ
ØØ25        DF = A1*D
ØØ26        EE = CMPLX(E,D)
ØØ27        EE1 = CMPLX(E,DF)
ØØ28        JC = CMPLX(ZX,Ø.)
ØØ29        LKK = CMPLX(XL,Ø.)
ØØ3Ø        UN = CMPLX(1.,Ø.)
ØØ31        PRINT 2Ø3
ØØ32        PRINT 1Ø1,(PFOR(I),I = 1,NTIME)
ØØ33        PRINT 1ØØ,(NOP(I),I = 1,3)
ØØ34        PRINT 2ØØ,E,DE,XL,ZX
ØØ35        PRINT 1ØØ,(NPP(I),I = 1,2)
ØØ36        PRINT 2ØØ,HM,STEP
ØØ37        PRINT 1ØØ,NELEM,NPOIN,NLOAD,NTIME
ØØ38        PRINT 2ØØ,AØ,A1,A2,A3,A4,A5,A6,A7,A8
ØØ39        DO 3 LL = 1,NPOIN
ØØ4Ø        PEF(LL) = Ø.
ØØ41        IF(LL.EQ.NLOAD) PEF(LL) = PFOR(1)
ØØ42      3 CONTINUE
ØØ43        NST = NTIME - 1
ØØ44        DO 1 J = 1,NST
ØØ45        IF(J.NE.1) GO TO 4
ØØ46        CALL COMPL(DISP,HM,PEF,LKK,EE,JC,UN,
          + NELEM,NPP,NOP,NPOIN,STEP,NLOAD)
ØØ47        PRINT 2Ø2
ØØ48        TIME = Ø.
ØØ49        PRINT 2Ø1,J,TIME,(DISP(NN),NN = 1,NPOIN)
ØØ5Ø        DO 2 I = 1,NPOIN
ØØ51        RYCH(I) = Ø.
ØØ52        ZRCH(I) = Ø.
ØØ53      2 CONTINUE
ØØ54      4 KK = J + 1
ØØ55        DO 6 I = 1,NPOIN
ØØ56        IF(I.EQ.NLOAD) PEF(I) = PFOR(J) + THE
```

```
          + *(PFOR(KK) - PFOR(J)) + HM*(A0*DISP(I)
          + + A2*RYCH(I) + 2.*ZRCH(I)) + DE/3.14*(A1
          + *DISP(I) + 2.*RYCH(I) + A3*ZRCH(I))
0057        IF(I.NE.NLOAD) PEF(I) = HM*(A0*DISP(I) +
          + A2*RYCH(I) + 2.*ZRCH(I)) + DE/3.14*(A1*
          + DISP(I) + 2.*RYCH(I) + A3*ZRCH(I))
0058      6 CONTINUE
0059        CALL COMPL(DISTHE,HM1,PEF,LKK,EE1,JC,
          + UN,NELEM,NPP,NOP,NPOIN,STEP,NLOAD)
0060        DO 5 I = 1,NPOIN
0061        VZRCH(I) = A4*(DISTHE(I) - DISP(I)) +
          + A5*RYCH(I) + A6*ZRCH(I)
0062        VRYCH(I) = RYCH(I) + A7*(VZRCH(I) + ZRCH(I))
0063        VDISP(I) = DISP(I) + STEP*RYCH(I) +
          + A8*(VZRCH(I) + 2.*ZRCH(I))
0064        DISP(I) = VDISP(I)
0065        RYCH(I) = VRYCH(I)
0066        ZRCH(I) = VZRCH(I)
0067      5 CONTINUE
0068        DISP(NPOIN) = 0.
0069        TIME = J*STEP
0070        JI = J + 1
0071        PRINT 201,JI,TIME,(DISP(NN),NN = 1,NPOIN)
0072        PRINT 201,JI,TIME,(RYCH(NN),NN = 1,NPOIN)
0073        PRINT 201,JI,TIME,(ZRCH(NN),NN = 1,NPOIN)
0074      1 CONTINUE
0075     10 CONTINUE
0076    100 FORMAT(10I5)
0077    101 FORMAT(7F10.0)
0078    102 FORMAT(4F20.6)
0079    200 FORMAT((5X,5E12.4))
0080    203 FORMAT(///,5X,'CONTROL DATA OUTPUT —
          + WILSON INTEGRATION',/,5X,42(1H - ),//)
0081    202 FORMAT(///,1X,'STEP',4X,'TIME',3X'STATE
          + VECTORS IN NODES',/,1X,42(1H - ),/)
0082    201 FORMAT(1X, I4,2X,F4.2,2X,5E10.3,(/,13X,5E10.2))
0083        STOP
0084        END
      C
      C     ________________________________________
      C
```

```
      C          TRANSFER MATRIX METHOD ALGORITHM
      C
ØØØ1             SUBROUTINE COMPL(DISP,HM,PEF,LKK,EE,JC
               + UN,
               + NELEM,NPP,NOP,NPOIN,STEP,NLOAD)
ØØØ2             COMPLEX BK,CK,BKK,BKJ,SR,VKO,VKK
ØØØ3             COMPLEX UN,LKK,EE,JC
ØØØ4             DIMENSION BK(5,5),CK(5,5),BKK(5,5), BKJ(5,5),
               + SR(2,3),NPP(2),NOP(3),A(4,4),B(4),BEA(2,2),
               + BER(2,2),VKO(5),VKK(5),DISP(NPOIN),
               + PEF(NPOIN)
ØØØ5             CALL BKMTRX(BK,LKK,EE,JC,UN)
ØØØ6             DO 1 I=1,5
ØØØ7             DO 1 II=1,5
ØØØ8             BKK(I,II)=BK(I,II)
ØØØ9           1 CONTINUE
ØØ1Ø             NS=NELEM-1
ØØ11             DO 2 I=1,NS
ØØ12             II=I+1
ØØ13             P=PEF(II)
ØØ14             CALL CMTRX(CK,UN,P,HM,STEP)
ØØ15             CALL DMULTC(BKK,CK,BKJ,5,5,5)
ØØ16             CALL DMULTC(BKJ,BK,BKK,5,5,5)
ØØ17           2 CONTINUE
ØØ18             CALL BOUND(BKK,SR,NPP,NOP,5,2,3)
ØØ19             CALL TRANSF(A,B,2,3,4,SR,BEA,BER)
ØØ2Ø             CALL SIMQ(A,B,4,KS)
ØØ21             IF(KS.EQ.1) PRINT 2ØØ
ØØ22         2ØØ FORMAT(//,5X,'FAILS STABILITY OF SYSTEM',/)
ØØ23             CALL NULMTR(VKO,5,1)
ØØ24             DO 3 I=1,2
ØØ25             II=NOP(I)
ØØ26             IJ=2*I
ØØ27             IK=2*I-1
ØØ28           3 VKØ(II)=CMPLX(B(IK),B(IJ))
ØØ29             VKØ(5)=UN
ØØ3Ø             DISP(1)=REAL(VKO(1))
ØØ31             DO 5 I=1,NELEM
ØØ32             II=I+1
ØØ33             CALL DMULTC(BK,VKO,VKK,5,5,1)
ØØ34             IF(I.EQ.NELEM) DISP(I)=REAL(VKK(1))
```

```
0035        IF(I.EQ.NELEM) GO TO 4
0036        P = PEF(II)
0037        CALL CMTRX(CK,UN,P,HM,STEP)
0038        CALL DMULTC(CK,VKK,VKO,5,5,1)
0039        DISP(II) = REAL(VKO(1))
0040      4 CONTINUE
0041      5 CONTINUE
0042        RETURN
0043        END
     C
     C      ------------------------------------------------
     C
     C      NODAL MATRIX
     C
0001        SUBROUTINE CMTRX(CK,UN,P,HM,STEP)
0002        COMPLEX CK,UN,PC,HMC
0003        DIMENSION CK(5,5)
0004        CALL NULMTR(CK,5,5)
0005        DO 1 I = 1,5
0006      1 CK(I,I) = UN
0007        PC = CMPLX(P,0.)
0008        HMC = CMPLX(HM,0.)
0009        CK(4,5) = -PC
0010        CK(4,1) = -HMC
0011        RETURN
0012        END
     C
     C      ------------------------------------------------
     C
     C      TRANSFER MATRIX
     C
0001        SUBROUTINE BKMTRX(BK,LKK,EE,JC,UN)
0002        COMPLEX BK,LKK,EE,JC,UN
0003        DIMENSION BK(5,5)
0004        CALL NULMTR(BK,5,5)
0005        DO 1 I = 1,5
0006      1 BK(I,I) = UN
0007        BK(1,2) = LKK
0008        BK(1,3) = -LKK*LKK/(CMPLX(2.,0.)*EE*JC)
0009        BK(1,4) = -LKK*LKK*LKK/(CMPLX(6.,0.)*EE*JC)
0010        BK(2,3) = -LKK/(EE*JC)
```

```
0011          BK(2,4)=BK(1,3)
0012          BK(3,4)=LKK
0013          RETURN
0014          END
      C
      C       ________________________________________
      C
      C       MATRIX OF BOUNDARY CONDITIONS
      C
0001          SUBROUTINE BOUND(BKK,SR,NPP,NOP,NB,NS,N1)
0002          COMPLEX BKK,SR
0003          DIMENSION BKK(NB,NB),SR(NS,N1),NPP(NS),
             +NOP(N1)
0004          CALL NULMTR(SR,NS,N1)
0005          DO 1 I=1,NS
0006          J=NPP(I)
0007          IF(J.EQ.0) STOP 2
0008          DO 1 II=1,N1
0009          JJ=NOP(II)
0010          IF(JJ.EQ.0) STOP 3
0011          SR(I,II)=BKK(J,JJ)
0012        1 CONTINUE
0013          RETURN
0014          END
      C
      C       ________________________________________
      C
      C       SOLUTION OF SYSTEM OF EQUATIONS
      C
0001          SUBROUTINE SIMQ(A,B,N,KS)
0002          DIMENSION A(1),B(1)
0003          TOL=0.
0004          KS=0
0005          JJ=-N
0006          DO 65 J=1,N
0007          JY=J+1
0008          JJ=JJ+N+1
0009          BIGA=0.
0010          IT=JJ-J
0011          DO 30 I=J,N
0012          IJ=IT+I
```

```
0013          IF(ABS(BIGA) - ABS(A(IJ))) 20,30,30
0014        2BIGA = A(IJ)
           0
0015          IMAX = I
0016       30 CONTINUE
0017          IF(ABS(BIGA) - TOL) 35,35,40
0018       35 KS = 1
0019          RETURN
0020       40 I1 = J + N*(J-2)
0021          IT = IMAX - J
0022          DO 50 K = J,N
0023          I1 = I1 + N
0024          I2 = I1 + IT
0025          SAVE = A(I1)
0026          A(I1) = A(I2)
0027          A(I2) = SAVE
0028       50 A(I1) = A(I1)/BIGA
0029          SAVE = B(IMAX)
0030          B(IMAX) = B(J)
0031          B(J) = SAVE/BIGA
0032          IF(J - N) 55,70,55
0033       55 IQS = N*(J - 1)
0034          DO 65 IX = JY,N
0035          IXJ = IQS + IX
0036          IT = J - IX
0037          DO 60 JX = JY,N
0038          IXJX = N*(JX - 1) + IX
0039          JJX = IXJX + IT
0040       60 A(IXJX) = A(IXJX) - (A(IXJ)*A(JJX))
0041       65 B(IX) = B(IX) - (B(J)*A(IXJ))
0042       70 NY = N - 1
0043          IT = N*N
0044          DO 80 J = 1,NY
0045          IA = IT - J
0046          IB = N - J
0047          IC = N
0048          DO 80 K = 1,J
0049          B(IB) = B(IB) - A(IA)*B(IC)
0050          IA = IA - N
0051       80 IC = IC - 1
0052          RETURN
```

```
ØØ53          END
        C
        C     ________________________________________
        C
        C     MATRIX ELIMINATION
        C
ØØØ1          SUBROUTINE NULMTR(MATRIX,M,N)
ØØØ2          COMPLEX MATRIX
ØØØ3          DIMENSION MATRIX(M,N)
ØØØ4          DO 1 I=1,M
ØØØ5          DO 1 II=1,N
ØØØ6        1 MATRIX(I,II)=CMPLX(Ø.,Ø.)
ØØØ7          RETURN
ØØØ8          END
        C
        C     ________________________________________
        C
        C     MATRIX MULTIPLICATION
        C
ØØØ1          SUBROUTINE DMULTC(A,B,C,N1,N2,N3)
ØØØ2          COMPLEX A(1),B(1),C(1),D
ØØØ3          DO 1 I=1,N1
ØØØ4          DO 1 III=1,N3
ØØØ5          D=CMPLX(Ø.,Ø.)
ØØØ6          DO 2 II=1,N2                DO 2 II = 1,N2
ØØØ7          K1=(II-1)*N1+I
ØØØ8          K2=(III-1)*N2+II
ØØØ9        2 D=D+A(K1)*B(K2)
ØØ1Ø          K3=(III-1)*N1+I
ØØ11        1 C(K3)=D
ØØ12          RETURN
ØØ13          END
        C
        C     ________________________________________
        C
        C     TRANSFORMATION MATRIX
        C
ØØØ1          SUBROUTINE TRANSF(A,B,NSYS,NPS,NCS,
            + SYSMAT,A1,A2)
ØØØ2          COMPLEX SYSMAT
ØØØ3          DIMENSION SYSMAT(NSYS,NPS),A1(NSYS,NSYS),
```

```
        + A2(NSYS,NSYS),A(NCS,NCS),B(NCS)
ØØØ4      DO 1 I=1,NSYS
ØØØ5      DO 1 II=1,NSYS
ØØØ6      A1(I,II)=AIMAG(SYSMAT(I,II))
ØØØ7      A2(I,II)=REAL(SYSMAT(I,II))
ØØØ8    1 CONTINUE
ØØØ9      DO 2 I=1,NSYS
ØØ1Ø      K=2*I-1
ØØ11      L=2*I
ØØ12      NHEL=NCS-1
ØØ13      DO 2 II=1,NHEL,2
ØØ14      KK=II+1
ØØ15      J=KK/2
ØØ16      B(II)= -REAL(SYSMAT(J,3))
ØØ17      B(KK)=AIMAG(SYSMAT(J,3))
ØØ18      A(K,II)=A2(I,J)
ØØ19      A(L,II)=A1(I,J)
ØØ2Ø      A(K,KK)= -A1(I,J)
ØØ21      A(L,KK)=A2(I,J)
ØØ22    2 CONTINUE
ØØ23      RETURN
ØØ24      END
```

A-3

THINWALLED BEAM SUBSTRUCTURE
AXIAL BEHAVIOUR

STATE VECTOR $k_K = (v^S_{Z,K}, N^S_{Z,K})^T$

TRANSFER MATRIX

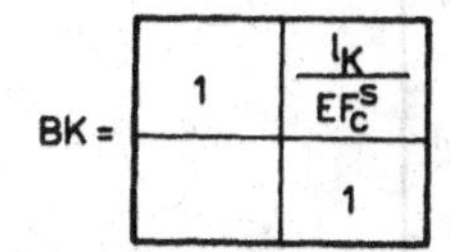

$$BK = \begin{bmatrix} 1 & \frac{l_K}{EF^S_C} \\ & 1 \end{bmatrix}$$

A-4

THINWALLED BEAM SUBSTRUCTURE
FLEXURAL BEHAVIOUR IN DIRECTION i (x OR y)

STATE VECTOR $k_K = (v^S_{i,K}, \varphi^S_{i,K}, M^S_{i,K}, Q^S_{i,K})^T$

TRANSFER MATRIX

$$BK = \begin{bmatrix} 1 & l_K & \frac{-l_K^2}{2EF^S_{ii}} & \frac{-l_K^3}{6EF^S_{ii}} \\ & 1 & \frac{-l_K}{EF^S_{ii}} & \frac{-l_K^2}{2EF^S_{ii}} \\ & & 1 & l_K \\ & & & 1 \end{bmatrix}$$

A-5

THINWALLED BEAM SUBSTRUCTURE
FLEXURAL BEHAVIOUR IN DIRECTION i
AND TENSION - TRANSFER MATRIX

STATE VECTOR $(v^S_{i,K}, \varphi^S_{i,K}, M^S_{i,K}, Q^S_{i,K})^T$ TENSION FORCE P_T

BK =

1	$l_K \frac{\sinh\gamma}{\gamma}$	$\frac{l_K^2}{EF^S_{ii}} \frac{\cosh\gamma - 1}{\gamma^2}$	$\frac{l_K^3}{EF^S_{ii}} \frac{\sinh\gamma - \gamma}{\gamma^3}$
	$\cosh\gamma$	$\frac{l_K}{EF^S_{ii}} \frac{\sinh\gamma}{\gamma}$	$\frac{l_K^2}{EF^S_{ii}} \frac{\cosh\gamma - 1}{\gamma^2}$
	$P_T l_K \frac{\sinh\gamma}{\gamma}$	$\cosh\gamma$	$l_K \frac{\sinh\gamma}{\gamma}$
			1

$$\gamma = \sqrt{\frac{P_T l_K^2}{EF^S_{ii}}}$$

A-6

THINWALLED BEAM SUBSTRUCTURE
FLEXURAL BEHAVIOUR IN DIRECTION i
AND COMPRESSION - TRANSFER MATRIX

STATE VECTOR $(v^S_{i,K}, \varphi^S_{i,K}, M^S_{i,K}, Q^S_{i,K})^T$ COMPRESSION FORCE P_C

BK =

1	$l_K \frac{\sin\gamma}{\gamma}$	$\frac{l_K^2}{EF^S_{ii}} \frac{1 - \cos\gamma}{\gamma^2}$	$\frac{l_K^3}{EF^S_{ii}} \frac{\gamma - \sin\gamma}{\gamma^3}$
	$\cos\gamma$	$\frac{l_K}{EF^S_{ii}} \frac{\sin\gamma}{\gamma}$	$\frac{l_K^2}{EF^S_{ii}} \frac{1 - \cos\gamma}{\gamma^2}$
	$-P_C l_K \frac{\sin\gamma}{\gamma}$	$\cos\gamma$	$l_K \frac{\sin\gamma}{\gamma}$
			1

$$\gamma = \sqrt{\frac{P_C l_K^2}{EF^S_{ii}}}$$

A-7

THINWALLED BEAM SUBSTRUCTURE
TORSION-BENDING BEHAVIOUR
TRANSFER MATRIX

STATE VECTOR $k_K = (\vartheta^s_{T,K}, \varphi^s_{T,K}, W^s_{T,K}, T^s_{T,K})^T$

$$\lambda^s_I = \sqrt{\frac{GJ_{TB}}{EF^s_{w_I w_I}}\,\varkappa_{w_I w_I}}$$

BK =

1	$\frac{\sinh(\lambda^s_I l_K)}{\lambda^s_I}$	$\frac{1-\cosh(\lambda^s_I l_K)}{(\lambda^s_I)^2 EF^s_{w_I w_I}}$	$\frac{\lambda^s_I l_K - \sinh(\lambda^s_I l_K)}{(\lambda^s_I)^3 EF^s_{w_I w_I}}$
	$\cosh(\lambda^s_I l_K)$	$-\frac{\sinh(\lambda^s_I l_K)}{\lambda^s_I EF^s_{w_I w}}$	$\frac{1-\cosh(\lambda^s_I l_K)}{EF^s_{w_I w_I}}$
	$-\lambda^s_I \sinh(\lambda^s_I l_K) EF^s_{w_I w_I}$	$\cosh(\lambda^s_I l_K)$	$\frac{\sinh(\lambda^s_I l_K)}{\lambda^s_I}$
			1

A-8

THINWALLED BEAM SUBSTRUCTURE
PURE TORSION BEHAVIOUR
TRANSFER MATRIX

STATE VECTOR $k_K = (\vartheta^s_{T,K}, T^s_{T,K})^T$

BK =

1	$\frac{l_K}{GJ_{TB}}$
	1

A-9

THINWALLED BEAM SUBSTRUCTURE

DISTORTIONAL BEHAVIOUR – TRANSFER MATRIX

STATE VECTOR $k_K = (\varphi^S_{V,K}, \beta^S_{V,K}, M^S_{W,K}, M^S_{VK})^T$

$$BK = \begin{bmatrix} \frac{2F_2ZW - F_1(Z^2-W^2)}{2ZW} & \frac{F_3W(3Z^2-W^2) - F_4Z(Z^2-3W^2)}{2ZW(Z^2+W^2)} & -\frac{F_1}{2ZW} & -\frac{F_4Z - F_3W}{2ZW(Z^2+W^2)} \\ & \frac{2F_2ZW(Z^2+W^2)+F_1(Z^4+W^4)}{2ZW(Z^2+W^2)} & -\frac{ZF_4+WF_3}{2ZW} & \frac{F_1}{2ZW} \\ & & \frac{F_1(Z^2-W^2)+2ZWF_2}{2ZW} & \frac{ZF_4+WF_3}{2ZW} \\ & & & \frac{F_1(Z^2-W^2)+2F_2ZW}{2ZW} \end{bmatrix}$$

$F_1 = \sinh(Zl_K)\sin(Wl_K)$

$F_2 = \cosh(Zl_K)\cos(Wl_K)$

$F_3 = \sinh(Zl_K)\cos(Wl_K)$

$F_4 = \cosh(Zl_K)\sin(Wl_K)$

$Z = ((J_{D_1}/(4F^S_{w_{II}w_{II}}))^{1/2}(1 + K_P(J_{D_1}/(4F^S_{w_{II}w_{II}}))^{1/2}))^{1/2}$ (K_P – see [27])

$W = ((J_{D_1}/(4F^S_{w_{II}w_{II}}))^{1/2}(1 - K_P(J_{D_1}/(4F^S_{w_{II}w_{II}}))^{1/2}))^{1/2}$

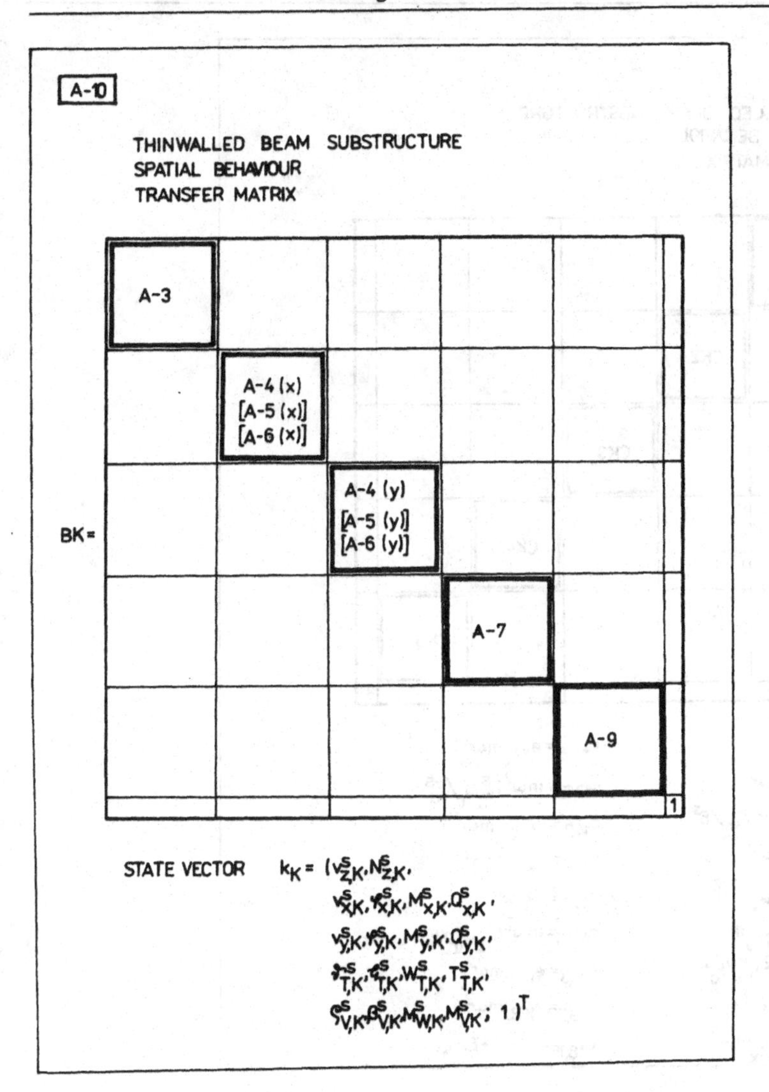

$k_K = (v^S_{z,K}, N^S_{z,K},$
$v^S_{x,K}, \varphi^S_{x,K}, M^S_{x,K}, Q^S_{x,K},$
$v^S_{y,K}, \varphi^S_{y,K}, M^S_{y,K}, Q^S_{y,K},$
$\vartheta^S_{T,K}, \varphi^S_{T,K}, W^S_{T,K}, T^S_{T,K},$
$\varphi^S_{V,K}, \beta^S_{V,K}, M^S_{W,K}, M^S_{V,K};\ 1)^T$

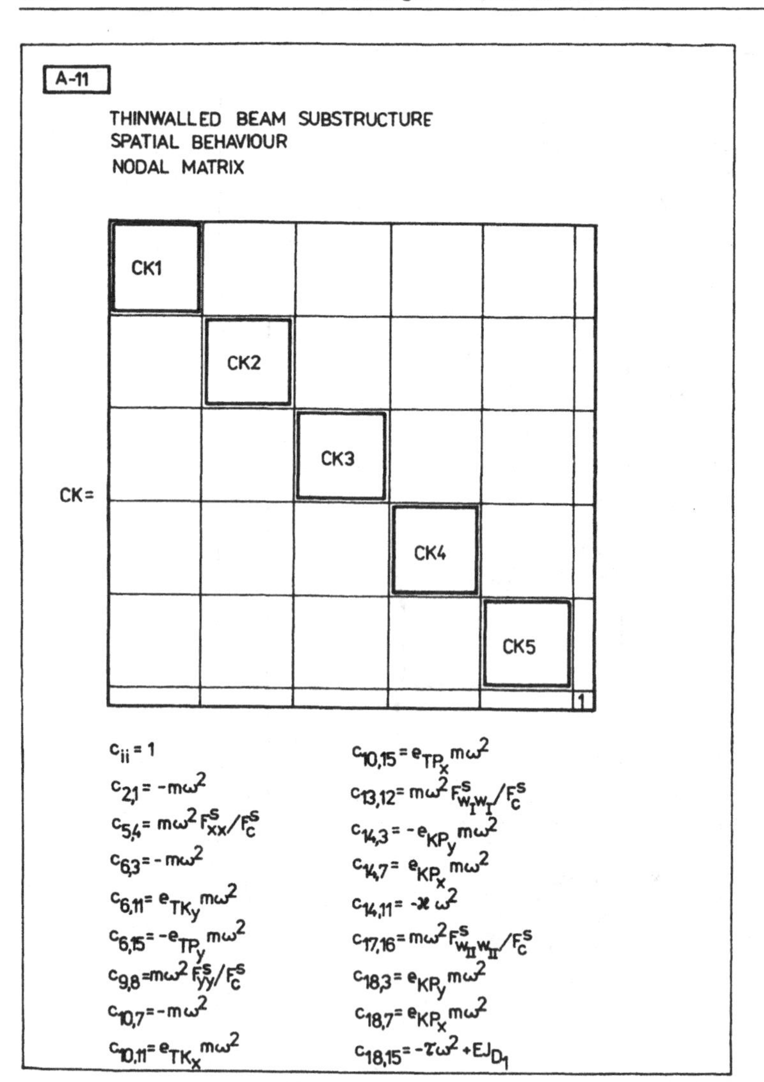
A-11
THINWALLED BEAM SUBSTRUCTURE
SPATIAL BEHAVIOUR
NODAL MATRIX
CK=
CK1
CK2
CK3
CK4
CK5
1
$c_{ii} = 1$
$c_{2,1} = -m\omega^2$
$c_{5,4} = m\omega^2 F^S_{xx}/F^S_c$
$c_{6,3} = -m\omega^2$
$c_{6,11} = e_{TK_y} m\omega^2$
$c_{6,15} = -e_{TP_y} m\omega^2$
$c_{9,8} = m\omega^2 F^S_{yy}/F^S_c$
$c_{10,7} = -m\omega^2$
$c_{10,11} = e_{TK_x} m\omega^2$
$c_{10,15} = e_{TP_x} m\omega^2$
$c_{13,12} = m\omega^2 F^S_{w_I w_I}/F^S_c$
$c_{14,3} = -e_{KP_y} m\omega^2$
$c_{14,7} = e_{KP_x} m\omega^2$
$c_{14,11} = -\varkappa \omega^2$
$c_{17,16} = m\omega^2 F^S_{w_{II} w_{II}}/F^S_c$
$c_{18,3} = e_{KP_y} m\omega^2$
$c_{18,7} = e_{KP_x} m\omega^2$
$c_{18,15} = -\tau\omega^2 + EJ_{D_1}$

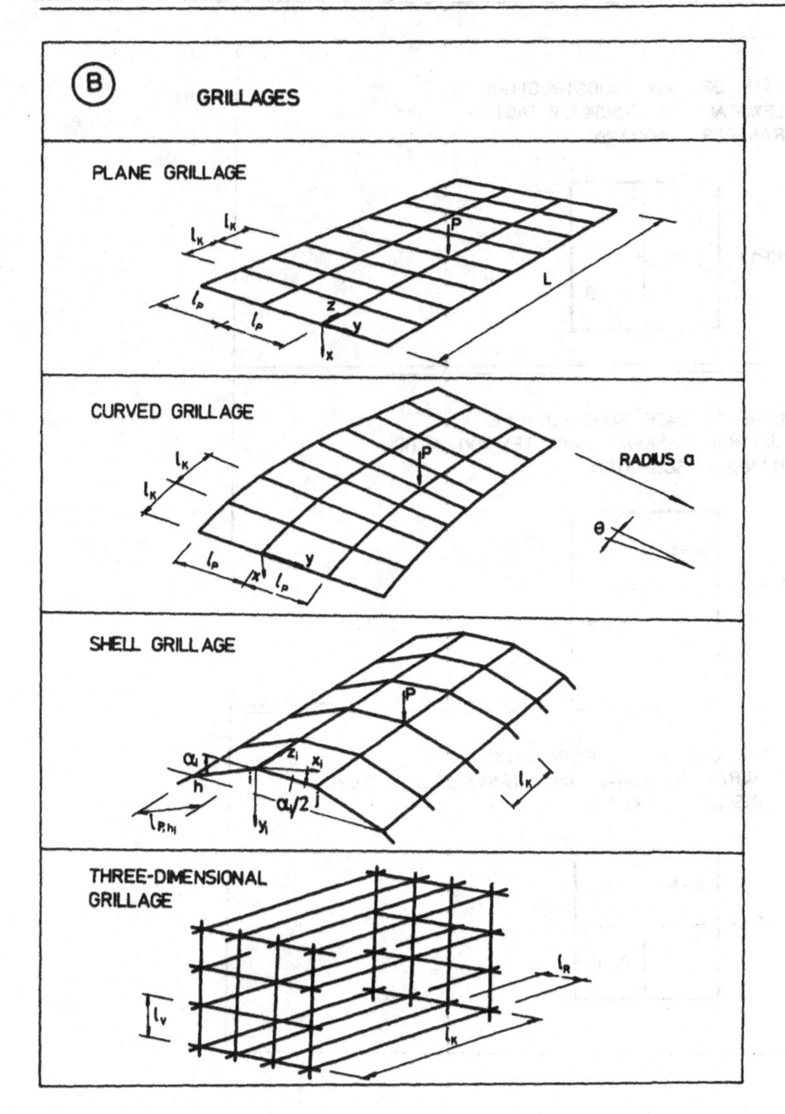
B
GRILLAGES
PLANE GRILLAGE
P
L
z
y
x
CURVED GRILLAGE
P
RADIUS a
θ
y
x
SHELL GRILLAGE
P
i
j
THREE-DIMENSIONAL
GRILLAGE

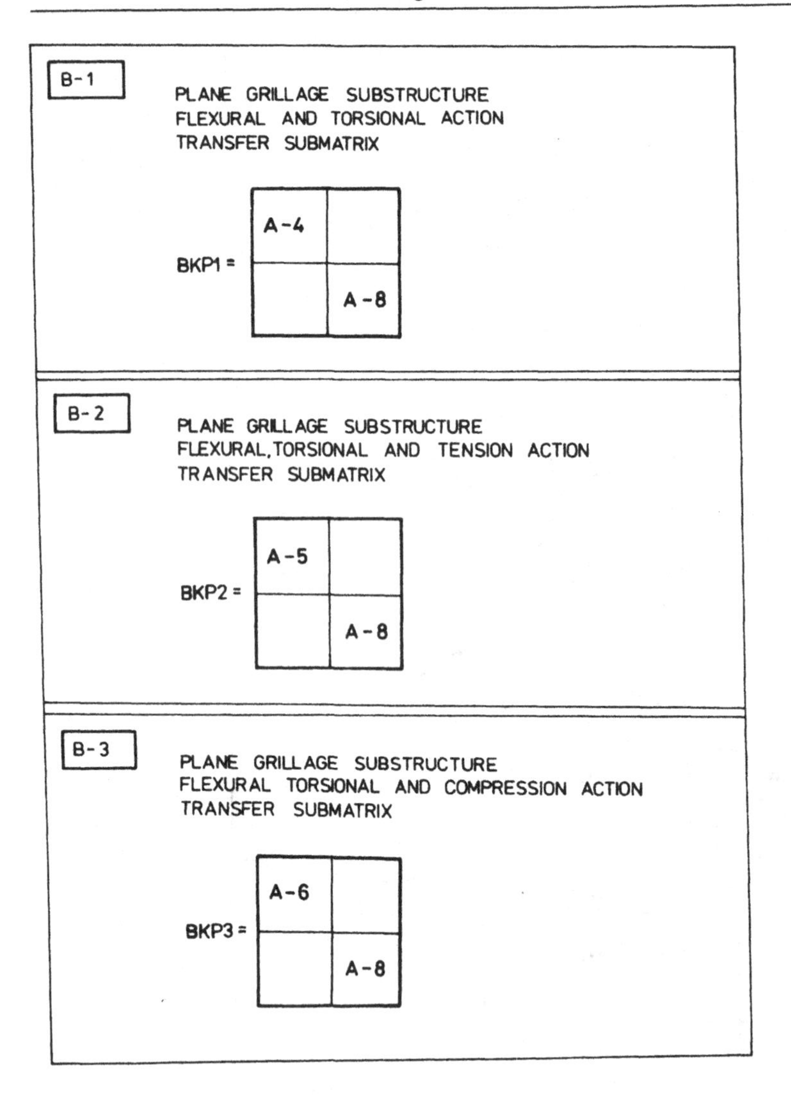
B-1
PLANE GRILLAGE SUBSTRUCTURE
FLEXURAL AND TORSIONAL ACTION
TRANSFER SUBMATRIX
BKP1 =
A-4
A-8
B-2
PLANE GRILLAGE SUBSTRUCTURE
FLEXURAL,TORSIONAL AND TENSION ACTION
TRANSFER SUBMATRIX
BKP2 =
A-5
A-8
B-3
PLANE GRILLAGE SUBSTRUCTURE
FLEXURAL TORSIONAL AND COMPRESSION ACTION
TRANSFER SUBMATRIX
BKP3 =
A-6
A-8

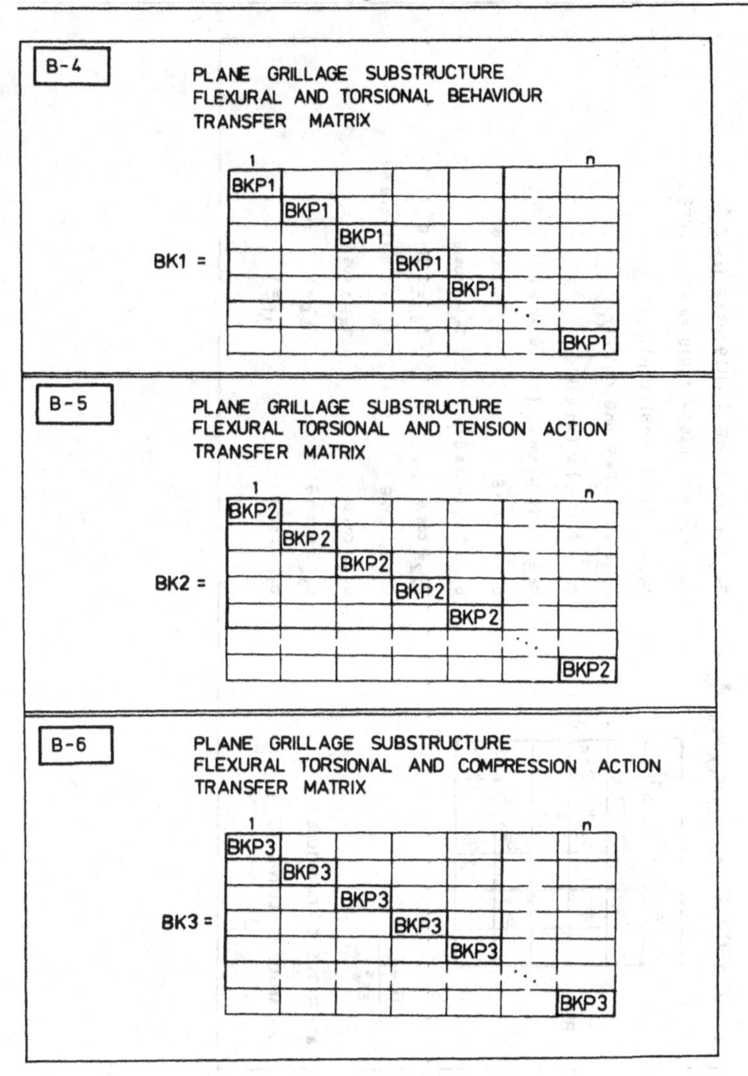
B-4
PLANE GRILLAGE SUBSTRUCTURE
FLEXURAL AND TORSIONAL BEHAVIOUR
TRANSFER MATRIX
1
n
BK1 =
BKP1
BKP1
BKP1
BKP1
BKP1
BKP1
B-5
PLANE GRILLAGE SUBSTRUCTURE
FLEXURAL TORSIONAL AND TENSION ACTION
TRANSFER MATRIX
1
n
BK2 =
BKP2
BKP2
BKP2
BKP2
BKP2
BKP2
B-6
PLANE GRILLAGE SUBSTRUCTURE
FLEXURAL TORSIONAL AND COMPRESSION ACTION
TRANSFER MATRIX
1
n
BK3 =
BKP3
BKP3
BKP3
BKP3
BKP3
BKP3

B - 7

CURVED GRILLAGE SUBSTRUCTURE
FLEXURAL AND TORSIONAL BEHAVIOUR
TRANSFER SUBMATRIX

STATE VECTOR $k_K = (v^s_{x,K}, \varphi^s_{x,K}, M^s_{x,K}, Q^s_{x,K})^T$

BKP =

$b_{1,1}$	$b_{1,2}$	$b_{1,3}$	$b_{1,4}$	$b_{1,5}$	$b_{1,6}$
	$b_{2,2}$	$b_{2,3}$	$b_{2,4}$	$b_{2,5}$	$b_{2,6}$
		$b_{3,3}$	$b_{3,4}$	$b_{3,5}$	$b_{3,6}$
			$b_{4,4}$		
	$b_{5,2}$		$b_{5,4}$	$b_{5,5}$	$b_{5,6}$
		$b_{6,3}$	$b_{6,4}$		$b_{6,6}$

$$\mu = \frac{GJ_{TB}}{EF^s_{w_I w_I}}$$

a - RADIUS OF CURVATURE

θ - ANGLE OF CURVATURE

$b_{1,3} = (a^2(2(1-\cos\theta)-(1+\mu)\theta\sin\theta))/(2\mu EF^s_{xx})$

$b_{1,4} = (a^3(2\theta-(3+\mu)\sin\theta + (1+\mu)\theta\cos\theta))/(2\mu EF^s_{xx})$

$b_{1,6} = \pm((1+\mu)\ a^2\ (\sin\theta - \theta\cos\theta))/(2\mu EF^s_{xx})$

$b_{2,3} = (a((1-\mu)\sin\theta - (1+\mu)\theta\cos\theta))/(2\mu EF^s_{xx})$

$b_{2,4} = (a^2(2(1-\cos\theta) - (1+\theta)\theta\sin\theta))/(2\mu EF^s_{xx})$

$b_{2,6} = \pm((1+\mu)\theta a \sin\theta)/(2\mu EF^s_{xx})$

$b_{5,3} = \pm((1+\mu) a\theta\sin\theta)/(2\mu EF^s_{xx})$

$b_{5,4} = \pm((1+\mu) a^2 (\sin\theta))$

$b_{5,6} = (a((1-\mu)\sin\theta + (1+\mu)\theta\cos\theta))/(2\mu EF^s_{xx})$

$b_{1,2} = a\ \sin\theta$	$b_{5,2} = \pm\sin\theta$
$b_{1,5} = \pm a(1-\cos\theta)$	$b_{5,5} = \cos\theta$
$b_{2,2} = \cos\theta$	$b_{6,3} = \pm\sin\theta$
$b_{2,5} = \pm\sin\theta$	$b_{6,4} = \pm a(1-\cos\theta)$
$b_{3,3} = \cos\theta$	$b_{6,6} = \cos\theta$
$b_{3,4} = a\ \sin\theta$	$b_{4,4} = 1$
$b_{3,6} = \pm\sin\theta$	$b_{1,1} = 1$

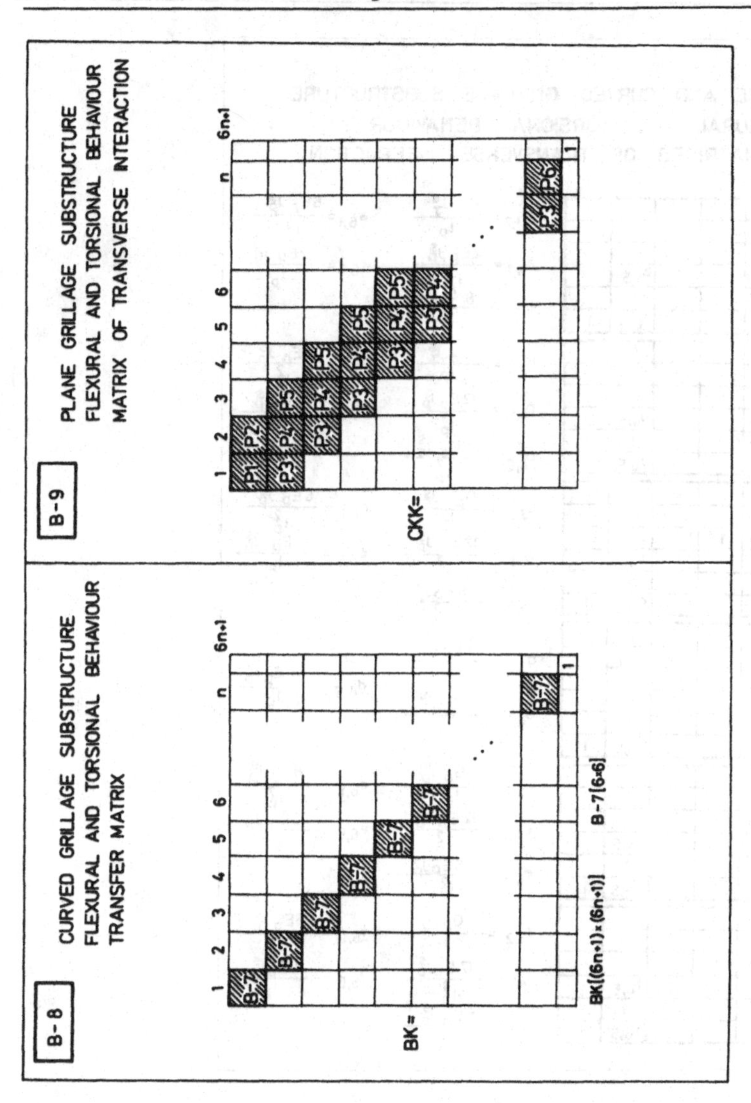
B-8
CURVED GRILLAGE SUBSTRUCTURE
FLEXURAL AND TORSIONAL BEHAVIOUR
TRANSFER MATRIX
BK=
1 2 3 4 5 6 n 6n+1
B-7
BK[(6n+1)x(6n+1)]
B-7[6x6]
B-9
PLANE GRILLAGE SUBSTRUCTURE
FLEXURAL AND TORSIONAL BEHAVIOUR
MATRIX OF TRANSVERSE INTERACTION
CKK=
1 2 3 4 5 6 n 6n+1
P1 P2
P3 P4 P5
P3 P6

B -10

PLANE AND CURVED GRILLAGE SUBSTRUCTURE FLEXURAL AND TORSIONAL BEHAVIOUR SUBMATRICES OF TRANSVERSE INTERACTION

$$P1 = \begin{bmatrix} 1 & & & & & \\ & 1 & & & & \\ & a_{3,2} & 1 & & & \\ a_{4,1} & & & 1 & a_{4,5} & \\ & & & & 1 & \\ a_{6,1} & & & & a_{6,5} & 1 \end{bmatrix}$$

$$a_{3,2} = -\frac{G_p J_T^s}{l_p} \qquad a_{6,1} = \frac{6 E_p J_p^s}{l_p^3}$$

$$a_{4,1} = \frac{12 E_p J_p^s}{l_p^3} \qquad a_{6,5} = -\frac{2 E_p J_p^s}{l_p}$$

$$a \;\; = -\frac{6 \; E_p J_p^s}{l_p^2}$$

$$P2 = \begin{bmatrix} 1 & & & & & \\ & 1 & & & & \\ & b_{3,2} & 1 & & & \\ b_{4,1} & & & 1 & b_{4,5} & \\ & & & & 1 & \\ b_{6,1} & & & & b_{6,5} & 1 \end{bmatrix}$$

$$b_{3,2} = \frac{G_p J_T^s}{l_p} \qquad b_{6,1} = -\frac{6 E_p J_p^s}{l_p^2}$$

$$b_{4,1} = -\frac{12 E_p J_p^s}{l_p^3} \qquad b_{6,5} = -\frac{2 E_p J_p^s}{l_p}$$

$$b_{4,5} = -\frac{6 E_p J_p^s}{l_p^2}$$

$$P3 = \begin{bmatrix} 1 & & & & & \\ & 1 & & & & \\ & c_{3,2} & 1 & & & \\ c_{4,1} & & & 1 & c_{4,5} & \\ & & & & 1 & \\ c_{6,1} & & & & c_{6,5} & 1 \end{bmatrix}$$

$$c_{3,2} = \frac{G_p J_p^s}{l_p} \qquad c_{6,1} = \frac{6 E_p J_p^s}{l_p^2}$$

$$c_{4,1} = -\frac{12 E_p J_p^s}{l_p^3} \qquad c_{6,5} = -\frac{2 E_p J_p^s}{l_p}$$

$$c_{4,5} = \frac{6 E_p J_p^s}{l_p^2}$$

$$P4 = \begin{bmatrix} 1 & & & & & \\ & 1 & & & & \\ & d_{3,2} & 1 & & & \\ d_{4,1} & & & 1 & & \\ & & & & 1 & \\ & & & & d_{6,5} & 1 \end{bmatrix}$$

$$d_{3,2} = \frac{2 G_p J_p^s}{l_p} \qquad d_{6,5} = -\frac{8 E_p J_p^s}{l_p}$$

$$d_{4,1} = \frac{24 E_p J_p^s}{l_p^3}$$

$$P5 = \begin{bmatrix} 1 & & & & & \\ & 1 & & & & \\ & e_{3,2} & 1 & & & \\ e_{4,1} & & & 1 & e_{4,5} & \\ & & & & 1 & \\ e_{6,1} & & & & e_{6,5} & 1 \end{bmatrix}$$

$$e_{3,2} = -\frac{G_p J_p^s}{l_p} \qquad e_{6,1} = -\frac{6 E_p J_p^s}{l_p^2}$$

$$e_{4,1} = -\frac{12 E_p J_p^s}{l_p^2} \qquad e_{6,5} = -\frac{2 E_p J_p^s}{l_p}$$

$$e_{4,5} = -\frac{6 E_p J_p^s}{l_p^2}$$

$$P6 = \begin{bmatrix} 1 & & & & & \\ & 1 & & & & \\ & f_{3,2} & 1 & & & \\ f_{4,1} & & & 1 & f_{4,5} & \\ & & & & 1 & \\ f_{6,1} & & & & f_{6,5} & 1 \end{bmatrix}$$

$$f_{3,2} = -\frac{G_p J_p^s}{l_p} \qquad f_{6,1} = -\frac{6 E_p J_p^s}{l_p^2}$$

$$f_{4,1} = \frac{12 E_p J_p^s}{l_p^3} \qquad f_{6,5} = -\frac{4 E_p J_p^s}{l_p}$$

$$f_{4,5} = \frac{6 E_p J_p^s}{l_p^2} .$$

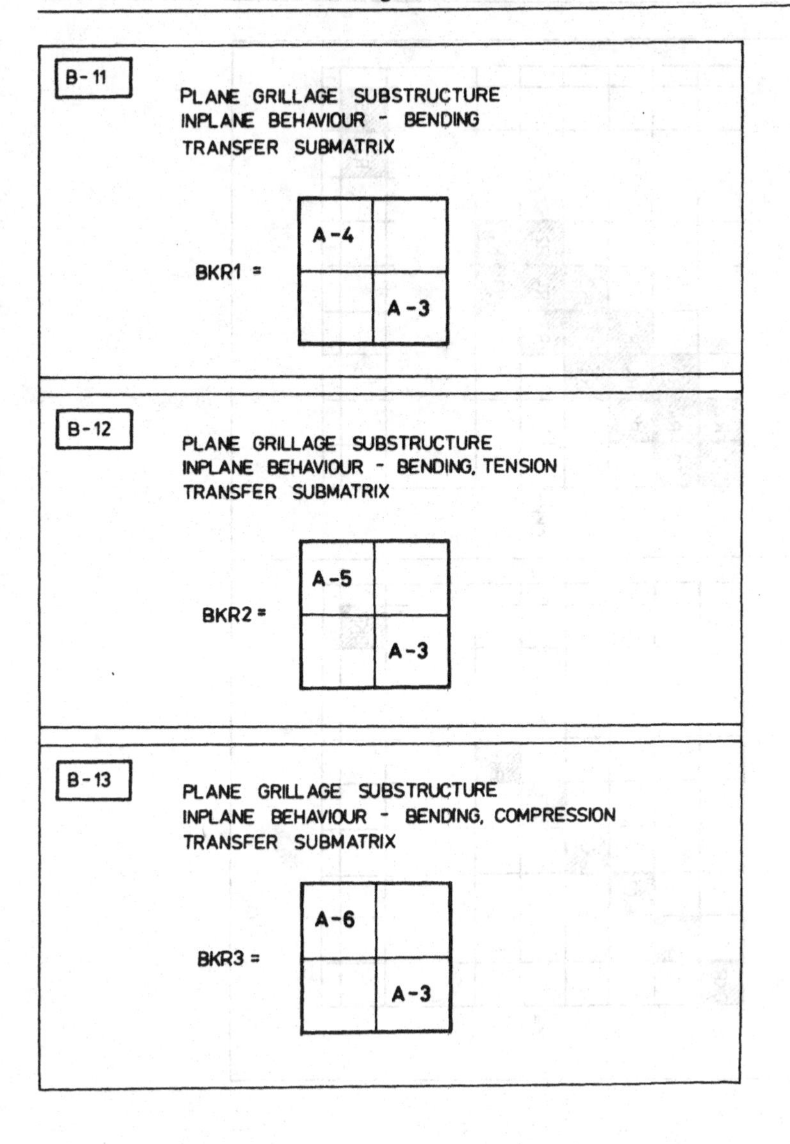
B-11
PLANE GRILLAGE SUBSTRUCTURE
INPLANE BEHAVIOUR - BENDING
TRANSFER SUBMATRIX
BKR1 =
A-4
A-3
B-12
PLANE GRILLAGE SUBSTRUCTURE
INPLANE BEHAVIOUR - BENDING, TENSION
TRANSFER SUBMATRIX
BKR2 =
A-5
A-3
B-13
PLANE GRILLAGE SUBSTRUCTURE
INPLANE BEHAVIOUR - BENDING, COMPRESSION
TRANSFER SUBMATRIX
BKR3 =
A-6
A-3

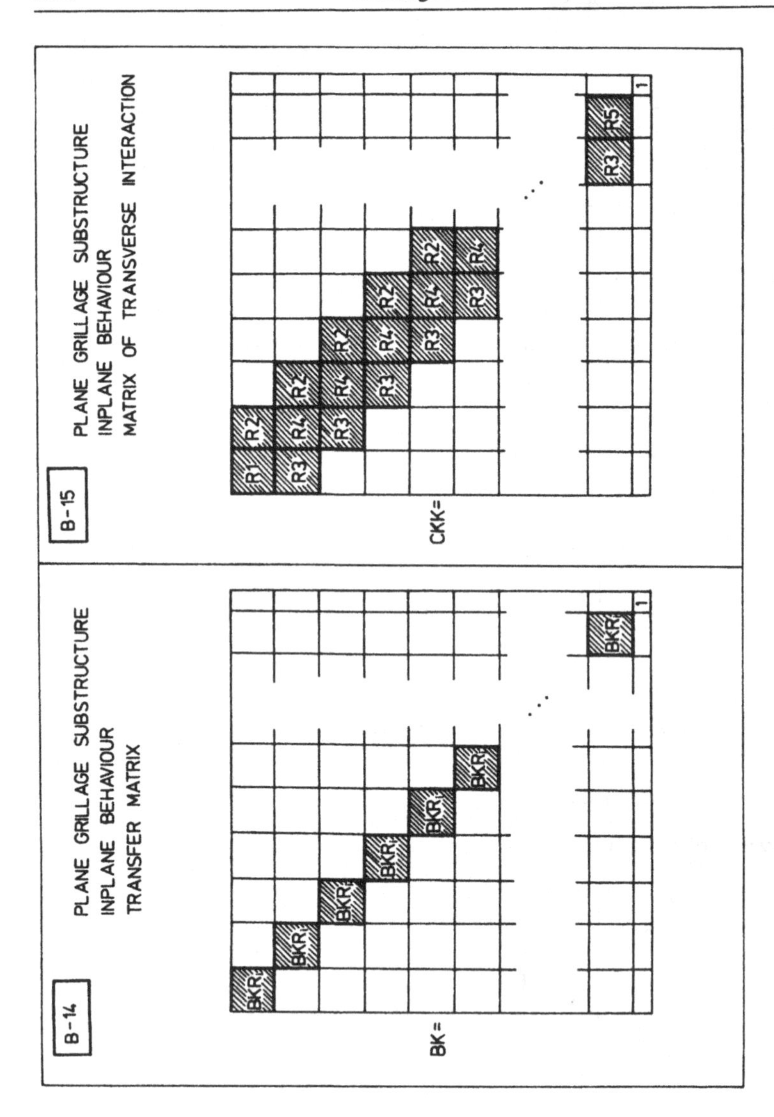
B-14
PLANE GRILLAGE SUBSTRUCTURE
INPLANE BEHAVIOUR
TRANSFER MATRIX
BK=
BKR
B-15
PLANE GRILLAGE SUBSTRUCTURE
INPLANE BEHAVIOUR
MATRIX OF TRANSVERSE INTERACTION
CKK=
R1
R2
R3
R4
R5

B - 16

PLANE GRILLAGE SUBSTRUCTURE INPLANE BEHAVIOUR - SUBMATRICES OF TRANSVERSE INTERACTION

$$R1 = \begin{bmatrix} 1 & & & & & \\ & 1 & & & & \\ & a_{3,2} & 1 & & a_{3,5} & \\ a_{4,1} & & & 1 & & \\ & & & & 1 & \\ & a_{6,2} & & & a_{6,5} & 1 \end{bmatrix}$$

$$a_{3,2} = -\frac{4E_p J_p^s}{l_p} \qquad a_{6,2} = -\frac{6E_p J_p^s}{l_p^2}$$

$$a_{3,5} = \frac{6E_p J_p^s}{l_p^2} \qquad a_{6,5} = \frac{12E_p J_p^s}{l_p^3}$$

$$a_{4,1} = \frac{E_p A}{l_p}$$

$$R2 = \begin{bmatrix} 1 & & & & & \\ & 1 & & & & \\ & b_{3,2} & 1 & & b_{3,5} & \\ b_{4,1} & & & 1 & & \\ & & & & 1 & \\ & b_{6,2} & & & b_{6,5} & 1 \end{bmatrix}$$

$$b_{3,2} = -\frac{2E_p J_p^s}{l_p} \qquad b_{6,2} = -\frac{6E_p J_p^s}{l_p^2}$$

$$b_{3,5} = -\frac{6E_p J_p^s}{l_p^2} \qquad b_{6,5} = -\frac{12E_p J_p^s}{l_p^3}$$

$$b_{4,1} = -\frac{E_p A}{l_p}$$

$$R3 = \begin{bmatrix} 1 & & & & & \\ & 1 & & & & \\ & c_{3,2} & 1 & & c_{3,5} & \\ c_{4,1} & & & 1 & & \\ & & & & 1 & \\ & c_{6,2} & & & c_{6,5} & 1 \end{bmatrix}$$

$$c_{3,2} = -\frac{2E_p J_p^s}{l_p} \qquad c_{6,2} = \frac{6E_p J_p^s}{l_p^2}$$

$$c_{3,5} = \frac{6E_p J_p^s}{l_p^2} \qquad c_{6,5} = -\frac{12E_p J_p^s}{l_p^3}$$

$$c_{1,4} = -\frac{E_p A}{l_p}$$

$$R4 = \begin{bmatrix} 1 & & & & & \\ & 1 & & & & \\ & d_{3,2} & 1 & & d_{3,5} & \\ d_{4,1} & & & 1 & & \\ & & & & 1 & \\ & d_{6,2} & & & d_{6,5} & 1 \end{bmatrix}$$

$$d_{3,2} = -\frac{8E_p J_p^s}{l_p} \qquad d_{3,5} = 0$$

$$d_{4,1} = \frac{2E_p A}{l_p} \qquad d_{6,2} = 0$$

$$d_{6,5} = \frac{24E_p J_p^s}{l_p^3}$$

$$R5 = \begin{bmatrix} 1 & & & & & \\ & 1 & & & & \\ & e_{3,2} & 1 & & e_{3,5} & \\ e_{4,1} & & & 1 & & \\ & & & & 1 & \\ & e_{6,2} & & & e_{6,5} & 1 \end{bmatrix}$$

$$e_{3,2} = -\frac{4E_p J_p^s}{l_p} \qquad e_{6,2} = \frac{6E_p J_p^s}{l_p^2}$$

$$e_{3,5} = -\frac{6E_p J_p^s}{l_p^2} \qquad e_{6,5} = \frac{12E_p J_p^s}{l_p^3}$$

$$e_{4,1} = \frac{E_p A}{l_p^2}$$

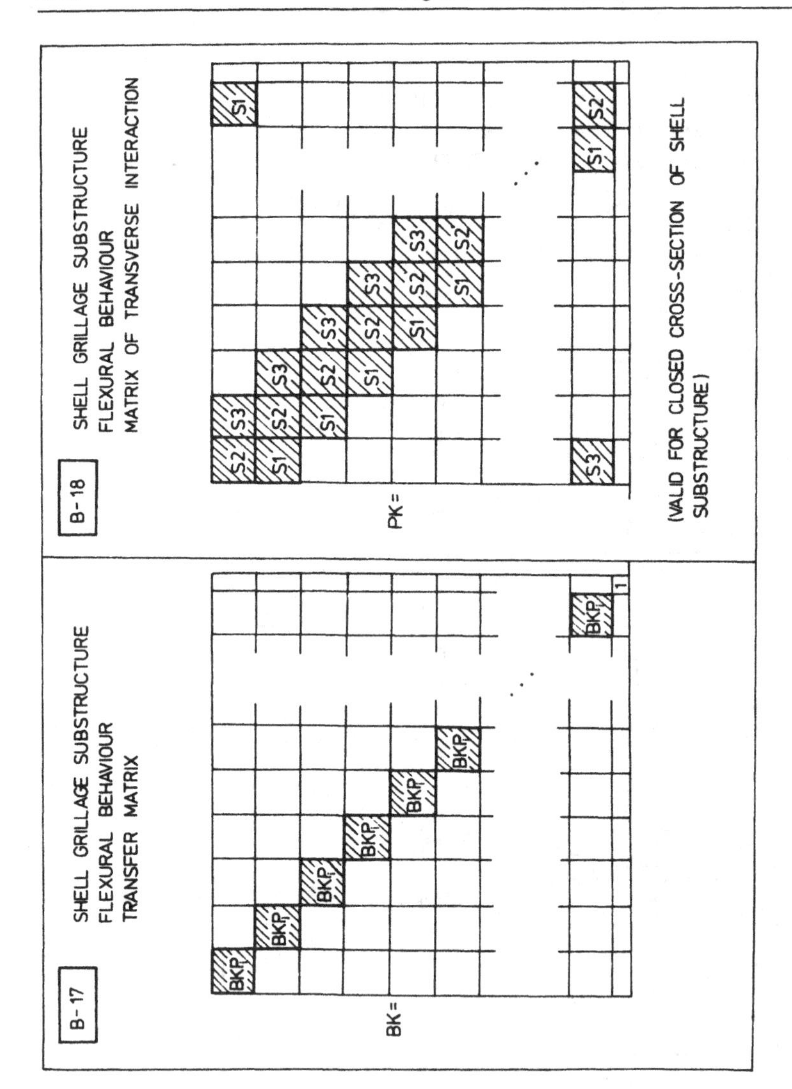
B-17
SHELL GRILLAGE SUBSTRUCTURE
FLEXURAL BEHAVIOUR
TRANSFER MATRIX
BK =
BKP
1
B-18
SHELL GRILLAGE SUBSTRUCTURE
FLEXURAL BEHAVIOUR
MATRIX OF TRANSVERSE INTERACTION
PK =
S1
S2
S3
(VALID FOR CLOSED CROSS-SECTION OF SHELL
SUBSTRUCTURE)

B - 19

SHELL GRILLAGE SUBSTRUCTURE - FLEXURAL AND TORSIONAL BEHAVIOUR - SUBMATRICES OF TRANSVERSE INTERACTION

S1 =

$v^s_{x,h}$	$\varphi^s_{x,h}$	$M^s_{x,h}$	$Q^s_{x,h}$	$\vartheta^s_{x,h}$	$T^s_{x,h}$
	$a_{3,2}$				
$a_{4,1}$				$a_{4,5}$	
$a_{6,1}$				$a_{6,5}$	

$$a_{3,2} = G_p J^s_{T,hi} \cos\left(\frac{\alpha_h+\alpha_i}{2}\right) / \left(l_{p,hi} \cos\frac{\alpha_i}{2}\right)$$

$$a_{4,1} = \left(-12E_p J^s_{hi} \cos\frac{\alpha_h}{2} \cos\frac{\alpha_i}{2}\right) / l^3_{p,hi}$$

$$a_{4,5} = \left(-6E_p J^s_{hi} \cos\frac{\alpha_i}{2}\right) / l^2_{p,hi}$$

$$a_{6,1} = \left(-6E_p J^s_{hi} \cos\frac{\alpha_h}{2}\right) / l^2_{p,hi}$$

$$a_{6,5} = -2E_p J^s_{hi} / l_{p,hi}$$

S2 =

$v^s_{x,i}$	$\varphi^s_{x,i}$	$M^s_{x,i}$	$Q^s_{x,i}$	$\vartheta^s_{x,i}$	$T^s_{x,i}$
1					
	1				
	$b_{3,2}$	1			
$b_{4,1}$			1		
				1	
$b_{6,1}$				$b_{6,5}$	1

$$b_{3,2} = -\frac{G_p}{\cos(\frac{\alpha_i}{2})} \left(\frac{J^s_{T,hi}}{l_{p,hi}} + \frac{J^s_{T,ij}}{l_{p,ij}}\right)$$

$$b_{4,1} = 12E_p \left(\frac{J^s_{hi}}{l^3_{p,hi}} + \frac{J^s_{ij}}{l^3_{p,ij}}\right) \cos^2\left(\frac{\alpha_i}{2}\right)$$

$$b_{6,1} = 6E_p \left(\frac{J^s_{hi}}{l^2_{p,hi}} - \frac{J^s_{ij}}{l^2_{p,ij}}\right) \cos\left(\frac{\alpha_i}{2}\right)$$

$$b_{6,5} = 4E_p \left(\frac{J^s_{hi}}{l_{p,hi}} + \frac{J^s_{ij}}{l_{p,ij}}\right)$$

S3 =

$v^s_{x,j}$	$\varphi^s_{x,j}$	$M^s_{x,j}$	$Q^s_{x,j}$	$\vartheta^s_{x,j}$	$T^s_{x,j}$
	$c_{3,2}$				
$c_{4,1}$				$c_{4,5}$	
$c_{6,1}$				$c_{6,5}$	

$$c_{3,2} = \frac{G_p (J^s_{T,ij} + J^s_{T,ij}) \cos(\frac{\alpha_i+\alpha_j}{2})}{l_{p,ij} \cos(\frac{\alpha_i}{2})}$$

$$c_{4,1} = \left(-12E_p J^s_{ij} \cos\left(\frac{\alpha_i}{2}\right) \cos\left(\frac{\alpha_j}{2}\right)\right) / l^2_{p,ij}$$

$$c_{4,5} = \left(6E_p J^s_{ij} \cos\left(\frac{\alpha_i}{2}\right)\right) / l^2_{p,ij}$$

$$c_{6,1} = \left(6E_p J^s_{ij} \cos\left(\frac{\alpha_i}{2}\right)\right) / l^2_{p,ij}$$

$$c_{6,5} = -2E_p J^s_{ij} / l_{p,ij}$$

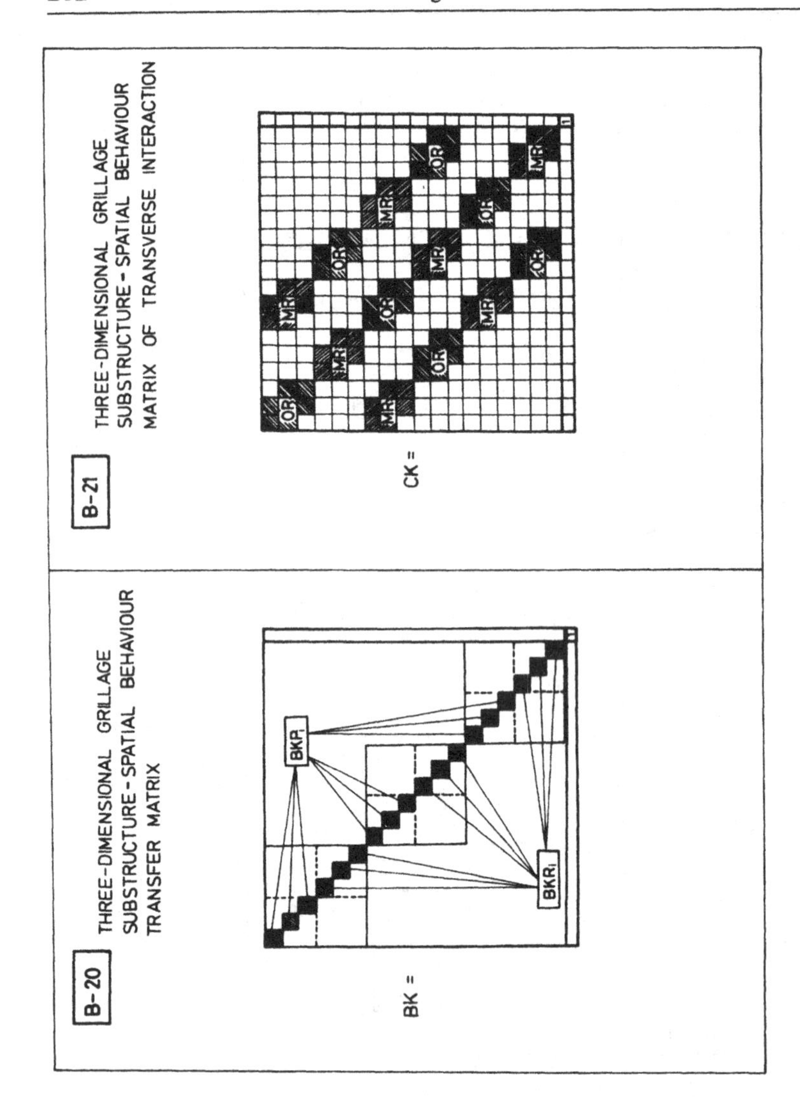
B-20
THREE-DIMENSIONAL GRILLAGE
SUBSTRUCTURE - SPATIAL BEHAVIOUR
TRANSFER MATRIX
BK =
BKP1
BKR1
B-21
THREE-DIMENSIONAL GRILLAGE
SUBSTRUCTURE - SPATIAL BEHAVIOUR
MATRIX OF TRANSVERSE INTERACTION
CK =
OR
MR

B-22

THREE-DIMENSIONAL GRILLAGE SUBSTRUCTURE
SPATIAL BEHAVIOUR
SUBMATRICES OF TRANSVERSE INTERACTION

OR =

P1	P2	
P3	P4	P5
	P3	P4

MR =

R1	R2	
R3	R4	R2
	R3	R4

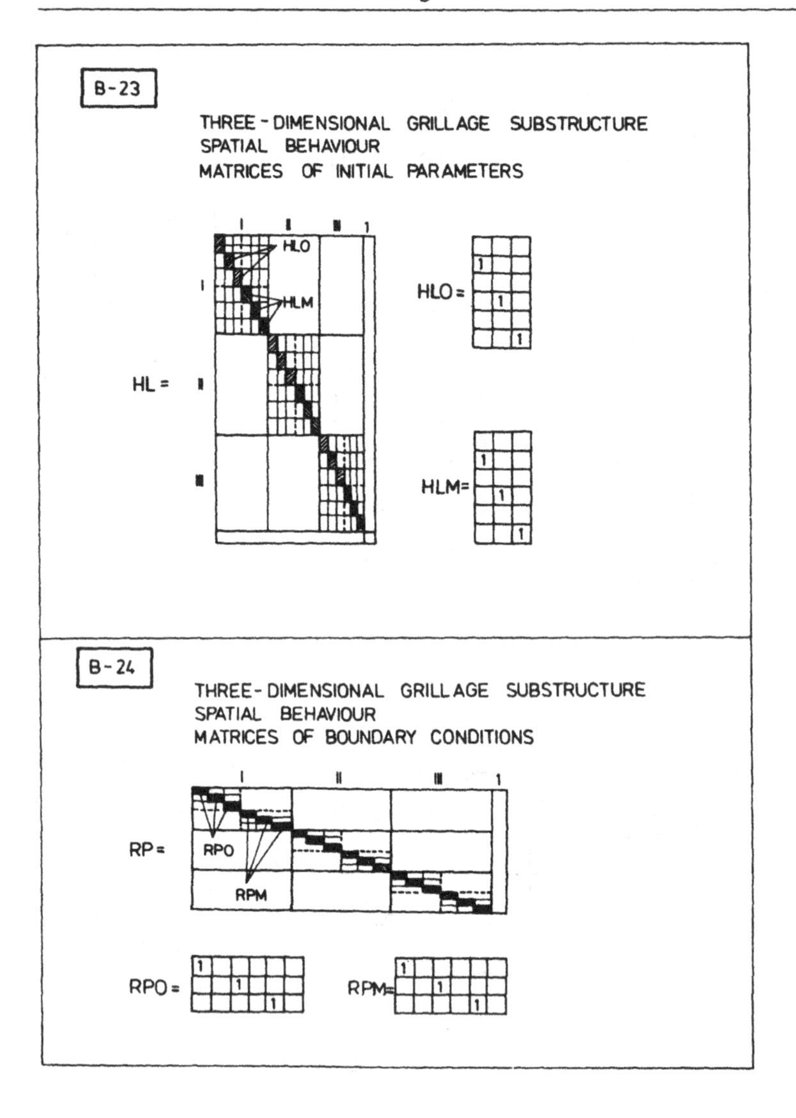
B-23
THREE-DIMENSIONAL GRILLAGE SUBSTRUCTURE
SPATIAL BEHAVIOUR
MATRICES OF INITIAL PARAMETERS
HL =
HLO
HLM
HLO =
HLM =
B-24
THREE-DIMENSIONAL GRILLAGE SUBSTRUCTURE
SPATIAL BEHAVIOUR
MATRICES OF BOUNDARY CONDITIONS
RP =
RPO
RPM
RPO =
RPM =

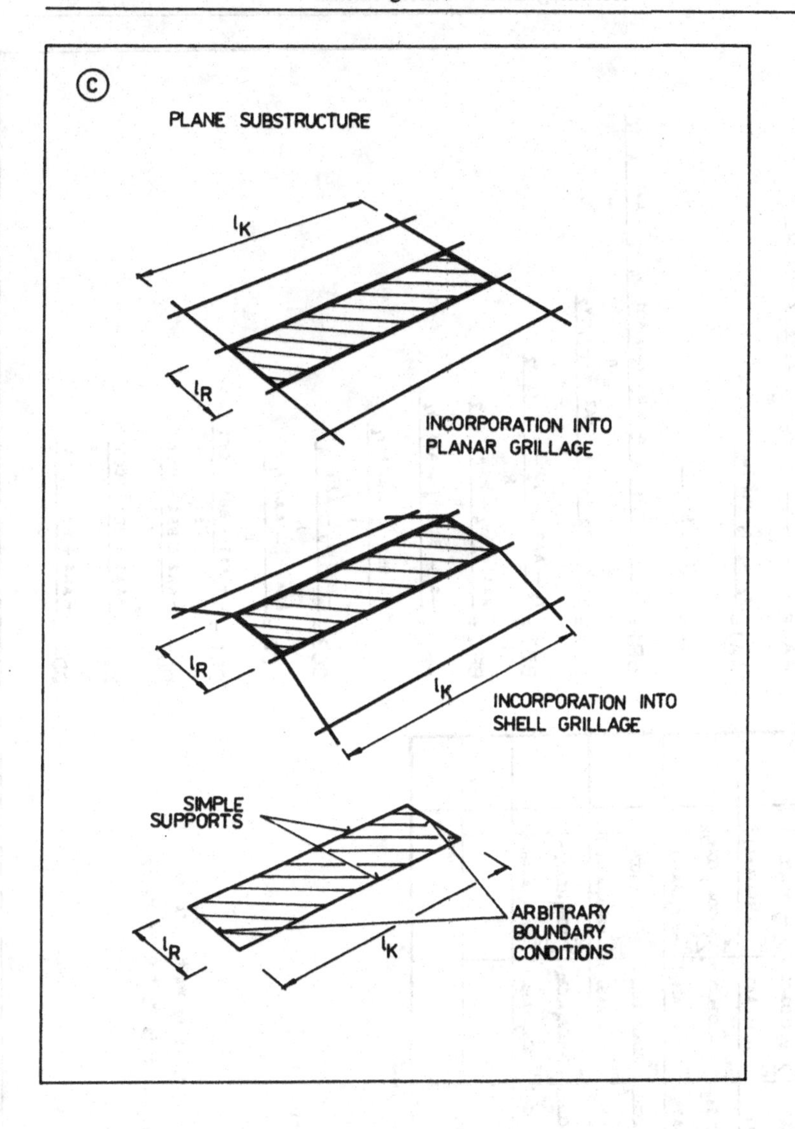
C
PLANE SUBSTRUCTURE
l_K
l_R
INCORPORATION INTO PLANAR GRILLAGE
l_R
l_K
INCORPORATION INTO SHELL GRILLAGE
SIMPLE SUPPORTS
l_R
l_K
ARBITRARY BOUNDARY CONDITIONS

C - 1 PLANE SUBSTRUCTURE, FLEXURAL - TORSIONAL BEHAVIOUR - TRANSFER SUBMATRIX

$BK^{(p)} =$

$\sum_{n=1}^{\infty}(DD1+CC1+BB1+AA1)$	$\sum_{n=1}^{\infty}(DD2+CC2-BB2+AA2)$	$\sum_{n=1}^{\infty}\frac{Y}{ZZ}(DD3+CC3+BB3+AA3)$	$\sum_{n=1}^{\infty}\frac{Y}{W}(DD4+CC4-BB4+AA4)$		
$\sum_{n=1}^{\infty}(DD1.s_1+CC1.s_2+BB1s_3-AA1s_4)$	$\sum_{n=1}^{\infty}(DD2s_1+CC2s_2-BB2s_3+AA2s_4)$	$\sum_{n=1}^{\infty}\frac{Y}{ZZ}(DD3s_1+CC3s_2+BB3s_3+AA3s_4)$	$\sum_{n=1}^{\infty}\frac{Y}{W}(DD4s_1+CC4s_2-BB4s_3+AA4s_4)$		
$\sum_{n=1}^{\infty}\frac{ZZ}{Y}(DD1u_1+CC1u_2+BB1u_3-AA1u_4)$	$\sum_{n=1}^{\infty}\frac{ZZ}{Y}(DD2u_1+CC2u_2-BB2u_3+AA2u_4)$	$\sum_{n=1}^{\infty}(DD3u_1+CC3u_2+BB3u_3+AA3u_4)$	$\sum_{n=1}^{\infty}\frac{ZZ}{W}(DD4u_1+CC4u_2-BB4u_3+AA4u_4)$		
$\sum_{n=1}^{\infty}\frac{W}{Y}(DD1p_1+CC1p_2+BB1p_3-AA1p_4)$	$\sum_{n=1}^{\infty}\frac{W}{Y}(DD1p_1+CC2p_2+BB2p_3-AA2p_4)$	$\sum_{n=1}^{\infty}\frac{W}{Z}(DD3p_1+CC3p_2+BB3p_3+AA3p_4)$	$\sum_{n=1}^{\infty}(DD4p_1+CC4p_2-BB4p_3+AA4p_4)$		
				1	$-\frac{l_K}{(GJ_T)(p)}$
					1

Substitutions

$t_i = s_i - 1$

$u_i = s_i \; t_i$

$p_i = s_i \; t_i - 2\Omega\lambda^2 t_i - \varepsilon(s_i - \lambda^2)$

$w_i = p_i - p_1$

$v_i = u_i - u_1$

$r_i = s_i - s_1$

$$m = (v_3 \; r_2^2 \; w_4 - v_3 \; r_2 \; r_4 \; w_2 - r_3 \; r_2 \; v_2 \; w_4 + r_3 \; r_4 \; w_2 \; v_2 - v_4 \; r_2^2 \; w_3 + v_4 \; r_2 \; w_3 + v_4 \; r_2 \; r_3 \; w_2 + r_4 \; v_2 \; w_3 \; r_2 - r_4 \; v_2 \; r_3 \; w_2) \; x_0^{s4}$$

$$h = (v_3 \; r_2 - r_3 \; v_2) \; x_0^{s3}$$

$$j = s_2 \; x_0^{s2}$$

$$d = x_0^{s1}$$

$$AA1 = \frac{p_1(v_3 \; r_2^2 - r_2 \; r_3 \; v_2) + s_1(r_3 \; v_2 \; w_2 - v_3 \; r_2 \; w_2)}{m} + \frac{u_1(r_2 \; w_2 \; r_3 - r_2^2 \; w_3) + s_1(v_2 \; r_2 \; w_3 - v_2 \; r_3 \; w_2)}{m}$$

$$AA2 = \frac{-v_3 \; r_2 \; w_2 + r_3 \; v_3 \; w_2 + v_2 \; w_3 \; r_2 - v_2 \; r_3 \; w_2}{m}$$

$$AA3 = \frac{r_2 \; w_2 \; r_3 - r_2^2 \; w_3}{m}$$

$$AA4 = \frac{v_3 \; r_2^2 - r_3 \; r_2 \; v_2}{m}$$

$$BB1 = \frac{s_1 \; u_2 \; r_2 - s_1 \; u_1 \; r_2 - u_1 \; r_2 + AA1 \; v_4 \; r_2 - AA1 \; r_4 \; v_2}{h}$$

$$BB2 = \frac{r_2 \; u_2 - r_2 \; u_1 + r_2 \; AA2 \; v_4 - r_4 \; v_2}{h}$$

$$BB3 = \frac{r_2 - AA3 \; v_4 \; r_2 - AA3 \; r_4 \; v_2}{h}$$

$$BB4 = \frac{AA4 \; r_2 \; v_4 - AA4 \; r_4 \; v_2}{h}$$

$$CC1 = \frac{-s_1 - BB1 \; r_3 + AA1 \; r_4}{j}$$

$$CC2 = \frac{1 + BB2 \; r_3 - AA2 \; r_4}{j}$$

$$CC3 = \frac{-BB3 \; r_3 - AA3 \; r_4}{j}$$

$$CC4 = \frac{BB4 \; r_3 - AA4 \; r_4}{j}$$

$$DD1 = \frac{1 + AA1 - BB1 - CC1}{d}$$

$$DD2 = \frac{-AA2 + BB2 - CC2}{d}$$

$$DD3 = \frac{-AA3 - BB3 - CC3}{d}$$

$$DD4 = \frac{-AA4 + BB4 - CC4}{d}$$

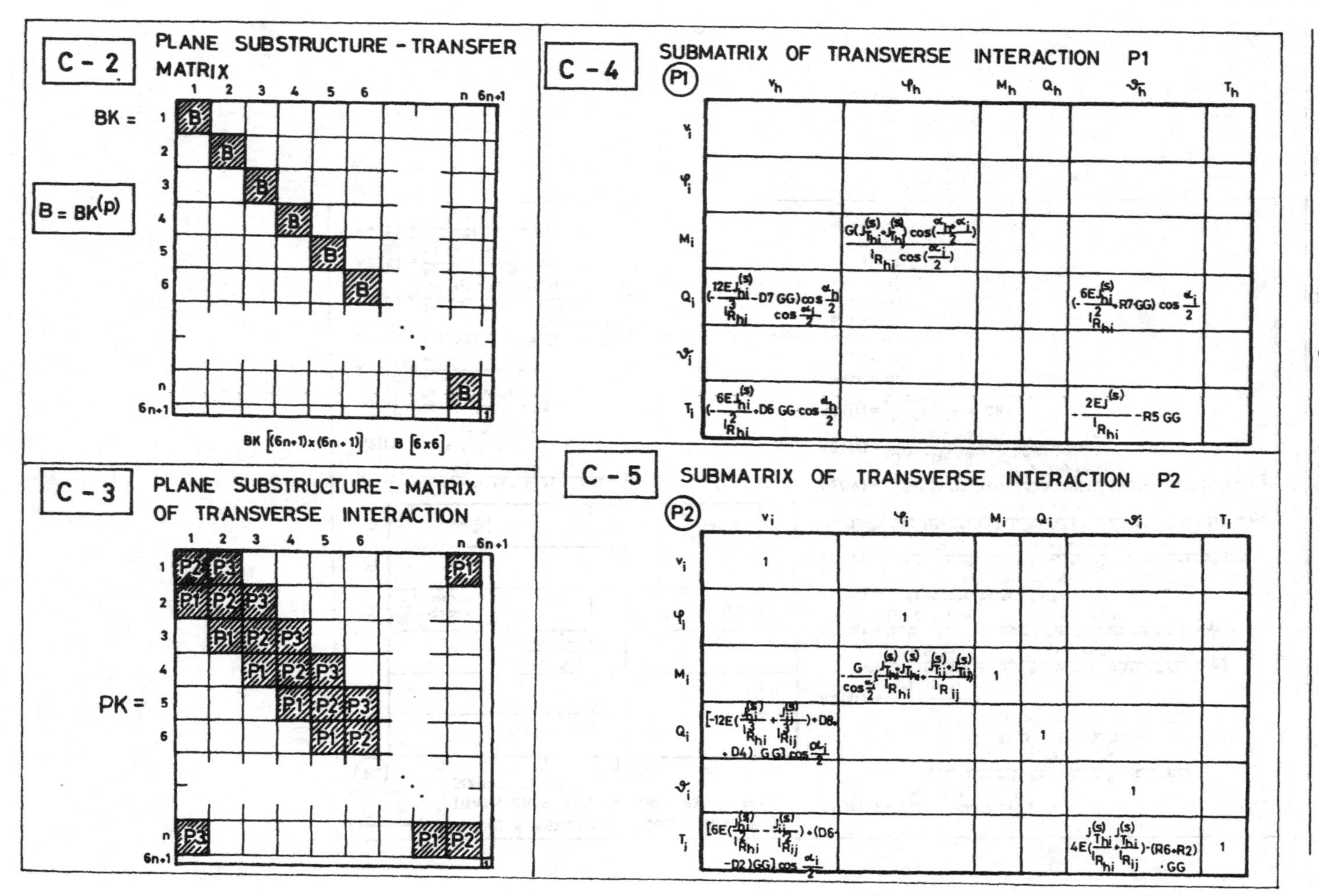
C - 2
PLANE SUBSTRUCTURE - TRANSFER MATRIX
BK =
B = BK(p)
BK [(6n+1)x(6n+1)] B [6x6]
C - 3
PLANE SUBSTRUCTURE - MATRIX OF TRANSVERSE INTERACTION
PK =
C - 4
SUBMATRIX OF TRANSVERSE INTERACTION P1
C - 5
SUBMATRIX OF TRANSVERSE INTERACTION P2

C - 6

PLANE SUBSTRUCTURE - SUBMATRIX OF TRANSVERSE INTERACTION P3 WITH SUBSTITUTIONS

(P3)

	v_j	φ_j	M_j	Q_j	ϑ_j	T_j
v_i						
φ_i						
M_i		$\frac{G(J_{ij}^{(s)}+J_{ij}^{(s)})\cos(\frac{\alpha_i+\alpha_j}{2})}{l_{R_{ij}}\cos\frac{\alpha_i}{2}}$				
Q_i	$\left(\frac{12EJ_{ij}^{(s)}}{l_{R_{ij}}^{3}}-D3\,GG\right)\cdot\cos\frac{\alpha_i}{2}\cos\frac{\alpha_j}{2}$				$\left(\frac{6EJ_{ij}^{(s)}}{l_{R_{ij}}^{2}}-R3\,GG\right)\cos\frac{\alpha_i}{2}$	
ϑ_i						
T_i	$\left(\frac{6EJ_{ij}^{(s)}}{l_{R_{ij}}^{2}}-D1\,GG\right)\cos\frac{\alpha_j}{2}$				$-\frac{2EJ_{ij}^{(s)}}{l_{R_{ij}}}-R1\,GG$	

$$A1(I) = \frac{h_3\,e_4 - h_4\,e_3 + h_4\,e_2 - h_3\,e_2}{(m_4\,h_3 - h_4\,m_3)\,e_2} \rightarrow R1$$

$$A2(I) = \frac{h_4\,e_3 - h_3\,e_4}{(m_4\,h_3 - h_4\,m_3)\,e_2} \rightarrow R2$$

$$A3(I) = \frac{-h_4}{m_4\,h_3 - h_4\,m_3} \rightarrow R3$$

$$A4(I) = \frac{h_3}{m_4\,h_3 - h_4\,m_3} \rightarrow R4$$

$$A1(II) = \frac{2m_3\,h_4(e_2-1) + h_3\,m_4\,e_3 + m_3\,e_4 - e_2\,h_3\,m_4 - e_2\,m_3}{e_2(h_3\,m_4 - m_3\,h_4)} \rightarrow R5$$

$$A2(II) = \frac{2m_3\,h_4 - h_3\,m_4\,e_3 - m_3\,e_4}{e_2(h_3\,m_4 - m_3\,h_4)} \rightarrow R6$$

$$A3(II) = \frac{h_3\,m_4 - 2m_3\,h_4}{h_3\,m_4 - h_4\,m_3} \rightarrow R7$$

$$A4(II) = \frac{m_3}{h_3\,m_4 - h_4\,m_3} \rightarrow R8$$

$$A1(III) = 1 - \frac{c_1}{e_2} + \frac{c_1\,f_2}{e_2\,h_3}\left(\frac{e_3}{e_2} - 1\right) + \frac{c_1\,g_2\,h_3 - a_1\,h_3\,e_2 + m_3\,b_1\,e_2 - m_3 c_1 f_2}{m_4\,h_3 - m_3\,h_4} \cdot \left(\frac{h_4 - 1}{e_2} - \frac{e_3\,h_4}{e_2\,e_2}\right) + \frac{h_1}{h_3} - \frac{b_1\,e_3}{h_3\,e_2} \rightarrow D1$$

$$A2(III) = -\frac{c_1}{e_2} - \frac{c_1\,f_2\,e_3}{e_2\,f_3\,e_2} + \frac{b_1\,e_3\,h_4}{h_3\,e_2\,e_2}\,\frac{c_1\,g_2\,h_3 - a_1\,h_3\,e_2 + m_3 b_1 e_2 - m_3 c_1 f_2}{m_4 h_3 - m_3 h_4} \rightarrow D2$$

$$A3(III) = \frac{c_1\,f_2}{e_2\,h_3} - \frac{b_1}{h_3} - \frac{h_4}{e_2}\,\frac{c_1 g_2 h_3 - a_1 h_3 e_2 + m_3 b_1 e_2 - m_3 c_1 f_2}{m_4\,h_3 - m_3\,h_4} \rightarrow D3$$

$$A4(III) = \frac{c_1\,g_2\,h_3 - a_1\,h_3\,e_2 + m_3\,b_1\,e_2 - m_3\,c_1\,f_2}{e_2(m_4\,h_3 - m_3\,h_4)} \rightarrow D4$$

$$A1(IV) = \frac{-e_2\,h_4\,m_3 - m_4\,e_2 + e_2\,h_4\,g_2\,e_3 - 2f_2\,h_4\,m_3\,e_2\,e_3 - f_2 m_4 e_2 e_3}{e_2\,e_2(h_4\,m_3 + m_4)} + \frac{f_2\,m_3\,m_4 - g_2 e_4 - e_2 h_4 g_2 e_2 + 2e_2 e_2 h_4 m_3 + e_2 f_2 m_4 e_2 - f_2 m_3 e_2 + g_2 e_2}{e_2\,e_2(h_4\,m_3 + m_4)} \rightarrow D5$$

$$A2(IV) = \frac{(h_4 m_3 + m_4)e_2 - e_2 h_4 g_2 e_3 + 2f_2 h_4 m_3 e_2 e_3 + f_2 m_4 e_2 e_3 - f_2 m_3 m_4 + g_2 e_4}{e_2\,e_2(h_4 m_3 + m_4)} \rightarrow D6$$

$$A3(IV) = \frac{e_2 h_4 g_2 - 2f_2 h_4 m_3 m_2 - f_2 m_4 e_2}{e_2(h_4\,m_3 + m_4)} \rightarrow D7$$

$$A4(IV) = \frac{f_2\,m_3 - g_2}{e_2(h_4 m_3 + m_4)} \rightarrow D8$$

Substitutions

$c_i = x^{a_i}$

$b_i = a_i\;x^{a_i - 1}$

$g_i = a_i - a_1$

$f_i = b_i - b_1$

$e_i = c_i - c_1$

$m_i = g_i - \frac{g_2\;e_i}{e_2}$

$h_i = f_i - \frac{f_2\;e_i}{e_2}$

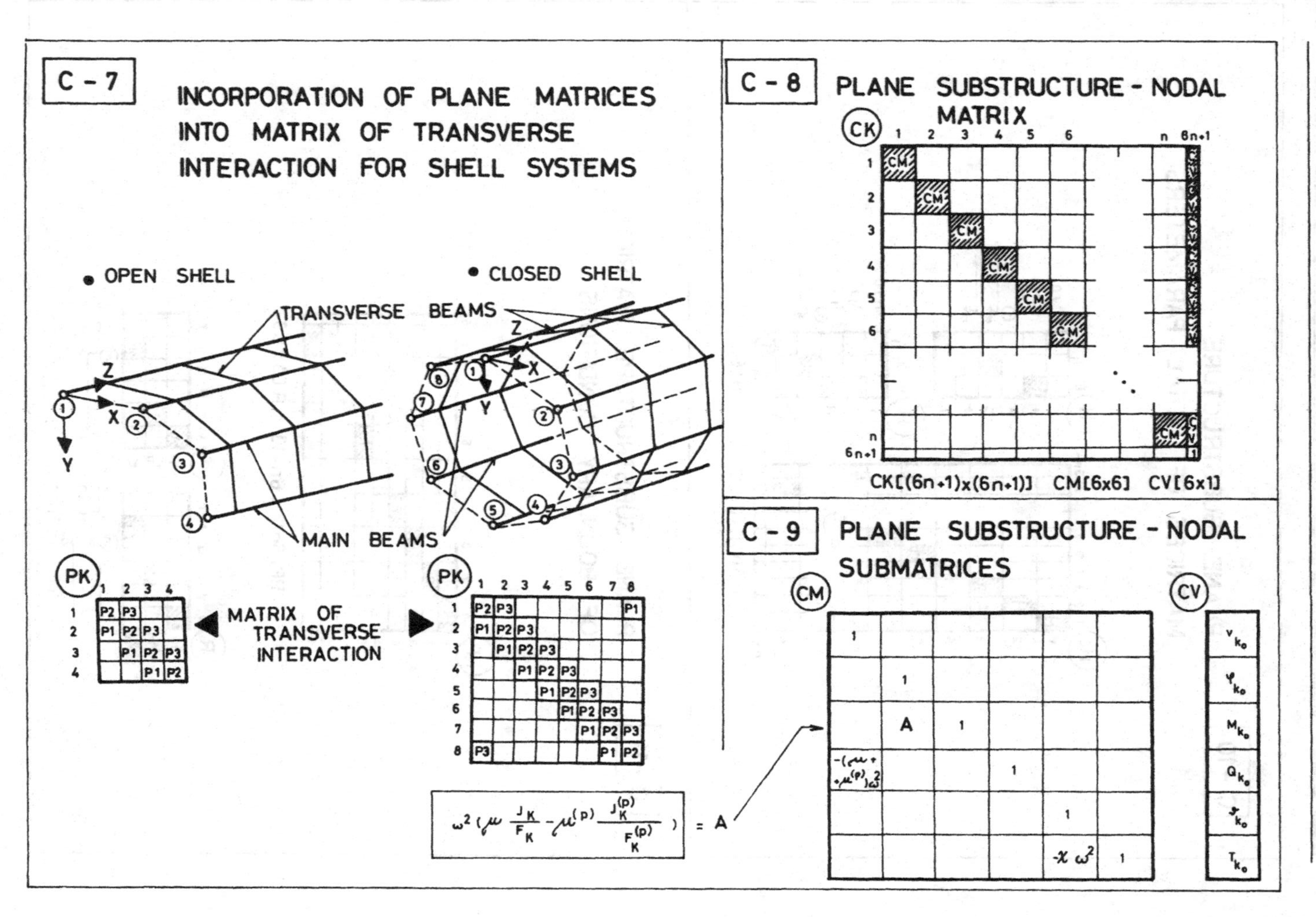
C – 7
INCORPORATION OF PLANE MATRICES INTO MATRIX OF TRANSVERSE INTERACTION FOR SHELL SYSTEMS
• OPEN SHELL
• CLOSED SHELL
TRANSVERSE BEAMS
MAIN BEAMS
X
Y
Z
PK
MATRIX OF TRANSVERSE INTERACTION
C – 8
PLANE SUBSTRUCTURE - NODAL MATRIX
CK
CM
CV
CK[(6n+1)x(6n+1)] CM[6x6] CV[6x1]
C – 9
PLANE SUBSTRUCTURE - NODAL SUBMATRICES
CM
CV
A
$\omega^2 (\mu \frac{J_K}{F_K} - \mu^{(p)} \frac{J_K^{(p)}}{F_K^{(p)}}) = A$
$-(\mu + \mu^{(p)})\omega^2$
$-\chi \omega^2$
v_{k_o}
φ_{k_o}
M_{k_o}
Q_{k_o}
ϑ_{k_o}
T_{k_o}

C - 10

PLANE SUBSTRUCTURE
MATRICES OF INITIAL PARAMETERS

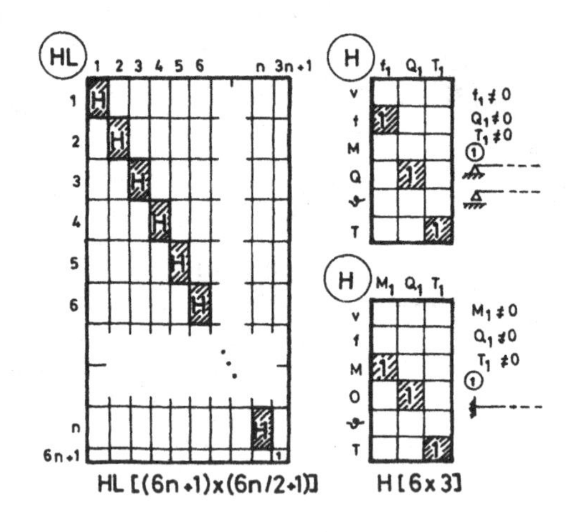

C - 11

PLANE SUBSTRUCTURE - MATRICES
OF BOUNDARY CONDITIONS

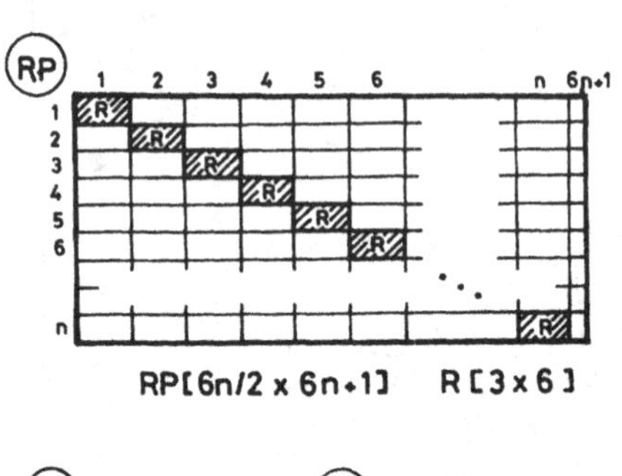

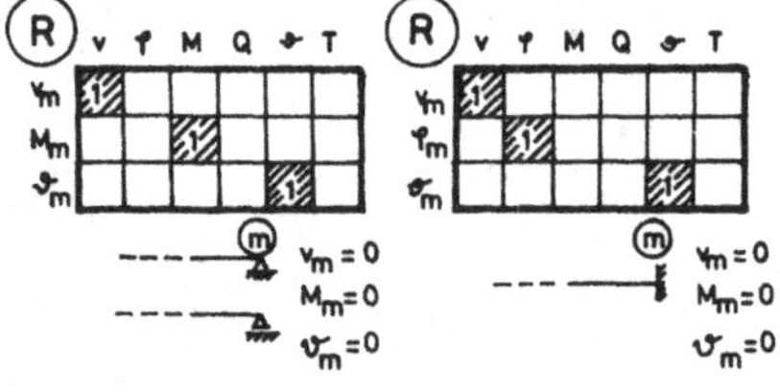

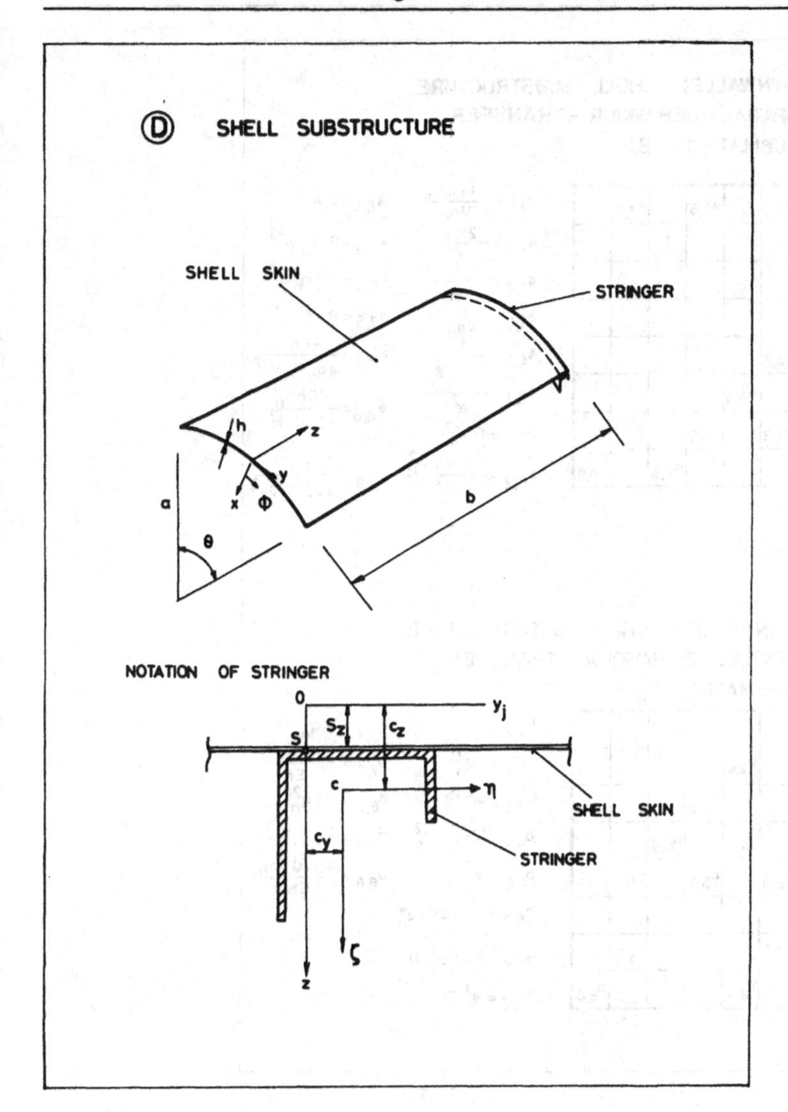

Ⓓ SHELL SUBSTRUCTURE
SHELL SKIN
STRINGER
h
z
y
x
Φ
a
θ
b
NOTATION OF STRINGER
0
y_j
s_z
c_z
S
c
η
SHELL SKIN
c_y
STRINGER
ζ
z

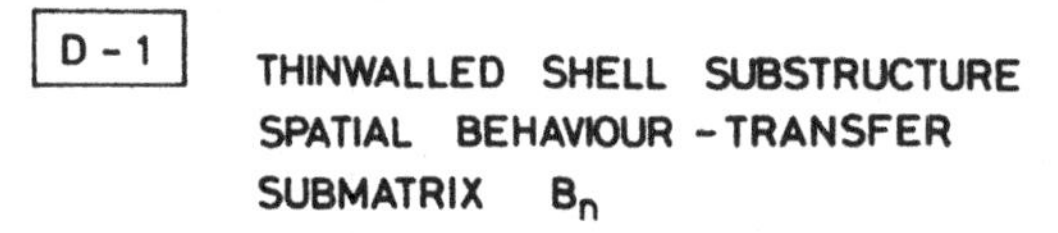

D - 1

THINWALLED SHELL SUBSTRUCTURE SPATIAL BEHAVIOUR - TRANSFER SUBMATRIX B_n

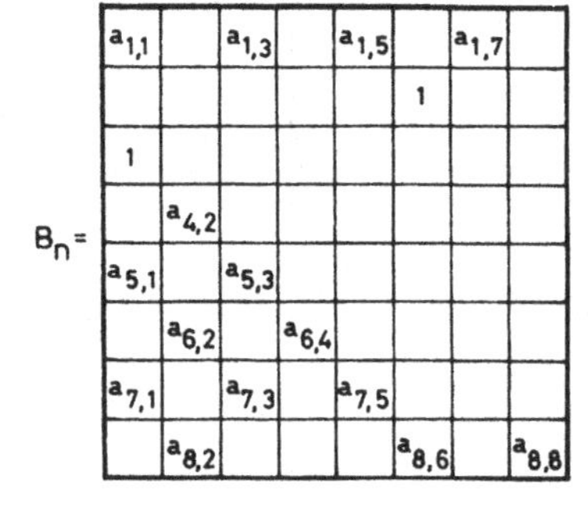

$$B_n = \begin{bmatrix} a_{1,1} & & a_{1,3} & & a_{1,5} & & a_{1,7} & \\ & & & & & 1 & & \\ 1 & & & & & & & \\ & a_{4,2} & & & & & & \\ a_{5,1} & & a_{5,3} & & & & & \\ & a_{6,2} & & a_{6,4} & & & & \\ a_{7,1} & & a_{7,3} & & a_{7,5} & & & \\ & a_{8,2} & & & & a_{8,6} & & a_{8,8} \end{bmatrix}$$

$a_{1,1} = -\frac{1+kq_n}{\nu q_n}$

$a_{1,3} = \frac{2kq_n}{\nu}$

$a_{1,5} = -\frac{k}{\nu q_n}$

$a_{1,7} = \frac{k}{\nu q_n}$

$a_{4,2} = a^{-1}$,

$a_{5,1} = -\frac{\nu D q_n^2}{a^2}$

$a_{5,3} = D\,a^{-2}$

$a_{6,2} = -\frac{(2-\nu)D\,q_n^2}{a^3}$

$a_{6,4} = D\,a^{-3}$

$a_{7,1} = D\,q_n^4\,a^{-3}$

$a_{7,3} = -2Dq_n^2 a^{-3}$

$a_{7,5} = D\,a^{-3}$

$a_{8,2} = \frac{-Eh}{a q_n (1+\nu)^2}$

$a_{8,6} = \frac{\nu E h q_n}{a(1+\nu)^2}$

$a_{8,8} = \frac{Eh}{a q_n (1+\nu)^2}$

D - 2

THINWALLED SHELL SUBSTRUCTURE SPATIAL BEHAVIOUR - TRANSFER SUBMATRIX B_n^{-1}

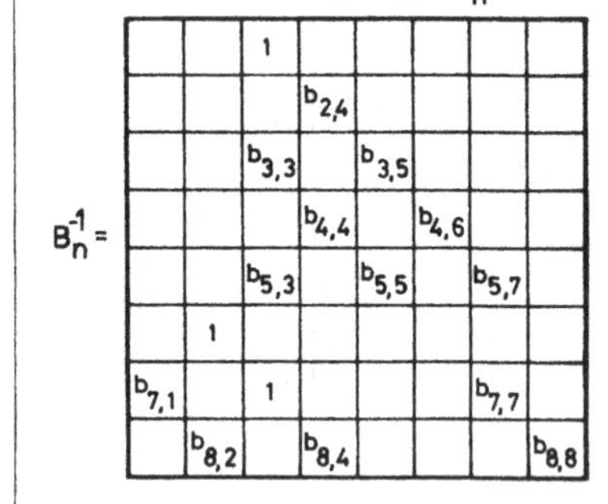

$$B_n^{-1} = \begin{bmatrix} & & 1 & & & & & \\ & & & b_{2,4} & & & & \\ & & b_{3,3} & & b_{3,5} & & & \\ & & & b_{4,4} & & b_{4,6} & & \\ & & b_{5,3} & & b_{5,5} & & b_{5,7} & \\ & 1 & & & & & & \\ b_{7,1} & & 1 & & & & b_{7,7} & \\ & b_{8,2} & & b_{8,4} & & & & b_{8,8} \end{bmatrix}$$

$b_{2,4} = a$

$b_{3,3} = q_n^2$

$b_{3,5} = a^2/D$

$b_{4,4} = (2-\nu)a q_n^2$

$b_{4,6} = a^3/D$

$b_{5,3} = -(1-2\nu)\,q_n^4$

$b_{5,5} = 2a^2 q_n^2/D$

$b_{5,7} = a^3/D$

$b_{7,1} = \nu\,q_n$

$b_{7,7} = \frac{a(1-\nu^2)}{Eh}$

$b_{8,2} = -\nu\,q_n^2$

$b_{8,4} = a$

$b_{8,8} = \frac{a(1+\nu)^2 q_n}{Eh}$

D - 3

THINWALLED SHELL SUBSTRUCTURE
SPATIAL BEHAVIOUR - NODAL
SUBMATRIX FOR STRINGER

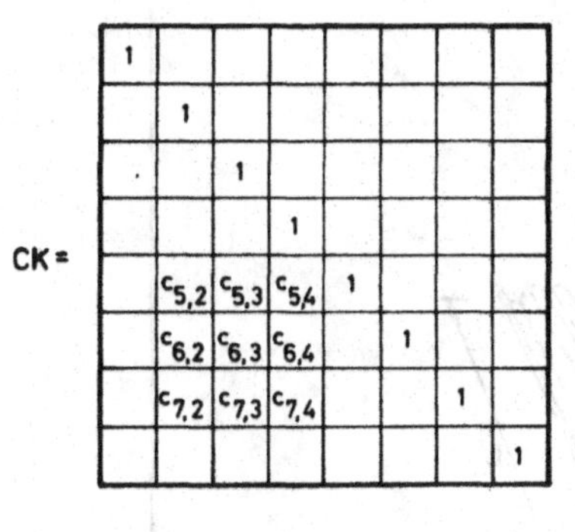

CK =

1							
	1						
		1					
			1				
	$c_{5,2}$	$c_{5,3}$	$c_{5,4}$	1			
	$c_{6,2}$	$c_{6,3}$	$c_{6,4}$		1		
	$c_{7,2}$	$c_{7,3}$	$c_{7,4}$			1	
							1

$c_{5,2} = \bar{f}$

$c_{5,3} = \bar{d}$

$c_{5,4} = \bar{c} - I\ \omega^2$

$c_{6,2} = -\bar{g}$

$c_{6,3} = -\bar{e} + m\ \omega^2$

$c_{6,4} = -\bar{d}$

$c_{7,2} = \bar{h} - m\ \omega^2$

$c_{7,3} = \bar{g}$

$c_{7,4} = \bar{f}$

SUBSTITUTIONS

$$\bar{f} = EJ_{\zeta} S_z \left(\frac{n\pi}{l}\right)^4 + \rho A (c_z - S_z)\ \omega^2$$

$$\bar{g} = EJ_{\eta\zeta} \left(\frac{n\pi}{l}\right)^4$$

$$\bar{h} = EJ_{\zeta} \left(\frac{n\pi}{l}\right)^4 - \rho A\ \omega^2$$

$$\bar{c} = EC_{WS} \left(\frac{n\pi}{l}\right)^4 + GC \left(\frac{n\pi}{l}\right)^2 - \omega^2 \rho J_S$$

$$\bar{d} = EJ_{\eta\zeta} S \left(\frac{n\pi}{l}\right)^4 - \omega^2 \rho A c_y$$

$$\bar{e} = EJ_{\eta} \left(\frac{n\pi}{l}\right)^4 - \omega^2 \rho A$$

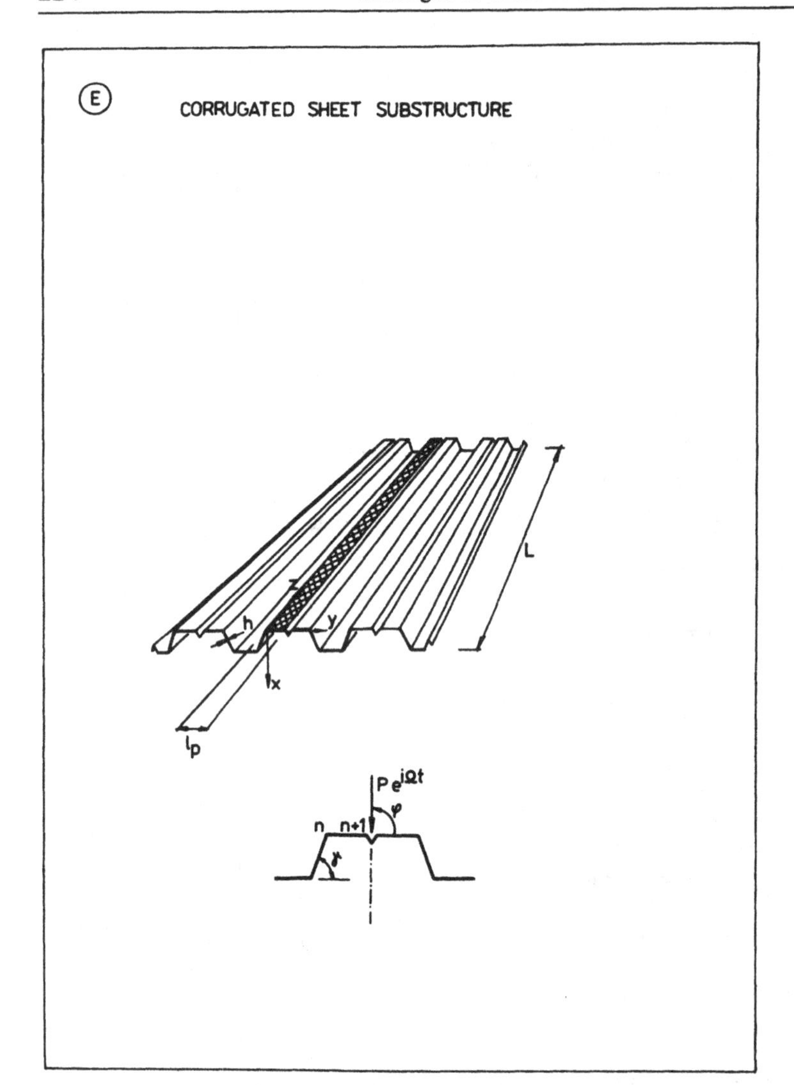
E
CORRUGATED SHEET SUBSTRUCTURE
L
h
z
y
x
l_p
$P e^{i\Omega t}$
φ
n
n+1
γ

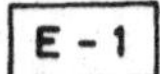

CORRUGATED SHEET SUBSTRUCTURE
FLEXURAL AND INPLANE BEHAVIOUR
TRANSFER SUBMATRIX

BKS=

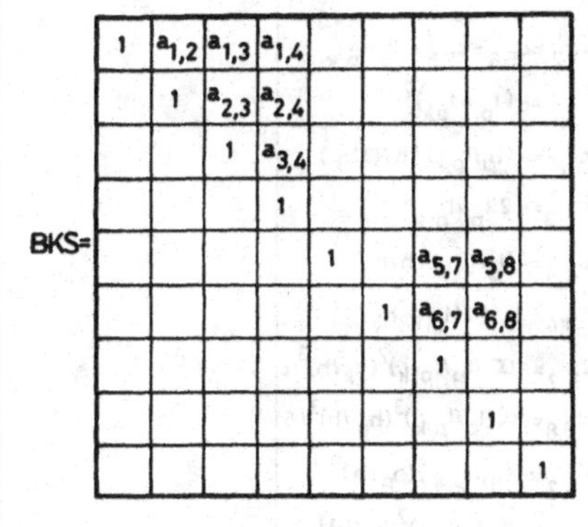

$a_{1,2} = 2\, l_p/L$

$a_{1,3} = -4\, l_p/L$

$a_{1,4} = -65\, E_p\, e\, l_p^3\,/(6L^4)$

$a_{2,3} = -\,4\, l_p/L$

$a_{2,4} = -96\, E_p\, h\, l_p^2\,/(5L^3)$

$a_{3,4} = 48\, E_p\, h\, l_p^2\,/(5L^2)$

$a_{5,7} = -\, l_p^2\,/(2\,E_p\, J_{p,x})$

$a_{5,8} = -\, l_p^3\,/(6E_p\, J_{p,x})$

$a_{6,7} = -\, l_p\,/(E_p\, J_{p,x})$

$a_{6,8} = -\, l_p^2\,/(2\,E_p\, J_{p,x})$

E - 2

CORRUGATED SHEET SUBSTRUCTURE
FLEXURAL AND INPLANE BEHAVIOUR
NODAL SUBMATRIX

CKS=

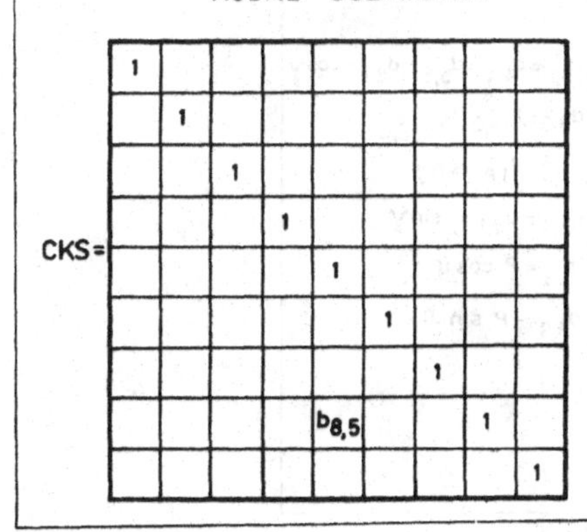

$b_{8,5} = -m\, \Omega^2$

E - 3

CORRUGATED SHEET SUBSTRUCTURE
FLEXURAL AND INPLANE BEHAVIOUR
MODIFIED TRANSFER SUBMATRIX

1	$c_{1,2}$	$c_{1,3}$	$c_{1,4}$					
	1	$c_{2,3}$	$c_{2,4}$					
		1	$c_{3,4}$					
			1					
				1	$c_{5,6}$	$c_{5,7}$	$c_{5,8}$	
					1	$c_{6,7}$	$c_{6,8}$	
						1	$c_{7,8}$	
							1	
								1

$c_{1,2} = c_{5,6} = c_{7,8} = l_p / l_{p,k}$

$c_{1,3} = -(l_p / l_{p,k})^2$

$c_{1,4} = -(l_p / l_{p,k})^3 \, h / (3 h_k)$

$c_{2,3} = -2 \, l_p / l_{p,k}$

$c_{2,4} = -(l_p / l_{p,k})^2 \, h / h_k$

$c_{3,4} = l_p \, h / (l_{p,k} \, h_k)$

$c_{5,7} = -\alpha \, (l_p / l_{p,k})^2 (h_k / h)^3 / 2$

$c_{5,8} = -\alpha (l_p / l_{p,k})^3 (h_p / h)^3 / 6$

$c_{6,7} = -(l_p / l_{p,k})(h_p / h)^3$

$c_{6,8} = -\alpha (l_p / l_{p,k})^2 (h_p / h)^3 / 2$

E - 4

CORRUGATED SHEET SUBSTRUCTURE
FLEXURAL AND INPLANE BEHAVIOUR
TRANSFORMATION SUBMATRIX

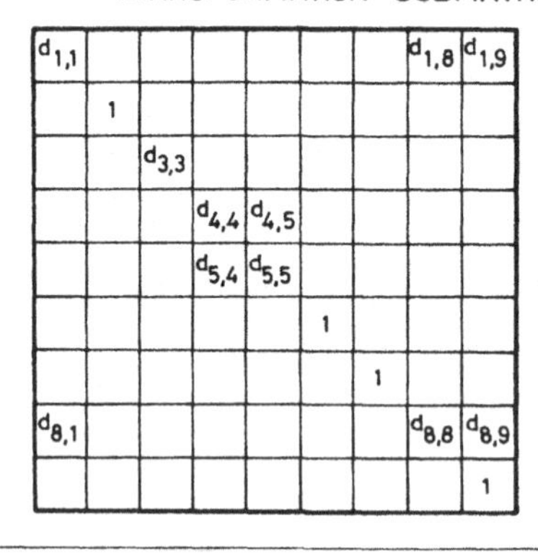

$d_{1,1}$							$d_{1,8}$	$d_{1,9}$
	1							
		$d_{3,3}$						
			$d_{4,4}$	$d_{4,5}$				
			$d_{5,4}$	$d_{5,5}$				
					1			
						1		
$d_{8,1}$							$d_{8,8}$	$d_{8,9}$
								1

$d_{1,1} = d_{4,4} = d_{5,5} = d_{8,8} = \cos\gamma$

$d_{3,3} = h_{n+2} / h_{n+1}$

$d_{4,5} = d_{1,8} = \sin\gamma$

$d_{5,4} = d_{8,1} = -\sin\gamma$

$d_{1,9} = P \cos\varphi$

$d_{8,9} = -P \sin\varphi$

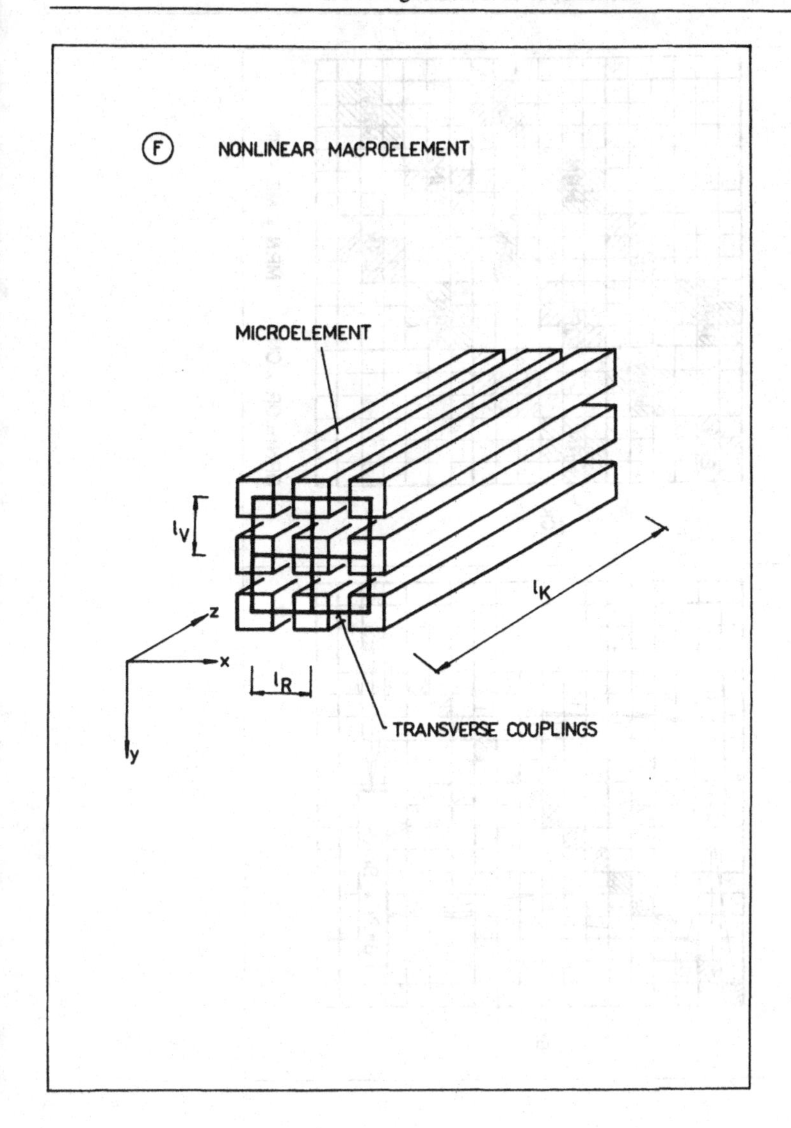
F
NONLINEAR MACROELEMENT
MICROELEMENT
l_V
l_K
z
x
y
l_R
TRANSVERSE COUPLINGS

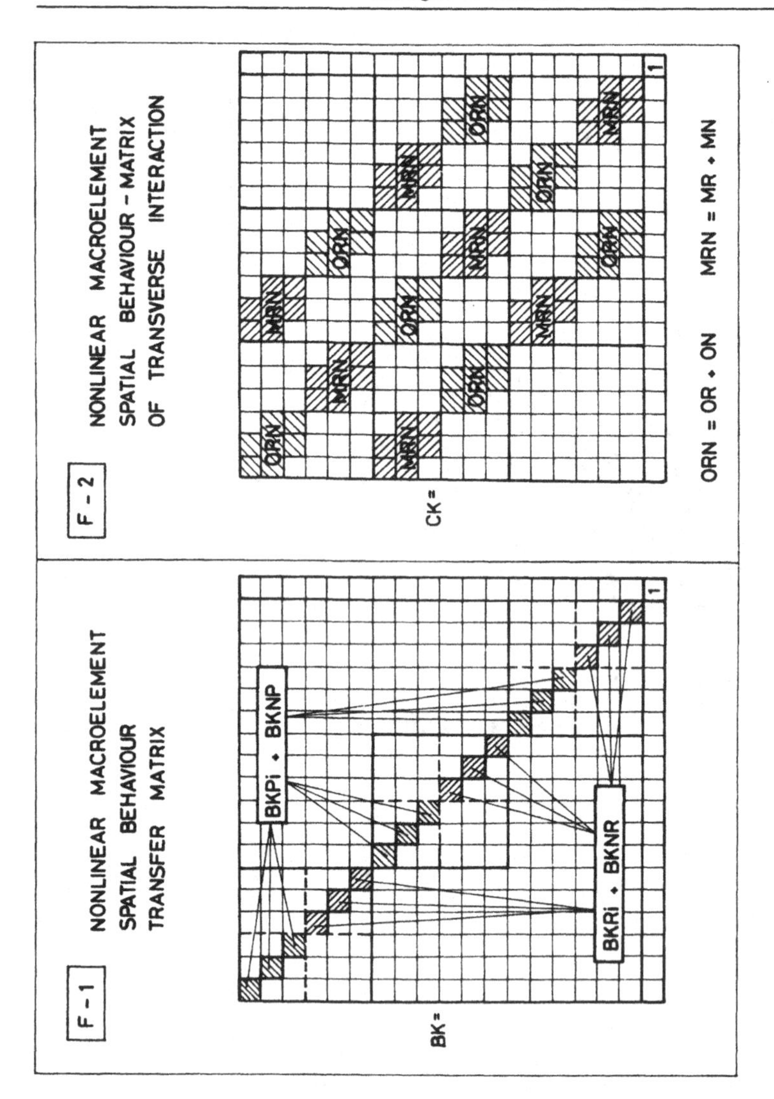
F - 1
NONLINEAR MACROELEMENT
SPATIAL BEHAVIOUR
TRANSFER MATRIX
BK =
BKPi + BKNP
BKRi + BKNR
1
F - 2
NONLINEAR MACROELEMENT
SPATIAL BEHAVIOUR - MATRIX
OF TRANSVERSE INTERACTION
CK =
ORN
MRN
1
ORN = OR + ON
MRN = MR + MN

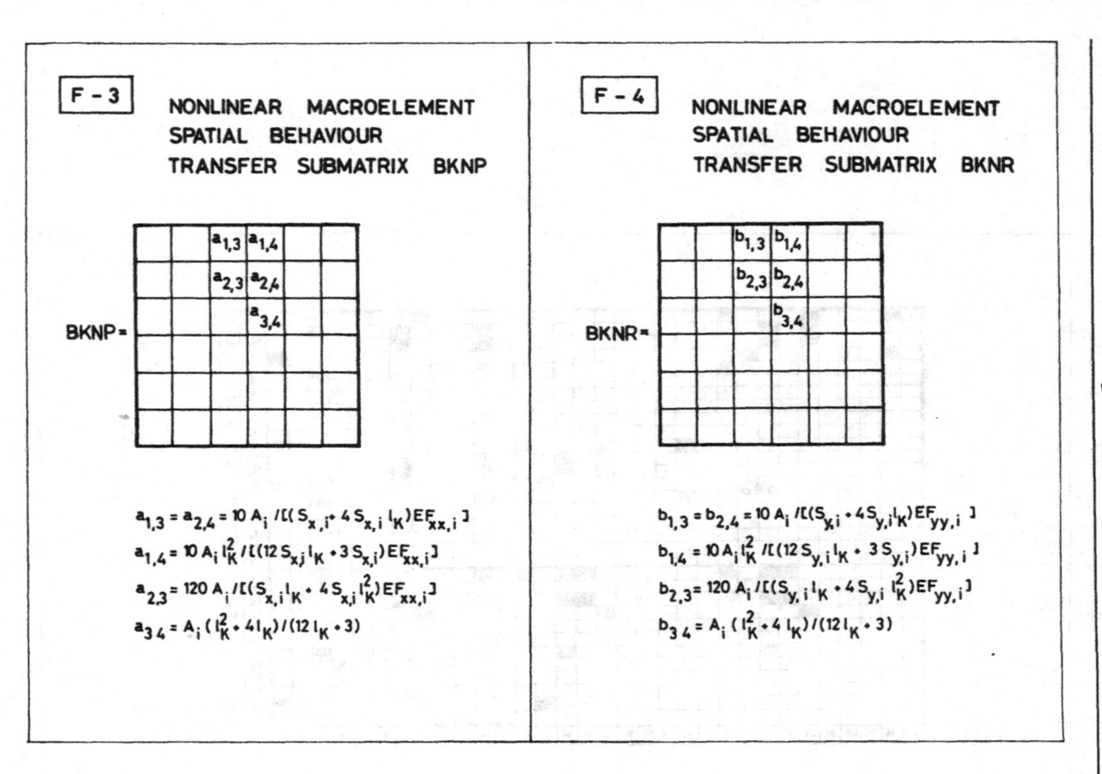
F - 3
NONLINEAR MACROELEMENT
SPATIAL BEHAVIOUR
TRANSFER SUBMATRIX BKNP
BKNP =
a1,3 a1,4
a2,3 a2,4
a3,4
a1,3 = a2,4 = 10 Ai /[(Sx,i + 4 Sx,i lK)EFxx,i]
a1,4 = 10 Ai lK2 /[(12 Sx,i lK + 3 Sx,i)EFxx,i]
a2,3 = 120 Ai /[(Sx,i lK + 4 Sx,i lK2)EFxx,i]
a3 4 = Ai (lK2 + 4 lK)/(12 lK + 3)
F - 4
NONLINEAR MACROELEMENT
SPATIAL BEHAVIOUR
TRANSFER SUBMATRIX BKNR
BKNR =
b1,3 b1,4
b2,3 b2,4
b3,4
b1,3 = b2,4 = 10 Ai /[(Sy,i + 4 Sy,i lK)EFyy,i]
b1,4 = 10 Ai lK2 /[(12 Sy,i lK + 3 Sy,i)EFyy,i]
b2,3 = 120 Ai /[(Sy,i lK + 4 Sy,i lK2)EFyy,i]
b3 4 = Ai (lK2 + 4 lK)/(12 lK + 3)

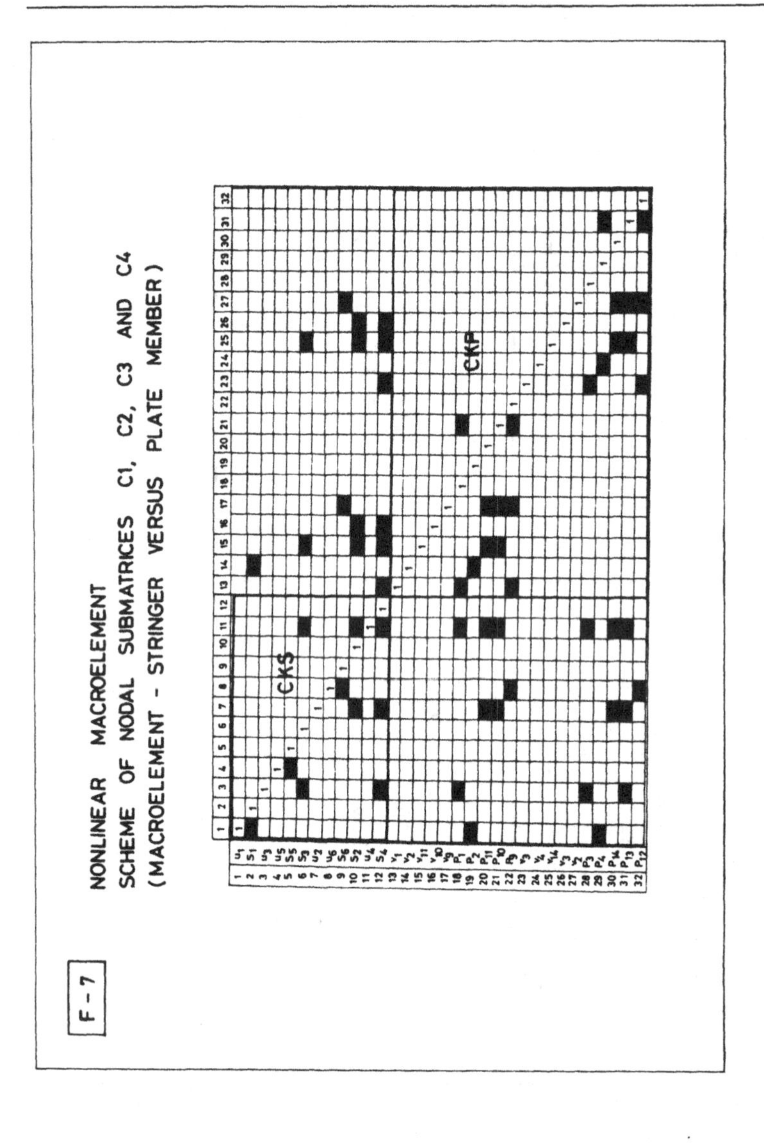
F - 7
NONLINEAR MACROELEMENT
SCHEME OF NODAL SUBMATRICES C1, C2, C3 AND C4
(MACROELEMENT - STRINGER VERSUS PLATE MEMBER)
CKS
CKP

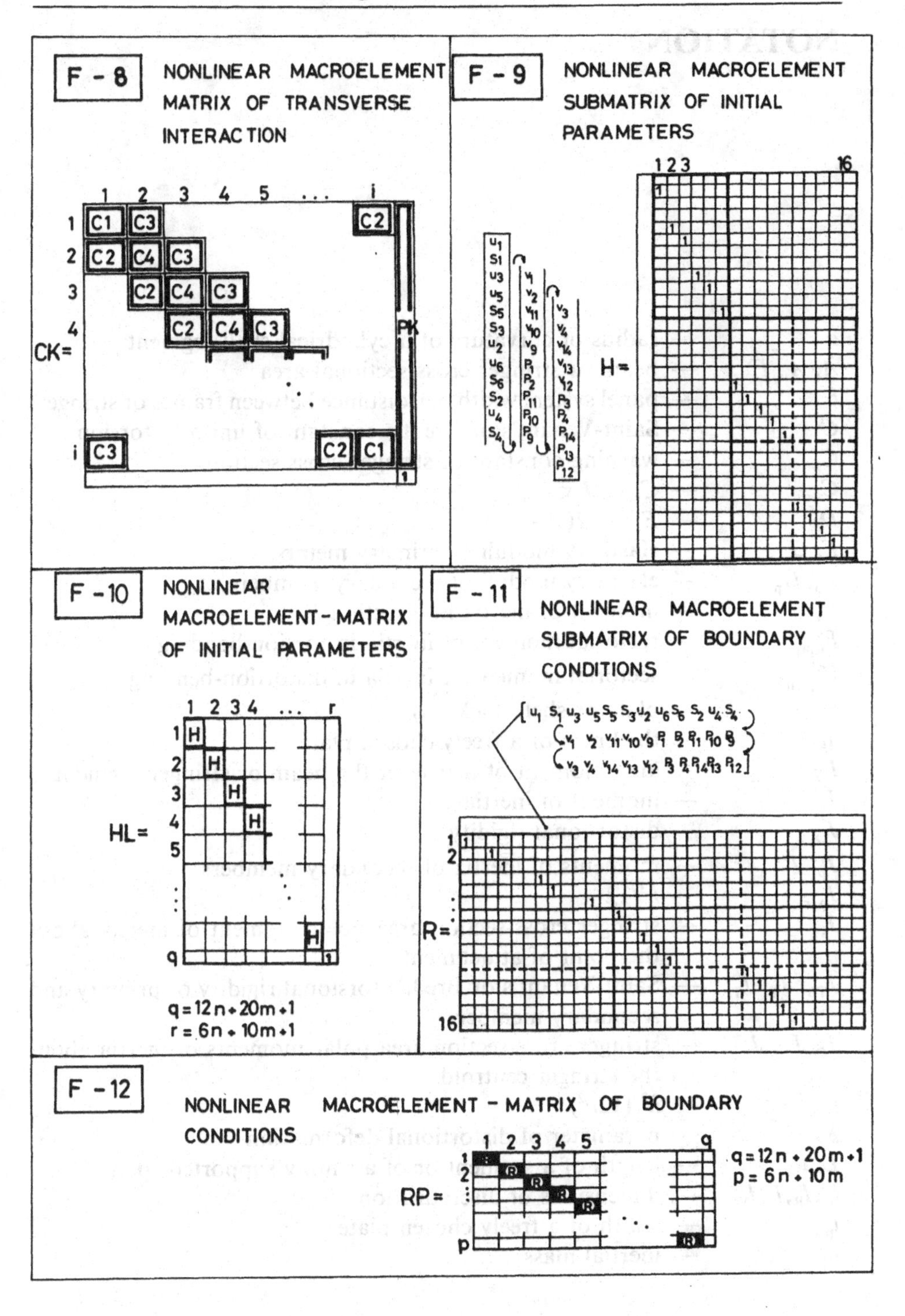
F - 8
NONLINEAR MACROELEMENT MATRIX OF TRANSVERSE INTERACTION
CK=
PK
F - 9
NONLINEAR MACROELEMENT SUBMATRIX OF INITIAL PARAMETERS
H=
F -10
NONLINEAR MACROELEMENT-MATRIX OF INITIAL PARAMETERS
HL=
q = 12n + 20m + 1
r = 6n + 10m + 1
F - 11
NONLINEAR MACROELEMENT SUBMATRIX OF BOUNDARY CONDITIONS
R=
F -12
NONLINEAR MACROELEMENT - MATRIX OF BOUNDARY CONDITIONS
RP=
q = 12n + 20m + 1
p = 6n + 10m

NOTATION

a	—	radius of curvature of a cylindrical shell segment
A, A_i, F_c^s	—	beam or stringer cross-sectional area
b	—	panel system width, i.e. distance between frames or stringers
C	—	Saint-Venant's or Bredt's constant of uniform torsion
C_w	—	warping constant of stringer cross-section
C_{ws}	—	$C_w + J_\zeta s_z^2$
D	—	$Eh^3/(12(1 - \nu^2))$
E, G	—	elasticity moduli of primary members
E_p, G_p	—	elasticity moduli of secondary members
F_{ii}^s	—	moment of inertia in direction i
$F_{w_I w_I}^s$	—	sectorial moment of inertia in torsion-bending
$F_{w_{II} w_{II}}^s$	—	sectorial moment of inertia in distortion-bending
h	—	plate or shell thickness
h_K	—	thickness of a freely chosen plate
I	—	area moment of inertia of the beam or stringer segment
J	—	moment of inertia
J_{D_I}	—	distortional rigidity
J_p^s, J^s	—	moments of inertia of secondary members
$J_{p,x}$	—	$h^3/12$
J_s	—	stringer cross-section area polar moment of inertia about the point of attachment
J_{TB}, J_T^s, J_T	—	Saint-Venant's or Bredt's torsional rigidity of primary and secondary members
J_ζ, $J_{\eta\zeta}$, J_η	—	stringer cross-section area polar moments of inertia about the stringer centroid
k	—	$h^2/(12a^2)$
K_p	—	parameter of distortional deformation
L, l	—	length of a segment or of a simply supported span
l_K, l_R, l_P, l_V	—	dimensions of discretization
$l_{p,K}$	—	width of a freely chosen plate
m	—	inertial mass

n — edge or mode number; index number of panel normal modes
P — inducing force
P_C, P_T — compressive and tensile force
q_n — $n\pi a/b$
s_i — parameter (see Section 2.6)
$S_{x,i}$, $S_{y,i}$ — pseudo-forces
v, φ, M, Q — state parameters in bending
x, y, z — coordinates
α, Ω, λ — parameters (see Section 2.6)
α_i — position angle
ϑ, T — state parameters in pure torsion
ω, Ω — circular frequency, forcing frequency
μ, $\mu^{(p)}$ — inertial mass of stringer and plate member
μ — $GJ_{TB}/(EF^{s}_{w_I w_I})$
γ — angle between two adjacent plates
ϱ — mass density
φ — angle between the load and the plate considered
Θ — angle of curvature
ν — Poisson's ratio
χ — inertial parameter in torsion
$\varkappa_{w_I w_I}$ — coefficient of shear deformation

REFERENCES

[1] Ahmad, S., Irons, B. M. and Zienkiewicz, O. C.: 'Analysis of Thick and Thin Shell Structures by Curved Elements', *Int. J. Num. Methods Engng* **2** (1970) 419—451.

[2] Argyris, J. H. and Dunne, P. C.: 'On the Application of the Natural Mode Technique to Small-Strain Large Displacement Problems', *Proc. World Congress on FEM in Structural Mechanics*, Bournemouth, 1975, Vol. 3, pp. 371—390.

[3] Argyris, J. H., Kelsey, S. and Kamel, W. H.: 'Matrix Methods of Structural Analysis', in F. de Veubeke (ed.), *Agardograph 72*, Pergamon Press, New York, 1964, pp. 105—120.

[4] Argyris, J. H. and Lochner, N.: 'On the Application of the SHEBA Shell Element', *Comp. Methods Appl. Mech. Engng* **1** (1972) 317—347.

[5] Bathe, K. J. and Wilson, E. L.: 'Stability and Accuracy of Direct Integration Methods', *Earthquake Engng Struct. Dynam.* **1** (1973) 107—118.

[6] Bathe, K. J. and Wilson, E. L.: *Numerical Methods in Finite Element Analysis*, Prentice-Hall International Inc., Englewood Cliffs, N. J., 1976.

[7] Belytschko, T. and Schoberle, F. D.: 'On the Unconditional Stability of an Implicit Algorithm for Nonlinear Structural Dynamics', *J. Appl. Mech.* (1975) 723—736.

[8] Bergan, P. G.: 'Nonlinear Analysis of Plates Considering Geometric and Material Effects', Rep. 72-1, The Norwegian Institute of Technology, Trondheim, 1972.

[9] Bergan, P. G., Clough, R. W. and Mojtahedi, S.: 'Analysis of Stiffened Plates Using the FEM', Rep. UCSESM 70-1, University of California, Berkeley, 1970.

[10] Bornscheuer, F. W.: 'Systematische Darstellung des Biege- und Verdrehvorganges unter besonderer Berücksichtigung der Wölbkrafttorsion', *Stahlbau* **21**, H.1 (1952) 6—27.

[11] Brebbia, A. C. and Deb Nath, J. M.: 'A Comparison of Recent Shallow Shell Finite Element Analyses', *Int. J. Mech. Sci.* **12** (1970) 849—857.

[12] Clough, R. W. and Felippa, C. A.: 'A Refined Quadrilateral Element for Analysis of Plate Bending', *Proc. 2nd Conference on Matrix Methods in Structural Mechanics*, 1968, pp. 399—440.

[13] Clough, R.W. and Wilson, E. L.: 'Dynamic Finite Element Analysis of Arbitrary Thin Shells', *Comp. Struct.* **1** (1971) 33—56.

[14] Dąbrowski, R.: 'Der Schubverformungseinfluß auf die Wölbkrafttorsion der Kastenträger mit verformbarem biegesteifem Profil', *Bauingenieur* **40** (1965) 117—140.

[15] Dawe, J. W.: 'Rigid Body Motions and Strain Displacement Equations of Curved Shell Finite Elements', *Int. J. Mech. Sci.* **14** (1972) 568—578.

[16] Denke, P. H.: 'Nonlinear and Thermal Effects on Elastic Vibrations', Tech. Rep. SM-30426, Douglas Aircraft Company Inc., Seattle, 1960.

[17] Dokainish, M. A.: 'A New Approach for Plate Vibrations: Combination of Transfer Matrix and Finite Element Techniques', *J. Engng Ind.* **B94** (1972) 121—137.

[18] Donnell, L. H.: 'A New Theory for the Buckling of Thin Cylinders under Axial Compression and Bending', *Trans. ASME* **56** (1934) 27—42.

[19] Falk, S.: 'Biegen, Knicken und Schwingen des mehrfeldrigen geraden Balkens', *Abh. Braunschweig. Wiss. Gesell.* **7** (1955) 30—41.
[20] Fiedler, M.: *Special Matrices and Their Utilization in Numerical Mathematics*, SNTL, Prague, 1970.
[21] Fletcher, R. and Powell, M. J. D.: 'A Rapidly Convergent Descent Method for Minimization', *Comp. J.* **6** (1936) 117—127.
[22] Fox, R. L.: *Optimization Methods for Engineering Design*, Addison-Wesley Publishing Group, Reading, Mass., 1971.
[23] Fuhrke, H.: 'Bestimmung von Balkenschwingungen mit Hilfe des Matrizenkalküls', *Ing. Arch.* **23**, H. 5 (1955).
[24] Fuhrke, H.: 'Bestimmung von Rahmenschwingungen mit Hilfe des Matrizenkalküls', *Ing. Arch.* **24**, H. 1 (1956).
[25] Gallagher, R. H.: 'Shell Elements', *Proc. World Congress on FEM in Structural Mechanics*, Bournemouth, 1975, Vol. 1, pp. E1—E35.
[26] Gallagher, R. H., Lien, S. and Mau, S. T.: 'A Procedure for Finite Element Plate and Shell Pre- and Post-Buckling Analysis', *Proc. 3rd Conference on Matrix Methods in Structural Mechanics*, Rep. AFFDL-TR-71-160, Wright-Patterson Air Force Base, 1971, pp. 857—879.
[27] Hees, G.: 'Querschnittsverformung des einzelligen Kastenträgers mit vier Wänden in einer zur Wölbkrafttorsion analogen Darstellung', *Bautechnik* 11 (1971).
[28] Heilig, R.: 'Beitrag zur Theorie der Kastenträger beliebiger Querschnittsform', *Stahlbau* **4** (1962).
[29] Henrici, P.: *Discrete Variable Methods in Ordinary Differential Equations*, John Wiley and Sons, New York, 1961 .
[30] Hildebrand, F. B.: *Introduction to Numerical Analysis*, McGraw-Hill Book Company, New York, 1956.
[31] Hill, R.: *The Mathematical Theory of Plasticity*, Oxford University Press, Oxford, 1950.
[32] Holand, I.: *Design of Circular Cylindrical Shells*, Oslo University Press, Oslo, 1957.
[33] Holand, I.: 'Simplified Calculation Methods of Shell Structures', *Proc. Colloquium on Simplified Calculation Methods*, Bruxelles, 1961, pp. 146—153.
[34] Horrigmoe, G.: 'Finite Element Instability Analysis of Free-Form Shells', Rep. 77-2, The Norwegian Institute of Technology, Trondheim, 1977.
[35] Hughes, T. J.: 'Stability, Convergence, Growth and Decay of Energy of the Average Acceleration Method in Nonlinear Structural Dynamics', *Comp. Struct.* **6** (1976) 1623—1635.
[36] Hughes, T. J.: 'A Note on the Stability of Newmark's Algorithm in Nonlinear Structural Dynamics', *Int. J. Num. Methods Engng* **11** (1977) 1711—1720.
[37] Idelsohn, S.: 'Analyses statique et dynamique des coques par la methode des elements finis', Ph. D. Thesis, University of Liège, Liège, 1974.
[38] Ilyushin, A. A.: 'Some Problems in the Theory of Plastic Deformations, *Prikl. Mat. Mekh.* **7** (1943) 92—101 (in Russian).
[39] Isaacson, E. and Keller, H. B.: *Analysis of Numerical Methods*, John Wiley and Sons, New York, 1966.
[40] Jones, R. F.: 'Shell and Plate Analysis by Finite Elements', *Proc. ASCE* **99** (1973) 889—902.
[41] Kamat, M. P. and Hayduk, R. J.: 'Energy Minimization Versus Pseudo-Force Technique for Nonlinear Structural Analysis', *Comp. Struct.* **11** (1980) 1231—1240.
[42] Kersten, R.: *Das Reduktionsverfahren der Baustatik*, Springer-Verlag, Berlin, 1962.
[43] Khojasteh-Bakht, M.: 'Analysis of Elastic-Plastic Shells of Revolution under Axisymmetric Loading by the FEM', Ph. D. Thesis, University of California, Berkeley, 1967.

[44] Kikuchi, F.: 'On the Validity of the FEM Analysis of Circular Arches Represented by an Assemblage of Beam Elements', *Comp. Methods Appl. Mech. Engng* **5** (1975) 253—270.
[45] Kirsch, U.: 'Synthesis of Structural Geometry Using Approximate Concepts', *Comp. Struct.* **15** (1982) 103—111.
[46] Knowles, N. C., Razzaque, A. and Spooner, J. B.: 'Some Practical Aspects of Nonlinear Analysis in Design', *Symposium on Structural Analysis*, Crawthorne, Berkshire, U. K., 1974, pp. 206—212.
[47] Lazan, B. J.: *Damping of Materials and Members in Structural Mechanics*, Pergamon Press, London, 1968.
[48] Leckie, A. and Pestel, E. C.: *Matrix Methods in Elastomechanics*, McGraw-Hill Book Company, New York, 1962.
[49] Lin, Y. K. and Donaldson, B. K.: 'A Brief Survey of Transfer Matrix Techniques with Special Reference to the Analysis of Aircraft Panels', *J. Sound Vibr.* **10** (1969) 817—824.
[50] Lock, A. C. and Sabir, A. B.: 'Algorithm for the Large Deflection of Geometrically Nonlinear Plane and Curved Structures', in J. R. Whiteman (ed.), *The Mathematics of Finite Elements and Applications*, Academic Press, New York—London, 1973, pp. 483—494.
[51] Marcal, P. V.: 'FEM Analysis with Material Nonlinearities', *Proc. Japan-U.S. Seminar on Matrix Methods of Structural Analysis and Design*, Tokyo, 1969, pp. 182—190.
[52] Marguerre, K. and Uhring, R.: 'Berechnung vielgliedriger Schwingerketten, I, Das Übertragungsverfahren und seine Grenzen', *ZAMM* **44** (1964) 91—98.
[53] Meskouris, K.: 'Elektronische Schwingungsberechnung ebener Rahmentragwerke mit Materialdämpfung', Thesis, Technical University, Munich, 1974.
[54] Moan, T.: 'A Note on the Convergence of FEM Approximations for Problems Formulated in Curvilinear Coordinate Systems', *Comp. Methods Appl. Mech. Engng* **3** (1974) 209—235.
[55] Naghdi, P. M.: 'Stress—Strain Relations in Plasticity and Thermoplasticity', in E. H. Lee and P. S. Symonds (eds.), *Proc. 2nd Symposium on Naval Structural Mechanics*, Pergamon Press, London, 1960, pp. 343—361.
[56] Newmark, N. M.: 'A Method of Computation for Structural Dynamics', *Proc. ASCE* **85** (1959) 512—527.
[57] Olson, M. D.: 'Analysis of Arbitrary Shells Using Shallow Shell Finite Elements', in Y.C. Young and E. E. Sechler (eds.), *Thin Shell Structures Theory Experiment and Design*, Prentice Hall International Inc., Englewood Cliffs, N. J., 1974.
[58] Pawsey, S. and Clough, R. W.: 'Improved Numerical Integration of Thick Shell Finite Elements', *Int. J. Num. Methods Engng* **3** (1971) 575—586.
[59] Pestel, E. and Mahrenholz, O.: 'Zum numerischen Problem der Eigenwertbestimmung mit Übertragungsmatrizen', *Ing. Arch.* **28** (1959) 733—742.
[60] Petersen, Ch.: 'Das Verfahren der Übertragungsmatrizen für gekrümmte Träger', *Bauingenieur* **41**, H. 3 (1966) 193—195.
[61] Petersen, Ch.: 'Entwicklung eines Böfaktors durch Time-History, Schwingungssimulation realer Windgeschwindigkeitsschriebe', *LKI-Arbeitsber. Sicherheitstheor.* H. 5 (1974) 273—282.
[62] Pian, T. H. and Tong, P.: 'Variational Formulations of Finite Displacement Analysis', in B. Fraeijs de Veubeke (ed.), *High Speed of Computing Elastic Structures*, University of Liège, Liège, 1971, pp. 43—63.
[63] Pilkey, W. D. and Chang, P. Y.: *Modern Formulas for Statics and Dynamics*, McGraw-Hill Book Company, New York, 1980.
[64] Pilkey, W. D. and Haviland, J. K.: 'A Method of Analysis of Line Structures by Transfer

Matrices Derived from Finite Elements', Tech. Rep. 74-1, University of Virginia, Charlottesville, Va., 1979.

[65] Prager, W.: 'The Theory of Plasticity: A Survey of Recent Achievements, James Clayton Lecture', *Proc. Inst. Mech. Engrs* **169** (1955) 41—57.

[66] Przemieniecki, J. S.: *Theory of Matrix Structural Analysis*, McGraw-Hill Book Company, New York, 1968.

[67] Rausch, E.: 'Einwirkung von Windstößen auf hohe Bauwerke, *VDI-Z* **77** (1933) 32—39.

[68] Razani, R.: 'Behaviour of Fully-Stressed Design of Structures and Its Relationship to Minimum-Weight Design', *J. Amer. Inst. Aeron. Astron.* **3** (1965) 2262—2268.

[69] Rechenberg, J.: *Evolutionsstrategie*, Reihe Problemata 15, F. Frommann Verlag, Stuttgart —Bad Cannstadt, 1977.

[70] Remseth, S. N.: 'Nonlinear Static and Dynamic Analysis of Space Structures', Rep. 78-2, The Norwegian Institute of Technology, Trondheim, 1977.

[71] Rheinboldt, W. C.: 'An Adaptive Continuation Process for Solving Systems of Nonlinear Equations', Rep. TR-393, University of Maryland, 1975.

[72] Schlaich, J.: 'Beitrag zur Frage der Wirkung von Windstößen auf Bauwerke', *Bauingenieur* **41** (1966) 815—827.

[73] Schmit, L. A. and Miura, H.: 'A New Structural Analysis/Synthesis Capability-Access 1', *Proc. AIAA/ASME/SAE 16th Conference*, Denver, Col., 1975, pp. 307—315.

[74] Schwefel, H. P.: *Numerische Optimierung von Computer Modellen mittels der Evolutionsstrategie*, Interdisziplinäre Systemforschung 26, Birkhäuser Verlag, Basel—Stuttgart, 1977.

[75] Sedlacek, G.: 'Systematische Darstellung des Biege- und Verdrehvorganges für prismatische Stäbe mit dünnwandigem Querschnitt unter Berücksichtigung der Profilverformung', Thesis, Technical University, Berlin, 1967.

[76] Sedlacek, G. and Roik, K.: 'Erweiterung der technischen Biege- und Verdrehtheorie unter Berücksichtigung von Schubverformungen', *Bautechnik* **47** (1970) 912—930.

[77] Sharifi, P. and Popov, E. P.: 'Nonlinear Buckling Analysis of Sandwich Arches', *Proc. ASCE* **97** (1971) 1397—1412.

[78] Söreide, T. H.: 'Collapse Behaviour of Stiffened Plates Using Alternative Finite Element Formulations', Rep. 77-3, The Norwegian Institute of Technology, Trondheim, 1977.

[79] Steinle, A.: 'Torsion und Profilverformung', Thesis, Technical University, Stuttgart, 1967.

[80] Stricklin, J. A. et al.: 'Nonlinear Dynamic Analysis of Shells by Matrix Displacement Method', *AIAA J.* **9** (1971) 629—636.

[81] Tesár, A.: 'Application of Transfer Matrix Method for Solution of Torsional Vibration of Thinwalled Beams', *Staveb. Čas.* **26** (1976) 707—718.

[82] Tesár, A.: 'Aerodynamic Stability of Thinwalled Box Beams', *Staveb. Čas.* **27** (1977) 56—70 (in Slovak).

[83] Tesár, A.: 'Vibration of Thinwalled Box Structures', *Proc. Czechoslovak Academy of Sciences*, No. 1, Academia, Prague, 1977.

[84] Tesár, A.: 'Aeroelastic Response of Transporter Shell Bridges in Smooth Air Flow', Tech. Rep. 78-1, The Norwegian Institute of Technology, Trondheim, 1978.

[85] Tesár, A.: 'Aeroelastic Response of Transporter Shell Bridges in Smooth Air Flow', *Acta Technica ČSAV* 1 (1979) 19—46.

[86] Tesár, A.: Ein Beitrag zur Spannungsermittlung von regelmäßigen, querausgesteiften Plattenbalkentragwerken, *Stahlbau* H. 5 (1981) 212—232 (in German).

[87] Tesár, A.: Dynamic Response of Curved Plate-Grider Bridges, *Acta Technica ČSAV* 4 (1982) 412—442.

[88] Tesár, A.: Nonlinear Dynamic Response of Thin Shells, *Acta Technica ČSAV* 6 (1983) 754—775.

[89] Tesár, A.: 'Nonlinear Vibration of Shell Structures', Rep. III-3-2/8.1, Institute of Structures and Architecture, Slovak Academy of Sciences, Bratislava, 1983 (in Slovak).
[90] Tesár, A.: 'Nonlinear Interactions in Resonance Response of Thin Shells', *Comp. Struct.* **18** (1984) 734—742.
[91] Tesár, A.: 'Post-Buckling Resonance Analysis of Thin Shells', *Int. J. Num. Methods Engng* **20** (1984) 1123—1132.
[92] Tesár, A.: 'Vibration of Thin Shell Structures', *Proc. Czechoslovak Academy of Sciences*, No. 2, Academia, Prague, 1984.
[93] Tesár, A.: 'Nonlinear Three-Dimensional Resonance Analysis of Shells', *Comp. Struct.* **41** (1985) 157—172.
[94] Tesár, A.: 'Nonlinear Vibration and Optimization of Shell Structures', Rep. III-3-2/8.1, Institute of Structures and Architecture, Slovak Academy of Sciences, Bratislava, 1985 (in Slovak).
[95] Tesár, A.: 'Resonance Response of Thin Shells', *Acta Technica ČSAV* 1 (1985) 88—103.
[96] Tesár, A.: 'Synthesis in Optimization of Shell Bridge Structures', *Staveb. Čas.* **35** (1985) 812—824 (in Slovak).
[97] Tesár, A.: 'Inelastic Stability Response of Thin Shells', *Proc. Int. Conference on Computational Mechanics*, Tokyo, 1986, pp. 812—818.
[98] Tesár, A.: 'Limit State Dynamics of Thin Shells', *Proc. Int. Conference on Steel Structures*, Budva, 1986, pp. 627—633.
[99] Tesár, A. and Fillo, Ľ.: 'Computer Calculation of Stress State of Prototype of Transporter Shell Bridge', Tech. Rep. Institute of Construction and Architecture, Bratislava — Research Institute of Metal Industry, Prešov, 1979 (in Slovak).
[100] Tesár, A. and Fillo, Ľ.: 'Typology of Transporter Shell Bridges', Tech. Rep. Institute of Construction and Architecture, Bratislava — Research Institute of Metal Industry, Prešov, 1979 (in Slovak).
[101] Tesár, A. and Fillo, Ľ.: 'Elastic-Plastic Analysis of Thinwalled Box Beams', *Acta Technica ČSAV* 5 (1980) 497—514.
[102] Tesár, A. and Fillo, Ľ.: 'Dynamic Behaviour of Thinwalled Corrugated Sheets', *Acta Technica ČSAV* 3 (1981) 347—376.
[103] Tesár, A. and Fillo, Ľ.: 'Spatial Analysis of Thinwalled Structures', *Acta Technica ČSAV* 2 (1982) 208—222.
[104] Tesár, A. and Martinček, G.: 'Application of Transfer Matrix Method for Solution of Nonperiodic Vibration', *Staveb. Čas.* **27** (1977) 25—36 (in Slovak).
[105] Tesár, A. and Rückschloss, J.: 'Solution of Dynamic Response Using Combination of Transfer Matrices with Wilson Method', *Staveb. Čas.* **34** (1984) 711—727 (in Slovak).
[106] Tesár, A. and Rückschloss, J.: 'Newmark—Wilson Combination of Integration Operators for Direct Solution of Time Response', *Staveb. Čas.* **35** (1985) 881—903 (in Slovak).
[107] Tesár, A. and Rückschloss, J.: Some Simulation Models of Wind Gusts on Structural Systems', *Staveb. Čas.* **35** (1985) 419—433 (in Slovak).
[108] Tesár, A. and Rückschloss, J.: 'Transfer Matrix Formulation for Solution of Torsion-Bending Vibration of Beams with Thinwalled Cross-Section, *Acta Technica ČSAV* 5 (1985) 562—573.
[109] Tesár, A. and Šolek, P.: 'Method FETM Applied for Solution of Dynamic Response of Mechanical Systems', *Proc. 3rd Conference on Dynamic and Elasticity Problems of Machine Structures*, Bratislava, 1983, pp. 293—299.
[110] Tesár, A. and Tesár, P.: 'Resonance Response of Thinwalled Elements of Carriage Bodies', *Strojn. Čas.* **35** (1984) 179—194 (in Slovak).
[111] Tesár, A. and Tesár, P.: 'Resonance Response of Stiffened Sandwich Panels', *Strojn. Čas.* **34** (1983) 509—522 (in Slovak).

[112] Tesár, A. and Wu Tan Khiem: Distortional Behaviour of Thinwalled Box Beams, *Acta Technica ČSAV* 2 (1978) 72—78.

[113] Tesár, A.: *Numerical Methods in Structural Dynamics*, Veda, Bratislava (in Slovak — in print).

[114] Tottenham, H.: 'The Linear Analysis of Thinwalled Space Structures', *Proc. Symposium on Use of Electronic Computers in Structural Engineering*, Southampton, 1959, pp. 97—110.

[115] Wagner, H.: *Verdrehung und Knickung von offenen Profilen*, Festschrift 25 Jahre TU Danzig, 1929.

[116] Weeks, G.: 'Temporal Operators for Nonlinear Dynamic Problems', *J. Engng Mech. Div., ASCE* **98** (1972) 893—902.

[117] Wempner, G.: 'Finite Elements, Finite Rotations and Small Strains of Flexible Shells', *Int. J. Solids Struct.* **5** (1969) 117—153.

[118] Willam, K.: 'FEM Analysis of Cellular Structures', Ph. D. Thesis, University of California, Berkeley, 1969.

[119] Wilson, E. L.: 'A Computer Program for the Dynamic Stress Analysis of Underground Structures', Rep. SESM 68-1, University of California, Berkeley, 1968.

[120] Wright, J. P.: 'Mixed Time Integration Schemes', *Comp. Struct.* **10** (1979) 1241—1250.

[121] Wu, L. and Witmer, E.: 'Nonlinear Transient Response of Structures by the Spatial FEM', *AIAA J. Appl. Mech.* (1975) 1072—1083.

[122] Wunderlich, W.: 'Differentialsystem und Übertragungsmatrizen der Biegetheorie allgemeiner Rotationsschalen', Thesis, Technical University, Hannover, 1966.

[123] Yang, T. Y.: 'High Order Rectangular Shallow Shell Finite Ellement', *Proc. ASCE* **99** (1973) 157—181.

[124] Zienkiewicz, O. C.: *The Finite Element Method in Structural and Continuum Mechanics*, McGraw-Hill Book Company, New York, 1967.

[125] Zurmühl, M.: *Matrizen*, Springer-Verlag, New York—Berlin—Heidelberg, 1962.

[126] Zurmühl, M.: 'Berechnung von Biegeschwingungen abgesetzter Wellen mit Zwischenbedingungen mittels Übertragungsmatrizen', *Ing. Arch.* **26**, H. 6 (1958) 282—297.

SUBJECT INDEX